PHYSIOLOGICAL CHEMISTRY
A Series Prepared under the General Editorship of
Edward J. Masoro, Ph.D.

I   PHYSIOLOGICAL CHEMISTRY OF LIPIDS IN MAMMALS

II   PHYSIOLOGICAL CHEMISTRY OF PROTEINS AND NUCLEIC ACIDS IN MAMMALS

III   ENERGY TRANSFORMATIONS IN MAMMALS: Regulatory Mechanisms

IV   ACID-BASE REGULATION: Its Physiology and Pathophysiology

V   REGULATION OF AMINO ACID METABOLISM IN MAMMALS

VI   PHYSIOLOGICAL CHEMISTRY OF CARBOHYDRATES IN MAMMALS

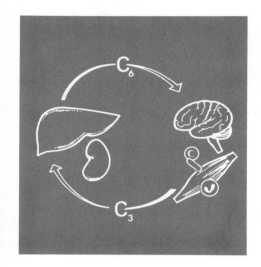

PHYSIOLOGICAL CHEMISTRY OF CARBOHYDRATES IN MAMMALS

Walton W. Shreeve, M.D., Ph.D.

Chief, Nuclear Medicine Service,
Veterans Administration Hospital,
Northport, New York;
Professor of Medicine (Nuclear Medicine)
Health Sciences Center,
State University of New York at Stony Brook,
Stony Brook, New York

1974

W. B. Saunders Company

Philadelphia · London · Toronto

W. B. Saunders Company: West Washington Square
                          Philadelphia, Pa. 19105

                          12 Dyott Street
                          London, WCIA   1DB

                          833 Oxford Street
                          Toronto, Ontario M8Z 5T9, Canada

Physiological Chemistry of Carbohydrates in Mammals   ISBN   0-7216-8265-0

© 1974 by W. B. Saunders Company. Copyright by W. B. Saunders Company. Copyright under the International Copyright Union. All rights reserved. This book is protected by copyright. No part of it may be reproduced, stored in a retrieval system, or transmitted in any form or by any means, electronic, mechanical, photocopying, recording, or otherwise, without written permission from the publisher. Made in the United States of America. Press of W. B. Saunders Company. Library of Congress Catalog card number 73-88265.

Print No.   9   8   7   6   5   4   3   2   1

*IN MEMORY OF
DONALD D. VAN SLYKE
AND
ALLEN R. HENNES*

EDITOR'S FOREWORD

The past three decades or so have seen biochemistry emerge as possibly the most vigorous of the biological sciences. This, in turn, has led to a level of autonomy that has cut the cord linking biochemistry with its historically most important parent, mammalian physiology. For investigators in the fields of both biochemistry and physiology, this vitality has been most useful. But because of the arbitrary separation of these two disciplines in most teaching programs and all textbooks, the vast majority of students do not see the intimate relationships between them. Consequently, the medical student and the beginning graduate student as well as the recently trained physician find it difficult, if not impossible, to utilize the principles of biochemistry as they apply to the physiological and pathological events they observe in man and other mammals.

Therefore, this series is designed not only to introduce the student to the fundamentals of biochemistry but also to show the student how these biochemical principles apply to various areas of mammalian physiology and pathology. It will consist of six monographs: (1) Physiological Chemistry of Lipids in Mammals; (2) Physiological Chemistry of Proteins and Nucleic Acids in Mammals: (3) Energy Transformations in Mammals; (4) Acid-Base Regulation: Its Physiology and Pathophysiology; (5) Physiological Chemistry of Carbohydrates in Mammals; and (6) Regulation of Amino-Acid Metabolism.

The series can be profitably used by undergraduate medical students. Recent medical graduates and physicians involved in areas of medicine related to metabolism should find that the series enables them to understand the theoretical basis for many of the problems they face in their daily work. Finally, the series should provide students in all areas of mammalian biology with a source of information on the biochemistry of the mammal that is not otherwise currently available in textbook form.

EDWARD J. MASORO

EDITOR'S
FOREWORD

PREFACE

Some years ago the editor of this series enlisted me to share his worthy purpose of opening to the minds of students a clearer understanding of the relationship of the trees of biochemistry to the forest of physiology. During this effort I myself have gained better appreciation of groves of exotic carbohydrates which give character, form, and distinctiveness to the forest. The relationships of these rare forms to the more common species were fascinating trails to follow. I became more thoroughly aware of the fine balances, the rich variety and adaptability of life in the forest, the place of each small and large species of carbohydrate in the ecological scheme. I must apologize that for the sake of narrative description the names of all the diligent, devoted naturalists cannot be marked beside the trail. Their works often have given me new inspiration in seeking more knowledge about the forest. I hope this summary may do so for others. For those who now or in the future will more directly care for the forest in its times of vulnerability or invasion, I hope this book will be of some general aid in husbandry.

For critical review of individual chapters I am indebted to several generous colleagues and friends, namely, Bernard N. Brodoff, Lewis K. Dahl, George C. Cotzias, A. Baird Hastings, Mitsuru Hoshi, Ezra Lamdin, Bernard R. Landau, Paul G. LeFevre, John A. Muntz, Edgar M. Neptune, Jr., John J. O'Neill, Edwin A. Popenoe, Merton F. Utter, the late Donald D. Van Slyke and Ching-Hui Wu. Very helpful throughout were the wise comments, perceptive corrections and vital encouragement of our editor, Edward J. Masoro. I thank my wife, Phyllis, and my children, Thomas, Daniel, James and Elizabeth, for their inspiration and their understanding.

WALTON W. SHREEVE

CONTENTS

CHAPTER 1

CHEMISTRY OF CARBOHYDRATES 1

    Monosaccharides.. 2
    Oligosaccharides 9
    Polysaccharides... 11

CHAPTER 2

DIGESTION AND ABSORPTION OF CARBOHYDRATES 38

    Hydrolysis of Polysaccharides 38
    Hydrolysis of Disaccharides 40
    Absorption and Transport of Monosaccharides 45
    Deficiencies of Digestion and Absorption.......................... 48

CHAPTER 3

CELLULAR TRANSLOCATION OF CARBOHYDRATES 53

    Concept of a Monosaccharide Carrier 54
    Stereospecificity of Monosaccharide Translocation 57
    Molecular Nature of the Membrane Binding Site............... 58
    Hormonal Effects on Glucose Transport........................... 61
    Monosaccharide Transport in Different Organs
        and Species.. 64
    Intracellular Translocation of Carbohydrates 76

CHAPTER 4

INTRACELLULAR UTILIZATION OF MONOSACCHARIDES 81

    Phosphorylation of Glucose 82
    Glycolysis 89
    Pentose Cycle 100
    Utilization of Pyruvate 110
    Tricarboxylic Acid Cycle 114
    Utilization of Fructose 125
    Utilization of Galactose 129
    Utilization of Mannose 134

CHAPTER 5

BIOSYNTHESIS OF MONOSACCHARIDES 139

    Gluconeogenesis 139
    Biosynthesis of Pentoses 158
    The Uronic Acid Pathway 164
    Biosynthesis of Hexosamines and Sialic Acids 172
    Biosynthesis of Fucose 178
    Biosynthesis of Galactose 180
    Biosynthesis of Fructose 180

CHAPTER 6

METABOLISM OF POLYSACCHARIDES 185

    Glycogen 185
    Proteoglycans 204
    Glycoproteins 219
    Oligosaccharides 230

CHAPTER 7

CHEMISTRY AND METABOLISM OF CARBOHYDRATES IN GLYCOLIPIDS 234

Chemistry 234
Occurrence and Function in Various Organs and
   Subcellular Structures 238
Metabolism 242
Pathological Conditions 249

CHAPTER 8

Turnover and Fate of Carbohydrates in the Circulation 253

Glucose 253
Lactic Acid and Pyruvic Acid 284
Glycerol 298
Galactose 304
Fructose 309

GLOSSARY 319

INDEX 321

CHAPTER

1

CHEMISTRY OF CARBOHYDRATES

To understand the physiological chemistry of carbohydrates one must understand their composition. The isomeric configuration of just one carbon atom in a monosaccharide can make a vast difference in its biological role, its susceptibility to hormone action, its utilization by different tissues, its fitness for linkage to other monosaccharides in formation of heteroglycans, and in the simple fact of its occurrence in nature. Among the possible myriad of stereoisomers of numerous oxygen functions on carbon atoms, of sequences of simple sugars within polysaccharides, and of location of glycosidic linkages, only a limited few have been selected by evolution to exist in nature and an even smaller number in mammals. Thanks to this situation, it is no large task to learn the alphabet and, indeed, the fundamental language of carbohydrate chemistry in mammals.

In this chapter are described first the monosaccharides with increasing numbers of carbon atoms, then the oligosaccharides and the polysaccharides with their single homogeneous type (glycogen) and their several heterogeneous types. In other chapters is contained the chemistry of disaccharides, of glycolipids, and of compounds in the tricarboxylic acid cycle. When the biochemical interactions of many of these carbohydrates with each other and with other classes of compounds are successively viewed, it is worth considering that still further functional aspects of the high stereospecificity, the "fine grammar," of the carbohydrates are yet to be revealed. In the future there is the possibility that elucidation of the more complete and higher-ordered architecture of the oligosaccharides, the glycolipids, and the polysaccharides will importantly signify various functional specificities, e.g., in the areas of immunology and even neuronal memory.

MONOSACCHARIDES

Monosaccharides may be defined as single straight chains of carbon atoms bearing multiple adjacent hydroxy groups, of which one is usually oxidized to a carbonyl group. Such compounds may have some further modification of the hydroxyl functions, e.g., oxidation of carbonyl to carboxyl groups, substitution of a hydroxyl by an amine group, or reduction of a hydroxyl group to a C—H bond.

Broadly, there can be included among the monosaccharides a few polyfunctional (mainly polycarboxylic) compounds which are formed in the complete catabolism of monosaccharides as well as in their synthesis from smaller units derived from amino acids, and so forth. These vital compounds, mainly members of the glycolytic scheme and the tricarboxylic (Krebs) cycle, will be considered in later chapters.

Simple carbohydrates (monosaccharides) occur in nature as molecules containing from two to nine carbon atoms. The most common and important monosaccharides found in living species generally, and particularly in higher species such as mammals, are the 5- and 6- carbon-membered sugars. With their polyfunctional nature and large number of asymmetric carbon atoms, monosaccharides afford the possibility of a rich variety of stereoisomers. Nevertheless, biological specificity of enzymes and transport mechanisms restricts natural monosaccharides to only a few of the possible isomers. An interesting challenge to biochemical investigators of the present era is the matter of defining for the polypeptide chains of the enzymes of carbohydrate metabolism the sequences and conformations of particular amino acids which determine the stereospecificity of the carbohydrate substrate. Most fundamental is the elucidation of the amino acid structure and sequence (and correspondingly the polynucleotide sequence in nucleic acids) which are complementary for the common monosaccharide, glucose.

Two-carbon Compounds

The simplest structure fitting the above definition of carbohydrates is glycolaldehyde (Fig. 1-1). This compound is not known to occur in

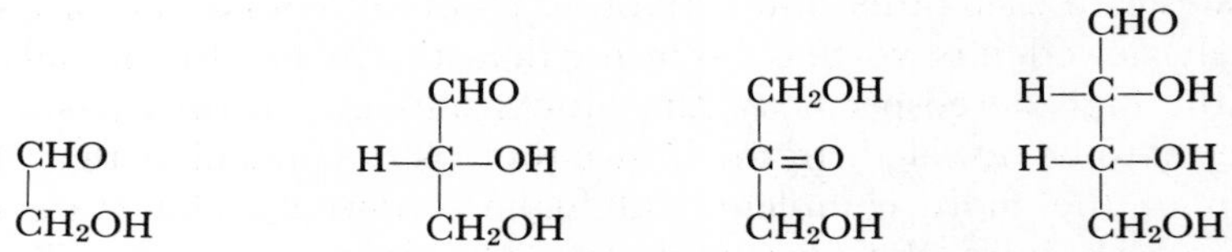

$$\begin{array}{ccccc}
& & \text{CHO} & \text{CH}_2\text{OH} & \text{CHO} \\
& & | & | & | \\
\text{CHO} & \text{H}-\text{C}-\text{OH} & \text{C}=\text{O} & \text{H}-\text{C}-\text{OH} \\
| & | & | & | \\
\text{CH}_2\text{OH} & \text{CH}_2\text{OH} & \text{CH}_2\text{OH} & \text{H}-\text{C}-\text{OH} \\
& & & & | \\
& & & & \text{CH}_2\text{OH} \\
\text{glycolaldehyde} & \text{D-glyceraldehyde} & \text{dihydroxy-} & \text{D-erythrose} \\
& & \text{acetone} &
\end{array}$$

Figure 1-1. Representative 2-, 3- and 4-carbon sugars.

free form in mammalian tissues, but it is considered to exist as a fragment derived from a larger monosaccharide and transferred to another in the action of an enzyme, transketolase, which is found in liver, red blood cells, and other tissues. This transfer is employed in the synthesis of certain sugars. Derivatives of glycolaldehyde, such as glycolic acid and glyoxylic acid, may occur as products in a pathway (minor for animals) of metabolism of the amino acid glycine, but are of little known significance in carbohydrate metabolism.

Three-carbon Compounds

A monosaccharide with three carbons represents the simplest structure which has an asymmetric carbon atom, i.e., the middle carbon. Of the two possible isomers the one commonly found in nature is termed D-glyceraldehyde. In the chemical formula commonly written as in Figure 1-1 the OH is placed on the right of the carbon to designate the D-isomer. L-glyceraldehyde is written with the hydroxyl group to the left of the carbon. This convention is employed for all higher monosaccharides, i.e., the position of the hydroxyl (right or left) on the highest-numbered* asymmetric carbon atom denotes whether the sugar has the D- or the L-stereospecific form.

D-glyceraldehyde occurs in the free form both unphosphorylated (derived, e.g., in the catabolism of fructose-1-phosphate) and, more commonly, as the derivative with a phosphate group on position C-3. In the latter form it is an intermediate in the "glycolytic" pathway common to many tissues for the breakdown of glycogen and glucose (Chap. 4).

Another 3-carbon monosaccharide important in metabolism is dihydroxyacetone, as seen also in Figure 1-1. This compound, as the monophosphorylated derivative, is readily interconvertible in the glycolytic pathway with D-glyceraldehyde-3-P. Furthermore, in a manner somewhat analogous to that for glycolaldehyde, the 3-carbon fragment of dihydroxyacetone monophosphate (DHAP) may be transferred by the enzyme transaldolase from certain 2-keto monosaccharides to condense with other aldol sugars. The polyalcoholic reduction product of DHAP, α-glycerol-P, is important as a precursor of the glycerol moiety of triglycerides and phospholipids. Also, α-glycerol-P may have a significant metabolic function in intracellular hydrogen transport.

Three-carbon compounds of lesser and indeed questionable significance are lactaldehyde and propanediol. Yet these compounds have been identified as occurring in mammalian tissues; propanediol (phosphate) may be present in rather high concentrations in brain.

---

*The carbons of carbohydrates are numbered beginning at that end of the chain with the most highly oxidized functional groups.

Four-carbon Compounds

The addition of one more asymmetric carbon atom provides the possibility of four stereoisomers of the tetrose series. Of these, the only one which is known in mammals is D-erythrose (Fig. 1-1). With phosphate attached to C-4 this compound occurs as a product of transketolations and transaldolations, as mentioned above, but it is found in only trace concentrations and occurs in metabolic pathways not considered of prime importance to mammalian organisms.

Five-carbon Compounds

Of the eight possible stereoisomeric pentoses (according to the rule of 2^n, where n is the number of asymmetric carbon atoms) by far the most important biologically is D-ribose (Fig. 1-2). This is well known as the isomer contained exclusively in ribonucleic acids (RNA) and in those mononucleotide derivatives (nicotinamide-adenine dinucleotides, the adenosine phosphates, uridine phosphates, and so forth) which are essential in transfers of electrons and substrates and in storage and transformations of energy. An important derivative of ribose is 2-deoxy-D-ribose (Fig. 1-2), which forms the structural carbohydrate component of deoxyribonucleic acids (DNA). In the polynucleotides the ribose (or deoxyribose) units are linked through phosphate bridges which are in diester combination with adjacent pentose units at the 3 and 5 positions, respectively.

A pentose which is much less common but occurs significantly in mammals is D-xylose (Fig. 1-2), which is a constituent of some heteropolysaccharides (polymers containing more than one kind of monosaccharide). A 2-keto pentose, L-xylulose (Fig. 1-2), as well as xylitol and D-xylulose, occurs as an intermediate in the glucuronic acid pathway (Chap. 5). A hereditary derangement in this pathway results in excessive urinary excretion of L-xylulose. This is known as essential pentosuria, a benign condition.

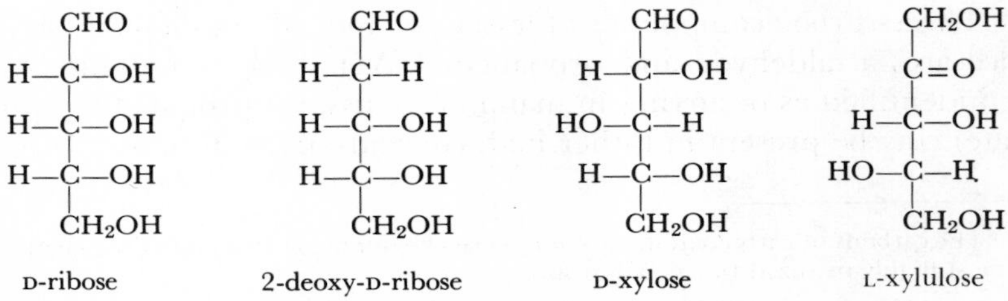

$$
\begin{array}{cccc}
\text{CHO} & \text{CHO} & \text{CHO} & \text{CH}_2\text{OH} \\
\text{H—C—OH} & \text{H—C—H} & \text{H—C—OH} & \text{C=O} \\
\text{H—C—OH} & \text{H—C—OH} & \text{HO—C—H} & \text{H—C—OH} \\
\text{H—C—OH} & \text{H—C—OH} & \text{H—C—OH} & \text{HO—C—H} \\
\text{CH}_2\text{OH} & \text{CH}_2\text{OH} & \text{CH}_2\text{OH} & \text{CH}_2\text{OH} \\
\text{D-ribose} & \text{2-deoxy-D-ribose} & \text{D-xylose} & \text{L-xylulose}
\end{array}
$$

Figure 1-2. Various pentoses in mammals.

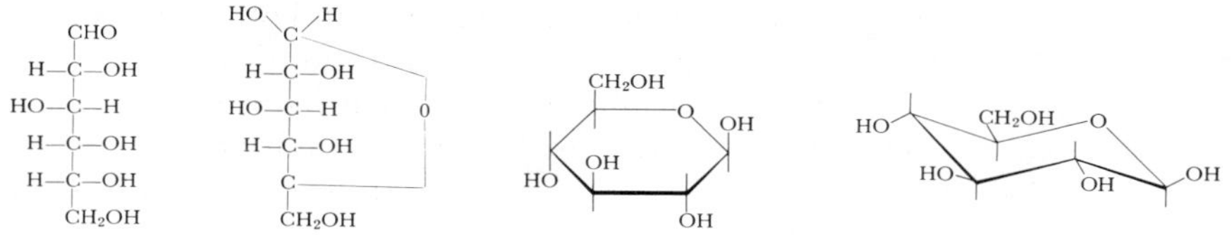

CHEMISTRY OF CARBOHYDRATES

Figure 1-3. Various representations of β-D-glucose.

Six-carbon Compounds

In the D-series there are eight possible isomers of hexose, and by far the most common and important biologically is D-glucose, the conventional open-chain two-dimension formula of which is shown in Figure 1-3a. Glucose is present in the blood in far higher concentration than any other monosaccharide. Its catabolism provides a vital source of energy for many tissues and an essential source for tissues of the central nervous system. Also, it is the exclusive structural unit of the polysaccharide glycogen, the storage form of sugar in mammalian liver, muscle, adipose tissue, and other organs. Various derivatives of glucose (but seldom glucose itself) are major components of the complex polysaccharides.

Whereas the open-chain form of D-glucose has four asymmetric carbon atoms, a fifth such center of asymmetry is introduced when the molecule undergoes an internal condensation leading to a ring formation through an oxygen bridge between the first and the fourth carbons (furanose ring) or the first and fifth carbons (pyranose ring). The latter is the more chemically stable and common. As seen in Figure 1-3b this structure may be included in the two-dimension formula, which shows the bridge written to the right of the chain with the hydroxyl on C-1 trans to that on C-2. This is called the β-form. It is the more active structure in metabolism of glucose in mammalian tissues, and it is also found in some glycosidic linkages (to be discussed). The α-form (bridge written to the right, but with hydroxyl of C-1 cis to that of C-2) exists in the glycosidic linkages of glycogen and elsewhere. The α and β isomers of ring forms of sugars are known as anomers. Of course, the formation of ring structures is also a property of pentoses and some other monosaccharides.

The cyclic forms of aldoses (and ketoses) are not immutable in aqueous solution, but rather are in equilibrium with a small amount of the open-chain form. When either the β-form or α-form is dissolved in water, an equilibrium mixture of the two gradually forms in solution. This phenomenon is called mutarotation, since the two forms separately often have quite different optical rotations.

The specificity of optical rotation has been much used in the past to identify glucose and other monosaccharides, or, when the type of pure sugar is known, to measure its concentration. This method is less often used today. Various chemical reactions of D-glucose and other sugars are more commonly used to measure their concentration. On treatment with strong acids aldohexoses form hydroxymethylfurfural (involving a double dehydration and anhydride formation with cyclization between C-2 and C-5), while aldopentoses form furfural. These compounds form colored products with certain reagents, e.g., anthrone, which is used to measure glucose and glycogen.

Another important reaction for isolation, characterization, and measurement of monosaccharides is that with phenylhydrazine to form "osazones," which are generally yellow crystalline compounds. Since this reaction involves complexing with, and loss of specificity at, carbons 1 and 2, D-glucose, another aldose, D-mannose, and the ketohexose, fructose, all form the same osazone. A further derivative of the osazone, the osatriazole, provides a colorless, more stable, and more easily purified crystalline product. It has proved advantageous for measuring the radioactivity of ^{14}C- or ^{3}H-labeled glucose by liquid scintillation spectrometry.

A common method for measuring glucose in blood, urine, and other solutions depends on the ready reduction by its free aldehyde group of certain metals, which, if appropriately complexed (as, for example, the iron in potassium ferricyanide), undergo a color development proportional to the concentration of glucose. Some other compounds in biological fluids also react, so the method is not highly specific. Of increasing use in recent years has been the very specific enzymatic oxidation of glucose to gluconic acid by an enzyme, glucose oxidase (notatin), obtained from various penicillin molds. This reaction can also be followed colorimetrically. A notable recent development is the capacity for measurement of very low concentrations of glucose and some of its main metabolic derivatives and products by enzymatic coupling of reactions of these compounds to a change in oxidation state of some coenzyme (e.g., pyridine nucleotide) which can be measured with great sensitivity fluorimetrically.

A more realistic way to represent the actual three-dimensional structure of β-D-glucose is the projection type of formulation shown in Figure 1-3c. Here it is seen that the angle of 109° between C—C bonds brings C-1 and C-5 (or C-4) relatively close together. This facilitates the migration of the proton of the hydroxyl of C-5 to associate with the oxygen of the aldehyde at C-1 with ring closure through the oxygen on C-5 (in the case of a pyranose ring). Such a structure is called an internal hemiacetal. Full acetals (or glycosides) arise from hemiacetals when the hydroxyl thus formed at C-1 condenses with the hydroxyl of another alcohol with the elimination of water. Glycosidic bonds between sugars are formed when the hydroxyl of the other alcohol is contained in another carbohydrate.

In transposing from the flat conventional formulations of Figures 1-

CHEMISTRY OF CARBOHYDRATES

3a and 1-3b to the projection formula of Figure 1-3c it should be noted that the functions written to the left and right of the carbons in Figures 1-3a and 1-3b are written, respectively, above and below the apparent plane of the ring in Figure 1-3c. A still more realistic depiction of β-D-glucose is Figure 1-3d, which indicates that a pyranose ring is not truly planar, but ordinarily assumes a "chair" conformation, as in Figure 1-3d, in order to maintain the normal valency angles. Other conformations, such as the "boat" and the "skew" forms, may also be assumed, but they have a higher energy content and are less stable. However, in the course of a reaction these other conformations may in some cases be favored or required.

Of the seven possible isomers of the D-hexose series besides D-glucose the only ones found in mammalian tissues are D-galactose and D-mannose. Their structures are given in Figure 1-4. D-Galactose is the hexose paired with D-glucose in the disaccharide lactose, found in milk. Also D-galactose is contained in certain of the heteroglycans (mucopolysaccharides and glycoproteins) and in some glycolipids. Another important hexose biologically is the 2-ketohexose, fructose (Fig. 1-4). Fructose in the free state is found in high concentration (about 0.5 per cent) in semen. Phosphorylated derivatives of this ketose are important as intermediates in the glycolytic sequence.

Another hexose which has great biological significance in mammals

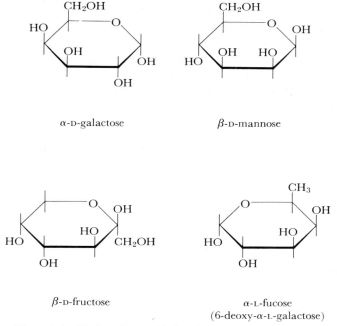

Figure 1-4. Various hexoses (other than glucose) in mammals.

is 6-deoxy-L-galactose or L-fucose (Fig. 1-4). This monosaccharide is a common constituent of the oligosaccharide components of glycoproteins and of the free oligosaccharides in human milk.

When glucose is oxidized at C-1 to the carboxylic acid it is known as D-gluconic acid. This compound may be formed by a minor reaction in mammalian tissues, but 6-phospho-D-gluconate occurs much more commonly as an intermediate in the pentose cycle pathway (Chap. 4). Oxidation of the primary alcohol group at C-6 to a carboxyl group forms the uronic acid derivative of a hexose. The term is derived from the analogue of glucose, glucuronic acid, which is excreted in the urine as a complex with various noxious agents. Formation of glucuronic acid and its detoxifying conjugates occurs in the liver. Glucuronic acid is also widely contained in a number of heteropolysaccharides. In at least one of the latter is found another uronic acid, L-iduronic acid.

Still another important type of derivative of hexoses are the amine sugars, e.g., 2-deoxy-2-amino-D-glucosamine (D-glucosamine) and 2-deoxy-2-amino-D-galactosamine (D-galactosamine). These hexosamines occur in heteroglycans and in gangliosides (Chap. 7) almost always in the form of their *N*-acetyl derivatives.

When fructose is reduced at its keto group there is formed the polyhydroxy alcohol, D-sorbitol, which is found in certain mammalian tissues and secretions. A group of cyclic hexahydric alcohols, the inositols, are found widely distributed in nature, including mammals. Myoinositol (Fig. 1-5) is contained in phosphatides and in semen. Scylloinositol (the C-2 epimer of myoinositol) is excreted in mammalian urine. A six-carbon compound which is also somewhat related to sugars is ascorbic acid or vitamin C (Fig. 1-5). The acidity derives from the activation of the protons of the C-2 and C-3 hydroxyls due to the adjacent double bond. There is a minimal formation (from hexoses) of ascorbic acid in mammalian tissues, but insufficient for body needs in certain species, including man, hence its classification as a vitamin.

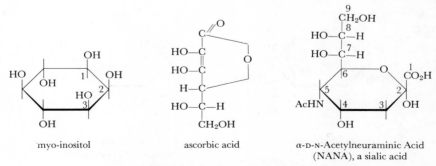

myo-inositol ascorbic acid α-D-N-Acetylneuraminic Acid (NANA), a sialic acid

Figure 1-5. Special derivatives of hexoses in mammals.

Seven-carbon Compounds

In mammalian tissues the occurrence of 7-carbon sugars is rare and seemingly of minor importance. However, sedo-heptulose-7-phosphate is a compound postulated to occur in the course of certain transketolation and transaldolation reactions, which form intermediates with varying numbers of carbon atoms in the process of metabolism of glucose via the pentose cycle pathway (Chap. 4).

Eight-carbon Compounds

Although an octulose has been isolated from the avocado and noted more recently in some bacteria, its occurrence has not been observed in or seriously postulated for mammalian species.

Nine-carbon Compounds

There is one unique example of the occurrence of a 9-carbon sugar found plentifully and of great physiological significance in heteroglycans and in glycolipids, i.e., sialic acid. The term "sialic acid" actually encompasses various derivatives of "neuraminic acid." Ths most common of these derivatives is N-acetylneuraminic acid (NANA) (Fig. 1-5); others are N-glycolyl-, O-acetyl-, and N,O-diacetyl-. The term neuraminic acid derives from the orginal identification of NANA as a component of glycolipids of brain. As seen in Figure 1-5, the molecule may be viewed as a condensation product between a hexosamine and pyruvic acid with a pyranose ring formation between the keto group at C-2 and hydroxyl at C-6. The free carboxyl group confers important properties on sialic acid.

OLIGOSACCHARIDES

Oligosaccharides contain approximately ten or fewer monosaccharide units joined in glycosidic linkage. The latter, as already noted, is a condensation between the hydroxyl group on C-1 of one (cyclized) unit and a hydroxyl group on one of the other carbon positions of another unit. Usually these linkages are -1,4- both in oligosaccharides and in polysaccharides, but they may be -1,2- or -1,3-. Branching occurs via various combinations, including -1,6- linkages.

The tissues of animals contain relatively few free oligosaccharides compared with those of plants. Only one seems to be of major occurrence, i.e., lactose in the milk of mammals. This disaccharide is composed of β-D-galactopyranosyl-1,4-D-glucopyranose (hereafter in nomenclature of glycosides the 6-membered pyranose ring may be assumed, unless stated otherwise). Whereas cow's milk contains only lactose in appreciable amount, human milk contains, in addition to lactose, several other oligosaccharides which often contain L-fucose and N-acetyl-D-glucosamine or both. The simplest and most common of these is the trisaccharide L-fucosyl-1,2-D-galactosyl-1,4-D-glucose or fucosyl-lactose. In the abbreviated notation of Figure 1-6, as given by Date, are the

2-fucosido-lactose

$$\text{Fu} \xrightarrow{\alpha-1,2} \text{Gal} \xrightarrow{\beta-1,4} \text{Gl} <$$

Lacto-difucotetraose

$$\text{Fu} \xrightarrow{\alpha-1,2} \text{Gal} \xrightarrow{\beta-1,4} \underset{\underset{\text{Fu}}{\uparrow \alpha-1,3}}{\text{Gl}} <$$

Lacto-N-tetraose

$$\text{Gal} \xrightarrow{\beta-1,3} \text{NAcGl} \xrightarrow{\beta-1,3} \text{Gal} \xrightarrow{\beta-1,4} \text{Gl} <$$

Lacto-N-fucopentaose I

$$\text{Fu} \xrightarrow{\alpha-1,2} \text{Gal} \xrightarrow{\beta-1,3} \text{NAcGl} \xrightarrow{\beta-1,3} \text{Gal} \xrightarrow{\beta-1,4} \text{Gl} <$$

Lacto-N-fucopentaose II

$$\text{Gal} \xrightarrow{\beta-1,3} \underset{\underset{\text{Fu}}{\uparrow \alpha-1,4}}{\text{NAcGl}} \xrightarrow{\beta-1,3} \text{Gal} \xrightarrow{\beta-1,4} \text{Gl} <$$

Lacto-N-fucohexaose I

$$\text{Fu} \xrightarrow{\alpha-1,2} \text{Gal} \xrightarrow{\beta-1,3} \text{NAcGl} \xrightarrow{\beta-1,3} \underset{\underset{\text{Fu}}{\uparrow \alpha-1,4}}{\text{Gal}} \xrightarrow{\beta-1,4} \text{Gl} <$$

Lacto-N-difucohexaose II

$$\text{Gal} \xrightarrow{\beta-1,3} \underset{\underset{\text{Fu}}{\uparrow \alpha-1,4}}{\text{NAcGl}} \xrightarrow{\beta-1,3} \text{Gal} \xrightarrow{\beta-1,4} \underset{\underset{\text{Fu}}{\uparrow \alpha-1,3}}{\text{Gl}} <$$

Figure 1-6. Names and structures of some oligosaccharides in human milk. Gal = D-galactose, Gl = D-glucose, Fu = L-fucose, NAcGl = N-acetyl-D-glucosamine, < = reducing group of glucose. (From Date, J. W.: Scand. J. Clin. and Lab. Invest. *16*:597, 1964.)

CHEMISTRY OF CARBOHYDRATES

structures of a few other oligosaccharides, besides lactose, found in human milk. The simplified names are used for easy reference. There are also some acidic oligosaccharides in human milk in which the acidic constituent is N-acetylneuraminic acid. The simplest of these is the trisaccharide neuramine-lactose. In the mammary gland are also found sulfate esters of lactose and NANA-lactose (sulfate attached at C-6 of galactose). The content of NANA in human milk is 40 times that in cow's milk.

Some of the oligosaccharides of human milk are excreted in the urine in small amounts during the latter half of pregnancy and in larger amounts (although still less than lactose) during lactation.

There are oligosaccharide units (also commonly containing fucose) which are attached to protein in the "glycoproteins," but since these are not free sugars they will be further considered in the section on complex polysaccharides.

POLYSACCHARIDES

Polysaccharides are polyglycosides composed of chains (linear and branching) of monosaccharides in which there are more than a few (i.e., ten or more) monosaccharide units. However, multiple oligosaccharide units which are attached to protein are often viewed as polysaccharides. Both in plants and in animals there are two general types of polysaccharides from a physiological standpoint, which may be termed nutrient and structural. There are also two major divisions from a chemical standpoint, i.e., homopolysaccharides or "homoglycans," which are composed almost entirely, if not exclusively, of one type of monosaccharide, and heteropolysaccharides or "heteroglycans," which contain as major components two or more different monosaccharides.

The nutrient or storage forms of polysaccharide are principally homoglycans both in plants (e.g., starch) and in animals (e.g., glycogen). Whereas structural polysaccharides in plants include both homoglycans (e.g., cellulose) and heteroglycans, in the animal kingdom the structural types are almost always varieties of heteroglycans. Moreover, the term "structural" is far too narrow a definition for the multiplicity of functions displayed by heteroglycans in higher animals such as mammals. Thus, heparin has anticoagulant and other properties, the "blood-group substances" are involved in immunospecificity, hormones and enzymes contain heteroglycans, and so forth.

Homoglycans

In mammals as in other animals there is only one chemical type of homoglycan, a polyglycoside of α-D-glucose units linked -1,4- with each other in linear chains and branched by means of α-1,6- linkages (Fig. 1-

7). Such compounds have the collective name of glycogen. The chemical structure of glycogen is very similar to that of starch, but the molecules of glycogen are more highly branched than those of starch. This appears to contribute to the higher water solubility of glycogen.

There are about 10 per cent as many -1,6- linkages in glycogen as -1,4- linkages (the degree of branching may be slightly variable depending on the source of glycogen). However, there is not regularity of branching—some linear chains may be as short as one to three glucose units or as long as 18 units. The use of the enzyme α-amylase to cleave glucose units from a chain down to a minimal stub length of one, two, or three residues beyond a branch point (so-called α-amylase limit dextrin) has afforded some calculations of chain length and degree of regularity and variability of branching.

Evidence indicates that α-1,6- linkages are thermodynamically more stable than α-1,4- linkages. Since the latter are ten times more numerous, it seems that some -1,4- linkages are sterically hindered from the action of branching enzyme. The latter enzyme probably acts by transferring a part of one chain to another chain rather than by making a branch in the same chain. Steric specificity of branching enzyme is thus probably a major factor in determining glycogen fine structure.

In order to account for the finding of glycogen molecules of relatively high molecular weight (exceeding 10×10^6) it must also be supposed that there is not only irregularity of branching but also a sizable fraction of internally terminating chains. These could very well be resistant to the action of degradative enzymes. A further implication from studies of successive degradation by amylase and by debranching enzyme is that the chains in the interior of the molecule are shorter than those on the periphery. On the other hand, statistical models require that some inner chains be much longer than the average value.

There has been some evidence that certain anomalous linkages or components may be present in glycogen. Thus nigerose, a disaccharide with two glucose units in -1,3- linkage, has been isolated from digests of glycogen, and isomaltotriose (containing successive -1,6- linkages) has also been found. It has been suggested, however, that these may represent "artifacts" of acid-catalyzed transfer reactions occurring during the process of analysis. Maltulose (α-D-glucosyl-1,4-D-fructose) has been identified in rabbit liver glycogen. Glucosamine has been identified in glycogen in trace amounts.

Viscosity and other physical measurements suggest that glycogen molecules are very nearly spherical, but more accurately have the shape of flattened ellipsoids. Molecular models of α-1,4-linked glucose chains show a tendency toward helix formation. Such helices are sterically capable of entwining with each other to form "double helices" in which the hydrophobic groups of adjacent helical chains are in contact. This configuration might facilitate acid-catalyzed (or enzyme-catalyzed) transfer reactions.

CHEMISTRY OF CARBOHYDRATES

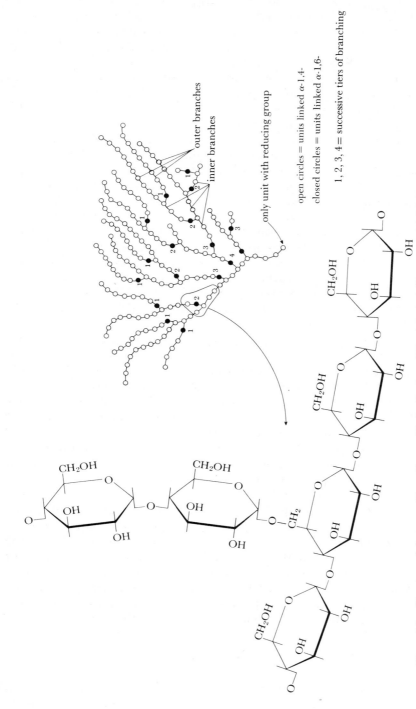

Figure 1-7. Structure of glycogen. (From Harper, H. A.: *Review of Physiological Chemistry*, 14th Ed., Lange Medical Publications, 1973.)

Samples of glycogen isolated by the more common methods (hot alkali, hot water–trichloracetic acid) provide a water-soluble product which for muscle glycogen generally shows a molecular weight of about 1×10^6 and for liver glycogen a molecular weight of about 5×10^6. However, more prolonged, careful extractions with cold water yield material with much higher molecular weight (150–200×10^6), sometimes combined with small amounts of protein. Other glycogen fractions are almost insoluble in water and have molecular weights up to 15–20×10^6. Not only are molecular weights higher after more gentle extraction but they show more specificity for given species, tissue, and physiological state. The metabolic activity of these super-molecules may be different from that of more common low-molecular-weight glycogens. Some extractions with cold water or dimethyl sulfoxide have provided glycogen samples with rather narrow molecular-weight ranges and therefore perhaps more distinct chemical as well as physiological entities.

Thus efforts are increasing to obtain glycogen fractions in relatively undegraded and pure condition and to distinguish molecules from aggregates of molecules. Electron micrography has recently aided such distinctions and suggested that some of the "giant" glycogens are indeed molecules rather than aggregates. Still, the effect of mild heating and lowered pH to cause changes to components of lower molecular weight suggests that other than covalent bonds are split.

Various data indicate that in the natural state glycogen molecules are hydrated to the extent of approximately 2 g. of water per g. of polysaccharide. This is one factor which makes such a "soggy" molecule a relatively burdensome form of storage of energy compared with fat. The other major factor is the thermodynamic yield of only 4 calories per g. of carbohydrate upon complete combustion compared with 9 calories per g. for fat. The combined factors make fat approximately 6 times more compact than carbohydrate for energy yield per unit of weight. The advantages that accrue to mobile living species (mammals in general and carnivores in particular) by the preferential storage of fat are obvious. Only in liver, where the percentage of glycogen is normally 3 to 5 per cent of wet weight, is there any sizable store of this carbohydrate. Owing to the occurrence of glucose-6-phosphatase (Chaps. 4 and 5), glycogen in liver and to a lesser extent in kidney is available to supply the blood with glucose. Rather than bulk energy storage the purpose of glycogen storage evidently is to maintain a crucial supply of glucose to the blood, brain, liver, and perhaps other tissues in the late postprandial period and in short-fasting periods when the mechanisms for gluconeogenesis are not fully mobilized to cope with the fasting condition.

Heteroglycans

In the heteroglycans or "glycosoaminoglycans" (including mucopolysaccharides), which are typically associated with more or less protein, one finds a fascinating menagerie of types of monosaccharides combined together. The biochemical variety reflects the versatility of functions which can be ascribed to these carbohydrate-protein complexes. Chemically there is a potentiality for polymorphism and microheterogeneity in the heteroglycans which is far greater than for proteins. That nature takes advantage of this fact is becoming more and more clear as biochemists unravel the polysaccharide chains to define such details as the sequence of particular monosaccharides, carbon positions of linkages among them, extent and regularity of branching, extent and position of acetylation and sulfation of amino and hydroxyl groups, and the types of linkage of polysaccharide chains to the protein backbones.

The chemical analysis of the complex polysaccharides is fraught with the hazard of producing artifacts. Many of the glycosidic linkages are quite labile to both acid and alkaline hydrolysis, which provide a means of linkage analysis, but also may lead to a puzzling assortment of degradation products. New applications of techniques of mass spectrometry and nuclear magnetic resonance spectroscopy give promise of aid in this direction. Probably the preparation of physiological entities can be ascertained only by the use of specific and sensitive biological tests, such as blood-group specificity in the case of the class of heteroglycans known as "blood-group substances."

Other heteroglycans may be involved in antigen-antibody specificity, as, for instance, the type-specific polysaccharides in the walls of bacteria. The effectiveness of heteroglycans as specific haptens, if not whole antigens, may be useful for the recognition, analysis, and classification of heteroglycans in mammalian tissues. (Consider, for comparison, the recent boon to endocrinology and medicine of the immunoassay of hormones and other proteins). It has been suggested that antibodies formed against the bacterial polysaccharide antigens may have properties of cross-reaction with the tissue mucopolysaccharides. This mechanism could account for some infectious manifestations in tissues containing mucopolysaccharides and possibly also some so-called "autoimmune" diseases.

Before the chemical content of carbohydrates in heteroglycans began to be elucidated there were classifications based on physical properties. Thus the true "mucins" in the secretions of the submaxillary gland and mucosa of the respiratory tract and other cavities were distinguished by their solubilities in acid, precipitability, and other properties from the "mucoids" isolated from ovarian cysts, ascites fluid, cartilage, cornea, egg white, and so forth. Subsequent chemical investigations led to separations partly on the basis of constituents and characteristics of polysaccharides and partly by the extent and firmness of binding to protein. Thereby two main groups, the "mucopolysaccharides" and the "mucoproteins" or "glycoproteins," were recognized. As indicated by the general summary of classifications in Table 1–1, investigators in the field

TABLE 1-1. *Classification of Heteroglycans*

GENERIC GROUP	GENERAL STRUCTURAL CHARACTERISTICS	TYPES AND SOURCES	INDIVIDUAL MEMBERS OR EXAMPLES	CONSTITUENT MONOSACCHARIDE UNITS*
Mucopolysaccharides (Meyer, Stacey) or polysaccharide-protein complexes (Jeanloz) or heteropolysaccharides (Gottschalk)	Relatively low protein content Easily split covalent or electrostatic linkages of carbohydrate to protein High degree of polymerization of carbohydrates Mostly linear structures with small repeating units Glycosidic linkages mostly of one kind in a given polysaccharide	Connective and collagenous tissues and mucus secretions	Hyaluronic acid Chondroitin-4-sulfate (A) Chondroitin-6-sulfate (C) Chondroitin Dermatan sulfate (B) Keratan sulfate	GlA, NAcGl GlA, NAcOSGal Gal, Xyl IdA, NAcOSGal (OS)Gal, NAcOSGl, Fu, NAcGal, SA
		Liver, lung, muscle, etc.	Heparin Heparan sulfate	OSGlA, NSOSGl (OS)GlA, NAc or NS(OS)Gl
Glycoproteins — including glycopolypeptides	Relatively high protein content Covalent linkages firmly bound Low polymers — usually multiple oligosaccharide prosthetic groups Often-branched structures without repeating units but with preferred sequences Variable types of linkages and greater variety of sugars in one polysaccharide	Blood-group substances	A,B,O(H),Le^a,Le^b	Gal, NAcGl, NAcGal, Fu, NANA
		Salivary gland glycoproteins	OSM, BSM, PSM, BSL, CSM **	NAcGl, NAcGal, Gal, Fu, SA (4 types)
		Plasma glycoproteins	α_1-acid glycoprotein Fetuin Zn- and Ba-α_2 glycoproteins Haptoglobin β_1-globulins Transferrin γ_1-Macroglobulins Immunoglobulins Prothrombin Fibrinogen	NAcGl, Gal, Man, Fu, SA (3 types) NAcGl, NAcGal, Gal, Man, SA NAcGl, Gal, Man, Fu, SA NAGl, Gal, Man, Gl, Fu, NANA NHex, Hex, Fu, SA NAcGl, Gal, Man, Fu, SA NHex, Hex, Fu, SA NAcGl, Gal, Man, Fu, SA NHex, Hex NAcGl, Gal, Man, SA

Urinary glycoproteins		T and H glycoprotein	NAcGl, NAcGal, Gal, Fu, NANA
	Hormones	Thyroglobulin	NAcGl, Gal, Man, Fu, SA
		Gonadotropins – ICSH, FSH, HCG, PMSG**	NAcGl, NAcGal, Gal, Man, Gl, Fu, SA
		TSH**	NAcGl, NAcGal, Man, Fu
		Erythropoietin	NAcGl, Gal, Man, SA
	Enzymes	Ribonuclease	NAcGl, NAcGal, Man, Gl, Gal, Fu, SA
		Ceruloplasmin	NHex, Gal, Man, Fu, SA
		Cholinesterase	SA, other
		γ-Glutamyl transpeptidase	SA, other
		Atropinesterase	SA, other
	Gastrointestinal tract	Intrinsic factor	NHex, Hex, Fu, SA
		Mucus secretions	NHex, Hex, Fu, SA
	Respiratory tract	Bronchial secretions	NHex, Hex, Fu, SA
	Female genital tract	Cervical mucus	NAcGl, NAcGal, Gal, Fu, SA
	Connective and collagenous tissues	Bone, cartilage, synovial fluid	NHex, Gal, Man, Gl, Fu, SA
		Arterial tissue	NAcGl, NAcGal, Gal, Man, Gl, Fu, SA
		Skin, tendon, reticulin	NAcGl, Gal, Man, Gl, NANA
		Vitreous body, cornea	NAcGl, NAcGal, Gal, Man, Gl, F, SA

*Abbreviations used: GlA=D-glucuronic acid, NAcGL=N-acetyl-D-glucosamine, NAcOSGl=N-acetyl,O-sulfyl-D-galactosamine, Gal=D-galactose, Xyl=D-xylulose, IdA=L-iduronic acid, NAcOSGl=N-acetyl, O-sulfyl-D-glucosamine, NAcGal=N-acetyl-D-galactosamine, Fu=L-fucose, SA=sialic acid, OSGlA=O-sulfyl-D-glucuronic acid, NSOSGl=N-sulfyl, O-sulfyl-D-glucosamine, NANA=N-acetylneuraminic acid, Man=D-mannose, Gl=D-glucose, NHex=hexosamine, Hex=hexose

**See text for full names.

Heteropolysaccharides

It was earlier supposed that all the "mucopolysaccharides" or heteropolysaccharides, if associated at all with protein, were combined loosely through electrostatic linkages. Later evidence indicated, however, that some of the chondroitin sulfate and heparin complexes are bound covalently by O-glycosidic linkages to the hydroxyl groups of serine or threonine. These glycosidic bonds are readily split by alkali. Thus, a characteristic of heteropolysaccharides is ease of separation from protein, but the binding may be covalent or electrostatic. In this group protein is usually a very minor component of the carbohydrate-protein complex, though probably not even hyaluronic acid is ever free of attachment. Also typical of this group is the high degree of polymerization. In hyaluronic acid several thousand monosaccharide residues are linked together, and there are about 160 in chondroitin sulfate of bovine nasal cartilage. In heteropolysaccharides the chains are for the most part linear and the carbon positions of linkages are relatively unvaried. The kinds of monosaccharides are few and usually there is a regular repetition in a chain of a small (usually disaccharide) unit.

In mammals all the heteropolysaccharides (except for keratan sulfate) are highly acidic owing to the charged carboxyl groups of uronic acids and the sulfate groups. They strongly bind cations, and the particular affinity of chondroitin sulfate for calcium has suggested a role for the polysaccharide in mineral deposition in bone and cartilage. The loose areolar stroma of subcutaneous tissues, fibrous ligaments, vascular walls, collagenous tissue in bone, cartilage, and skin, synovial sac and fluid of joint surfaces are among those structures grouped as connective tissues, all of which typically contain various heteropolysaccharides.

Hyaluronic Acid

Hyaluronic acid was first isolated from bovine vitreous humor. Its name derives from hyaloid (vitreous) + uronic acid. The basic chemical unit is the disaccharide called "acetylhyalobiuronic acid," which is composed of 3-0-β-D-glucuronyl-1,4-N-acetyl-D-glucosamine (Fig. 1-8). The linkages are thus alternating β-1,4- and β-1,3-. The repeating unit is strung in very long chains composing highly asymmetric molecules. A wide spectrum of molecular weights ranging from 5×10^4 to 3×10^7 has been found, differing with the source, the technique of preparation, and

CHEMISTRY OF CARBOHYDRATES

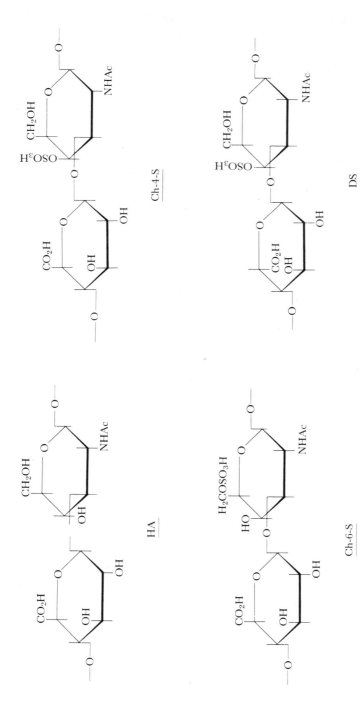

Figure 1-8. Repeating disaccharide units of hyaluronic acid (HA), chondroitin-4-sulfate (Ch-4-S), chondroitin-6-sulfate (Ch-6-S), and dermatan sulfate (DS).

the method of determination. Hyaluronic acid obtained from vitreous body has a molecular weight markedly lower than that isolated from synovial fluid, umbilical cord, skin, or rooster comb. Hyaluronic acid has been found also in cornea, bone, ligamentum nuchae, aorta and heart valve, tumor fluids, joint fluids, and fibroblasts. It may represent up to 5 per cent of the dry weight in some tissues. Those tissues with the highest concentrations include umbilical cord, vitreous humor, joint fluids, and skin. A component of hyaluronic acid in ox synovial fluid exists as a complex with a protein which comprises 25 to 30 per cent of the total complex. On the other hand, hyaluronic acid in a seemingly natural state and with less than 1 per cent protein content has been isolated also from ox synovial fluid.

Among heteropolysaccharides the most extensive shape studies have been carried out on hyaluronic acid. Light scattering and other measurements indicate that in physiological salt solution, hyaluronic acid has a somewhat stiff random coil structure. The latter may be flexible, however, depending on ionic strength, and so forth. If there are cross-links in the molecule (which might occur, for example, in ester bonds between carboxyl groups of glucuronic acid and hydroxyl groups of neighboring residues), then a cage structure may be inferred.

The radius of gyration of the chain molecule is very large, so that the chain extends over a solvent domain which is 10^3 to 10^4 times larger than the volume of the polysaccharide chain itself. Only small amounts of water appear to be actually bound to the chains of hyaluronic acid, but water is nevertheless "immobilized" to a certain extent in the polysaccharide domain owing to osmotic as well as frictional forces. Even at very low concentrations the chains of hyaluronic acid molecules begin to entangle according to measurements such as viscosity, sedimentation, equilibrium dialysis, and gel filtration. It is the properties of the resultant network which are physiologically important. Large molecules, such as proteins, diffuse slowly through such a network and osmotic pressure of dissolved proteins (e.g., albumin) is enhanced. Detailed studies of the lubricative properties of hyaluronic acid have been made.

Chondroitin-4-sulfate

A major constituent of cartilage (40 per cent of the dry weight of pig nasal septa) is chondroitin-4-sulfate, formerly designated chondroitin sulfate A. The contained sugars are mainly D-glucuronic acid and N-acetyl-D-galactosamine alternating in a repeating disaccharide unit, chondrosin, which was isolated early (1914), though its structure was not established until 40 years later. The structure of chondrosin (sulfate) is given in Figure 1-8. Glucuronic acid is linked β-1,3- to the galactosamine, while the latter is linked β-1,4- to glucuronic acid. The sulfate is on the galactosamine residue. There are smaller amounts of two other monosaccharides in the region of linkage to the protein. Xylose is in O-glycosidic attachment to a serine component of the peptide; two galactose units are linked serially to xylose with the outer galactose attached to glucuronic acid. As much as 20 per cent of the polysaccharide chains of chondroitin-4-sulfate may be linked via amide bonds to the protein rather than via serine.

Besides being in cartilage, chondroitin-4- sulfate is found in skin, bone, cornea, sclera, umbilical cord, ligamentum nuchae, aorta, urine, and chondrosarcoma. It is abundant in granulation tissue. The viscosity of chondroitin-4-sulfate is lower than that of hyaluronic acid, but the high dissociation of the sulfate groups gives it stronger ion-binding capacities than for hyaluronic acid. Presumably the chondroitin sulfates play a role similar to that of hyaluronic acid in connective tissues.

Measurement of light scattering and osmotic pressure indicates average molecular weights for various chondroitin sulfate–protein complexes of about 1 to 5×10^6. There is a protein core (about 25 per cent of the total weight) in a rod of length about 4,000 Å along which are constituted about 25 to 50 polysaccharide chains with molecular weights of about 3 to 5×10^4.

Chondroitin-6-sulfate

This compound (also known as chondroitin sulfate C) differs from chondroitin-4-sulfate only in having its sulfate located at C-6 of galactosamine (Fig. 1-8). Its molecular size and shape, viscosity, and electrophoretic mobility are practically identical to those of the 4-sulfate, but the optical rotation is slightly lower. The two chondroitin sulfates are usually isolated together from cartilage, bone, skin, tendon, umbilical cord, and heart valve and are separated with difficulty. Chondroitin-6-sulfate from saliva and from chordoma appears to be a pure polysaccharide.

Chondroitin

A product isolated from cornea appears to consist of the same repeating disaccharide unit of chondrosin as found in the chondroitin sulfates, but it has a very small content of sulfate. It is degraded by hyaluronidase, as are desulfated chondroitin sulfates.

Dermatan Sulfate

This heteropolysaccharide, formerly designated chondroitin sulfate B, differs in chemical structure from the chondroitin sulfates by containing, instead of D-glucuronic acid, the C-5 epimer of the latter, L-iduronic acid (Fig. 1-8). The sulfate is typically at C-4 of galactosamine. However, hypersulfation (additional sulfyl groups at C-2 or C-3) is common. Strongly negative optical rotation suggests that the anomery at C-1 of the L-iduronic acid is α, rather than β as depicted in Figure 1-8. Solubility, viscosity, and electrophoretic mobility are similar to the same properties of the chondroitin sulfates, but the molecular weight appears to be about half of the latter.

Dermatan sulfate is associated in connective tissue of skin, tendons, heart valve, and aorta with the other glycosaminoglycuronoglycans. The amount of dermatan sulfate in tissues relative to hyaluronic acid appears to increase with age. In Hurler's syndrome (gargoylism), a systemic disease of connective tissue in humans, there are large intracellular accumulations of dermatan sulfate, particularly in the liver, and the compound is excreted in the urine. In dermatan sulfate there are clusters along the polysaccharide chain of disaccharide units containing D-glucuronic acid instead of L-iduronic acid; such a form is a sort of copolymer of dermatan sulfate and chondroitin sulfates. The region of linkage to protein in dermatan sulfates is similar structurally to that for chondroitin sulfates. Dermatan sulfate appears to be identical with β-heparin, a by-product of preparation of heparin from beef lung with some anti-coagulant properties.

Keratan Sulfate

This polysaccharide was first isolated from cornea, hence its name. It was later found in nucleus pulposus and cartilage. In cartilage it may be linked to the same protein core as is chondroitin sulfate. Its tissue source and the presence of O-sulfyl groups seem to place it in a group with other heteropolysaccharides, as does the fact that it has a repeating disaccharide unit. However, the composition of this disaccharide and other chemical characteristics seem to resemble the carbohydrate moieties of glycoproteins. The disaccharide is composed of D-galactose and N-acetyl-D-glucosamine, and, besides, there are significant amounts of galactosamine, xylose, fucose, and sialic acid. The polysaccharide seems to be highly branched—this is a characteristic of glycoproteins. Keratan sulfate is cleaved by enzymes which degrade blood-group substances, but not by hyaluronidase or chondrosulfatases, and it cross-reacts with anti-blood-group sera after desulfation.

The majority of sulfate groups are positioned at C-6 of the hexosamine; however, some sulfates may be found also on the C-6 of galactose. The extent of the latter in keratan sulfate is said to increase with age. Also the proportion of keratan sulfate to chondroitin sulfates in cartilage generally increases with age. Keratan sulfate contains β-1,4-galactosidic and β-1,3- glucosaminidic linkages. Keratan sulfate of cornea displays glycosylamine linkage to asparagine or glutamine, whereas that of cartilage contains the polysaccharide unit linked to the hydroxy group of serine or threonine.

Heparin

Notable among the heteropolysaccharides for its anticoagulant properties, heparin, as indicated by its name, was first discovered in the liver. It is abundant in liver, lung and muscle (0.6 g. per kg. of beef muscle) and present in lesser amounts in heart, kidneys, thymus, spleen, blood, stomach and intestinal wall, and connective tissues in general. In fact, the ubiquitous mast cells in these various organs appear to be the sites of production, storage, and secretion of heparin.

The disaccharide repeating unit of heparin is composed of glucosamine and glucuronic acid with occasional presence of iduronic acid. Not only are hydroxyls of both types of monosaccharide sulfated, but so also are almost all the amino groups of glucosamine. Paradoxically, although sulfate is not added to its name, heparin is much more heavily sulfated than the chondroitin sulfates or heparan sulfate. Approximately seven out of eight glucosamine residues are N-sulfated and the remainder have free amino groups. To a similar extent the glucuronic acid residues are sulfated at C-2 and the glucosamine at C-3 or C-6. There may be a variability of heparin samples depending on the degree of N- and O- sulfation, but whether this is true biological variation or the result of desulfated artifacts arising during preparation is not clear. Most heparin contains few acetyl groups, which if present may be found near the protein linkage region. The glycosidic linkages are α-1,4- for both the glucuronidic and the glucosaminidic units. However, a small proportion of -1,6- linkages to the glucosamine units is present, which suggests some branching. Thus, in the positioning of glucuronidic linkages heparin differs from other polysaccharides containing glucuronic acid.

The molecular weight of various preparations of heparin lies between 15,000 and 20,000. It has a lower intrinsic viscosity than any of the other heteropolysaccharides composed of the same sugars. Heparin readily combines with basic substances such as histamine and protamines and with proteins at their isoelectric point. The resultant change at a tissue site, e.g., blood vessel wall, toward electronegativity is now thought to be significant in the inhibition of blood coagulation and possibly in the prevention of thrombus formation. It may be that this electro-negativity is an aspect of another important physiological effect of heparin, i.e., on the transport and metabolism of circulating triglycerides. Heparin has long been identified as the "clearing factor" of blood which acts as a kind of coenzyme or facilitating agent for the action of lipoprotein lipase.

Heparan Sulfate

This heteropolysaccharide (also called heparitin sulfate and heparin monosulfate) derives its name from the fact that it, like β-heparin, was found as a side-fraction in the isolation of heparin from liver and lung.

It does not have the anti-coagulant properties of heparin, but it resembles the latter in having D-glucosamine and D-glucuronic acid as constituent monosaccharides (except for a small amount of iduronic acid) and in possessing N-sulfyl and O-sulfyl groups. However, heparan sulfate is not as highly sulfated as is heparin and a larger proportion of its amino groups are acetylated. Heparan sulfate may indeed be not one but a family of substances with varying degrees of N- and O-sulfation. There are -1,4- hexosaminidic linkages as in other glycosylaminoglycuronoglycans, but the uronidic linkages are not -1,3- as in most heteropolysaccharides. It has been proposed that the glucuronic acid residues are linked -1,6- to the D-glucosamine units. There is various evidence for a branched structure. There is a xylosyl-serine bond in the linkage of glycan to proteins.

The molecular weights of some of the heparan sulfates are small; values between 1240 and 2075 have been reported in the case of samples isolated from the livers of patients with Hurler's syndrome; such livers are a rich source of this polysaccharide. The lower molecular weight may be peculiar to the abnormal Hurler's type.

Glycoproteins

Glycoproteins, in contrast to the polysaccharide-protein complexes (heteropolysaccharides), contain a relatively high amount of protein to which multiple oligosaccharide units are relatively firmly attached as "prosthetic groups." The biochemical properties and functions of the carbohydrate moieties are therefore closely connected with those of the protein portions. The latter have a variety of biological functions in which no common denominator seems to correspond to the carbohydrate content (but see Chap. 6). In general, it would appear that the carbohydrate moiety in all cases captures and entrains water in the vicinity of the glycoprotein, which may be important in the solubility and the movement of other small molecules.

There is a greater variety of monosaccharides in the glycoproteins than in the heteropolysaccharides. The same hexosamines (D-glucosamine and D-galactosamine) are found and they are almost always acetylated at the amino group. Galactose and mannose are the two hexoses most commonly found; glucose is occasionally present. Also, the special hexose, L-fucose (6-deoxy-L-galactose), is contained in glycoproteins, usually as a terminal unit on the non-reducing (free) end of a chain or side-chain. The unique nine-carbon compound, sialic acid, is yet another type of sugar present in glycoproteins but not heteropolysaccharides.

The carboxyl group of sialic acid (usually N-acetyl-neuraminic acid or NANA) is generally free and the NANA located at the terminal non-reducing end of an oligosaccharide of a glycoprotein. These two facts would predict important biological properties of sialic acid and this is indeed supported by certain observations of the effect of cleaving sialic acid by dilute acid or enzymes from various glycoproteins. From such studies it now appears that sialic acid contained in blood-group substances, submaxillary gland protein, urinary mucoprotein, and glycolipids is vital for the property common to all these substances — inhibition of hemagglutination by influenza virus. This plus other evidence has suggested that attachment of the virus to the red cell wall depends on a configuration in the virus complementary to and specific for neuraminic acid. The virus structure has an activity which has been termed "neuraminidase," since upon its action NANA is released without the appearance of any other sugar. Such findings provide important indications as to the possible mechanisms of influenza (and perhaps other) infections and relevant functions of some glycoproteins. From similar studies the presence of sialic acid appears to be necessary for the viscosity of salivary glycoproteins, the biological activity of gonadotropins and other hormones, and the proper function of Castle's intrinsic factor (see below).

Blood-Group Substances

The carbohydrate nature of the blood-group substances was recognized early and has been very extensively studied. Because carbohydrate constitutes 80 to 90 per cent of the molecule and the polypeptide moiety is relatively small (11 to 12 amino acids), earlier classifications identified blood-group substances as neutral mucopolysaccharides. However, subsequent knowledge of the monosaccharide constituents and of the branching structure of multiple oligosaccharide chains has led to their placement in the category of glycoproteins. More accurately they may be termed glycopeptides.

A clue to the discovery of the carbohydrate content of the A, B, and O (H) blood-group substances was the finding that these substances are secreted in a water-soluble form in saliva, gastric juice, semen, and other secretions. Hitherto they had been known to occur only in an insoluble form at the surface of red blood cells, where they constitute some of the agglutinogens responsible for clumping of red cells in incompatible serum of a different blood-group type. Blood-group substances may be found in high concentration in ovarian cyst fluid and indeed to some extent in a large variety of secretions, fluids, and tissues with the exception of nerve tissue, epithelium, skin appendages, bone and cartilage. However, only 75 per cent of persons of A, B, or O (H) blood-group types are secretors of A, B, or O (H) substances; the other 25 per cent secrete a different (Le[a]) specific substance.

The sugars contained in the blood-group substances are D-galactose, N-acetyl-D-glucosamine, N-acetyl-D-galactosamine, L-fucose, and

NANA. These are arranged in relatively short multiple branched carbohydrate chains attached at frequent intervals to a peptide backbone probably by linkage to serine or threonine residues. Sequence and linkage analysis of the carbohydrate chains has indicated a pattern and evolution of structure which can be correlated to a considerable extent with blood-group specificity (Fig. 1-9). Structures specific for B substance (not in Fig. 1-9) contain terminal units of α-D-galactose instead of those of α-N-acetyl-D-galactosamine found in A substance. The locations of sialic acid (NANA) units are not definitely known, although some evidence suggests that these may be attached to terminal β-galactosidic or to subterminal N-acetyl-glucosaminidic units of the carbohydrate chains of Le^a substance. From the existence of the chemical specificities, as well as other evidence, e.g., inhibition of hemagglutination or precipitation by simple sugars, it can be inferred that it is the sugar sequence, stereochemistry, and position of the glycosidic linkages at the non-reducing end of the carbohydrate chains which determine the high degree of immunological specificity of the molecules. Antigenicity of the ABH and Lewis type substances does not depend on sialic acid, but that of M and N blood-group antigens does.

Glycoproteins that carry blood-group specificity may have protective and lubricative functions as components of mucinous secretions. However, the physiological differences among the different blood-group substances remain obscure. There are no pathological conditions specifically associated with each, but there are some differences in predilection to certain diseases, e.g., to duodenal ulcer and toxemia of pregnancy in individuals of group O blood and to pernicious anemia and gastric carcinoma for individuals of group A. These clinical observations suggest that physiological differences will eventually be found.

Salivary Gland Glycoproteins

The "mucin clot" from the secretion of submaxillary glands was early recognized to contain a glycoprotein. It has been intensively studied and characterized, particularly since interest in this glycoprotein was heightened by discovery of its very strong inhibition of influenza virus hemagglutination. This, as discussed above, led to recognition of its content of sialic acid which is present at the non-reducing end of a disaccharide unit and linked to an N-acetyl-D-hexosamine by an α-2,6-glycosidic bond. About 800 of these disaccharide units are distributed along the polypeptide chain. The entire molecule has a molecular weight of about 1×10^6 and a high intrinsic viscosity. The latter property largely accounts for the viscous nature of saliva—particularly in ruminants, in which the submaxillary glands are the main mucus-secreting salivary glands. The property of viscosity in aqueous solution resides largely with the sialic acid because the mutual repulsion of the carboxyl groups stiffens the random-coil shape of the molecule.

CHEMISTRY OF CARBOHYDRATES

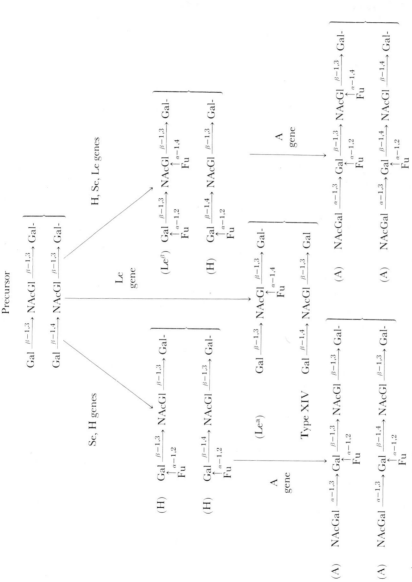

Figure 1-9. Structure of oligosaccharide units in blood-group specific substances (abbreviations as in Fig. 1-6). (From Watkins, W. A.: in *Glycoproteins, Their Composition, Structure and Function*, Elsevier Publishing Co., 1966).

Ovine submaxillary gland glycoprotein (OSM) may have part of its disaccharide units linked to protein through a glycosidic-ester linkage between C-1 of acetylgalactosamine and the free carboxyl groups of aspartyl and glutamyl residues. Other disaccharides are bound glycosidically to the β-carbon atoms of serine and threonine. The relative proportion of the latter type of linkage may be greater in mature sheep than in lambs. Bovine submaxillary gland glycoprotein (BSM) has the majority of its carbohydrate linked to serine or threonine.

The sialic acid of OSM appears to be entirely NANA. In BSM there are several sialic acids, i.e., all three of the more common ones (NANA, O-acetylneuraminic acid, and N-glycolylneuraminic acid) plus N-acetyl-O-diacetyl-neuraminic acid (one acetyl at C-7, the other at C-8 or C-9). Also a small amount of glucosamine is present in addition to galactosamine and there are trace amounts of galactose and fucose. Some other salivary gland glycoproteins have been less thoroughly investigated. Porcine submaxillary gland glycoprotein (PSM) and those of bovine sublingual gland (BSL) and canine submaxillary gland (CSM) all contain more galactose and fucose and less sialic acid than OSM and BSM. BSL and CSM contain as much acetylglucosamine as acetylgalactosamine. The sialic acid of PSM is almost exclusively N-glycolylneuraminic acid.

Plasma Glycoproteins

In plasma there is a great multiplicity of glycoproteins. Certain fractions have been identified with the major electrophoretic components, but the process of subfractionation and separation of individual glycoproteins by paper and gel electrophoresis, immunoelectrophoresis, and so forth is a continuing effort which should help eventually to clarify the physiological or pathological significance of many serum glycoproteins. An obvious spur to this effort is the fact that in various diseases all of the known components may rise in concentration and there may be significantly altered ratios of them. The α_2- globulin fraction is very flexible, rising particularly in acute systemic reactions. It has been suggested that the α_2- globulin fraction has something to do with detoxification of products in the serum. Such a function for the sialic acid in glycoproteins (and for the glucuronic acid and sulfuric acid in heteropolysaccharides) could be analogous to the formation of glucuronate and sulfate complexes with noxious agents in the liver.

Normally bound to the total (alcohol-precipitable) proteins in 100 ml. of plasma are about 120 mg. of hexose, 80 mg. of hexosamine, 60 mg. of sialic acid, and 90 mg. of fucose. The hexose portion consists of approximately half mannose and half galactose. There are no uronic acids present. Among the electrophoretic components the α-globulins are particularly rich in carbohydrate. However, the β- and γ-globulins also contain appreciable amounts. The various types of monosaccharides are found in similar proportions among the electrophoretic globulins. However, γ-globulin is particularly rich in fucose. Also, in γ-globulin the molar ratio of mannose to galactose is about 3 : 1 in contrast to the usual 1 : 1 ratio in serum glycoproteins, and the content of hexosamine is somewhat lower than for other glycoproteins. It is said that the fucose:hexose ratio in various globulin fractions can be quite an individual characteristic in healthy persons as well as in those with certain diseases of the reticuloendothelial system.

Certain fairly well-characterized glycoproteins of plasma are listed in Table 1–1 and are discussed below. Some other glycoproteins of hormone or enzyme nature are also found circulating in plasma, but these are listed and discussed separately.

α_1- Acid-Glycoprotein.   About 10 per cent of the total hexose in glycoproteins is contained in a fraction called "seromucoid," which is precipitated after removal of the large bulk of serum proteins. The major component of seromucoid in a more purified fraction is known as α_1-acid-glycoprotein. A very similar, perhaps identical, glycoprotein isolated by a different technique has been known as "orosomucoid." Refinements of technique of separation continue to demonstrate subfractions or components of α_1-acid-glycoproteins. Six fractions identified by cellulose chromatography show successively increasing ratios of sialic acid to glucosamine and of fucose or galactose to mannose.

Human plasma has proved to be a rich source of α_1-acid-glycoproteins. Lesser amounts (in order of decreasing concentration) are found in plasma of ox, rabbit, horse, guinea pig, and rat. From the urine of nephrotic patients large amounts may be obtained. Outstanding characteristics of this protein are its high solubility and its heat stability. It has a very low isoelectric point, which increases upon removal of the sialic acid by neuraminidase. Because its strong anodal mobility (at low pH) is considerably more than that of albumin, this fraction and certain changes of it in disease are readily demonstrated in the electrophoretic pattern.

α_1-acid-glycoprotein is composed of about 40 per cent carbohydrate and has a rather low molecular weight (40 to 45 × 10^3) compared with other plasma proteins. For each molecule it is estimated there are about 20 residues of galactose, 20 of mannose, 30 of N-acetyl-glucosamine, 16 to 17 of sialic acid, and three to four of fucose. However, the various monosaccharides are found in varying proportion, depending on the source of α_1-acid-glycoprotein. For instance, whereas mannose and galactose are each found in amounts of about 6 to 8 per cent in human, bovine, and equine plasma, in rat plasma and Yoshida ascites fluid the α_1-acid-glycoprotein contains 3 per cent or less of total hexose, although the sialic acid concentrations (11 to 13 per cent) are similar to those in other species. All three main sialic acids can be found—in human plasma NANA and possibly N,O-diacetylneuraminic acid, and in bovine and horse plasma NANA and N-glycolylneuraminic acid. Electrophoretic polymorphism may be due to positional isomerism in attachment of sialic acid to galactose.

There appear to be several (5 to 8) polysaccharide units or chains each attached by one linkage to the protein core and each consisting of 8 to 15 monosaccharide units. Detailed investigation by enzymatic cleavage, periodate oxidation, and so forth has suggested that a prevailing monosaccharide sequence is sialylgalactosyl-glucosaminoylmannose. A branched octasaccharide unit, the sequential structure of which is known, has been obtained by graded acid hydrolysis. Some details of linkage characteristics are known, but others are still in the process of elucidation. The linkage of carbohydrate to protein has long been known to be very resistant to alkaline hydrolysis, and the linkage is probably a glycosidic one from C-1 of N-acetylglucosamine to the β-amide group of asparagine.

FETUIN. Fetuin is an α-globulin found in high concentration in fetal calf serum. It contains about 20–30 per cent carbohydrate, which includes 4 to 5 per cent galactose, 2 to 3 per cent mannose, 4 to 6 per cent hexosamine (largely N-acetylglucosamine with a trace of N-acetylgalactosamine) and 6 to 9 per cent sialic acid (mostly NANA, but also a small amount of N-glycolylneuraminic acid). As with glycoproteins generally, the sialic acid is in a terminal position and its free carboxyl group is largely responsible for the low isoelectric point of the protein. Various chemical degradation studies suggest that a trisaccharide unit, sialyl-2,3-galactosyl-1, 4-N-acetylglucosamine, occurs as a terminal branched structure on an inner core consisting of residues of mannose and N-acetylhexosamine. Some of the hexosamine in this glycoprotein is not acetylated. Fetuin (or a closely related glycoprotein associated with it) has the interesting biological property of promoting growth when added to HeLa cells in tissue culture.

ZN-α_2-GLYCOPROTEIN AND βA-α_2-GLYCOPROTEIN. In the α_2-globulin fraction of plasma proteins there are a homogeneous glycoprotein precipitable by Zn ions and two glycoproteins precipitable by Ba ions. The molecular weights are, respectively, 41×10^3 and 49×10^3 for Zn-α_2-glycoprotein and βa-α_2-glycoproteins. Both of these types contain N-acetylglucosamine, galactose, mannose, fucose, and sialic acid with slightly different proportions of these sugars but a total carbohydrate content of about 16 to 18 per cent for each.

HAPTOGLOBIN. A group of hemoglobin-binding proteins in serum, which migrate electrophoretically in the α_2-globulin region and have a molecular weight of about 85×10^3, contain about 20 per cent carbohydrate (5 per cent NANA, 5.5 per cent glucosamine, 1 per cent fucose, about 0.5 per cent glucose, and 8.5 per cent galactose and mannose). Certain arrangements of monosaccharide sequence (with sialic acid terminal, as usual) are suggested by graded acid hydrolysis. Various haptoglobin components are genetically determined; it would be of interest to know whether these different components (or those of the genetic types of transferrin mentioned below) correspond to any differences in carbohydrate content or sequence.

β_1-GLOBULINS. In the β_1-globulin fraction of serum there are two glycoproteins, designated β_{1c}-globulin and β_{1a}-globulin, which are related to the serum complement complex. The carbohydrate content is very low, but includes hexosamine, hexose, fucose, and sialic acid.

TRANSFERRIN. A fraction of plasma proteins in the β_1-globulin electrophoretic component normally binds iron and is known as transferrin (siderophilin or β_1-metal-combining globulin). This glycoprotein is probably synthesized in the liver (like most others in plasma), and, as its name implies, has the function of transporting iron in the serum from sites of absorption and hemolysis to sites of hemoglobin synthesis. It may also serve as a secondary defense mechanism against infection. A further interesting property of transferrin is its molecular variation according to genetic, and particularly racial, types.

Transferrin has a molecular weight of about 9×10^4 and a carbohydrate content of about 5.5 per cent, including residues of galactose (eight), glucosamine (eight), mannose (four), sialic acid (four), and fucose (one). These appear to be divided into two branched-chain polysaccharides with terminal sialic acid units. The chains seem to be attached to asparagine in the protein structure.

γ_1-MACROGLOBULINS. In this fraction of normal human serum (as well as that of patients with macroglobulinemia) there is a mixture of glycoproteins with an average carbohydrate content of about 10 per cent, including about 4.5 per cent each of hexosamine and hexoses, 0.7 per cent fucose, and 1.5 per cent sialic acid. In the γ_1-macroglobulin of normal human serum, galactose and mannose occur in a ratio of about 1:2.

IMMUNOGLOBULINS. Immunoglobulins of mammalian serum include three main members, which may be designated IgG, IgM, and IgA, although other synonyms are used. Although all are glycoproteins, their antibody activity is not dependent on the carbohydrate moiety for fundamental reactivity of antigen with antibody, but various associated phenomena (e.g., complement fixation) could be functions of the carbohydrate content.

In the serum of all species examined, IgG constitutes about 90 per cent of the total immunoglobulin. Its molecular weight is about 15×10^4. In human IgG the hexose and hexosamine are each present in about 1 per cent concentration, while fucose and sialic acid each constitute 0.2 to 0.3 per cent. In various species, including human, IgM and IgA contain somewhat higher amounts of each of the carbohydrates. The ratio of galactose:mannose appears to vary among different species (man, horse, rabbit, ox). There may be two or three oligosaccharide units per molecule in immunoglobulins and each of them is highly branched. Glucosamine, fucose, and sialic acid are all included among non-reducing end groups in IgG. The linkage of oligosaccharide to protein is primarily via glycosidic attachment of glucosamine to the β-amide group of asparagine, but some serine-type linkages are found.

Of particular interest is the fact that pathological immunoglobulins show considerable variation in carbohydrate content. The urinary proteins related to immunoglobulins are well known as "Bence-Jones protein."

PROTHROMBIN. The proenzyme prothrombin contains 6.5 per cent hexose and 1.7 per cent hexosamine. Interest has centered on the partial release of free carbohydrate when prothrombin is converted to thrombin by certain agents, but the nature of the carbohydrate released is in dispute.

FIBRINOGEN. The substrate of thrombin in the plasma, i.e., fibrinogen, also contains carbohydrate attached to the protein (probably through asparagine). According to a molecular weight of 33×10^4 there are seven residues of galactose, 14 of mannose, 14 of hexosamine, and seven of sialic acid. Studies of glycopeptides obtained by enzymatic digestion suggest that six or seven oligosaccharides are present on a protein core. All contain galactose, mannose, and glucosamine in the same proportion; but the sialic acid content varies. A minor fraction of the carbohydrate is released to the medium when fibrinogen is converted to fibrin.

Urinary Glycoproteins

In normal human urine there are at least 20 different glycoproteins which have been fractionated from each other. They range in total carbohydrate content from 18 to 95 per cent, which is relatively high compared with most glycoproteins. Some of these urinary proteins can be identified with plasma proteins.

T AND H GLYCOPROTEIN. One urinary glycoprotein, which is called T and H glycoprotein for two of its investigators, Tamm and Horsfall, has been particularly well characterized. It comprises about 10 per cent of the urinary proteins. T and H glycoprotein is a long, fibrous molecule; with a relatively large molecular weight, 7×10^6, it appears to be a rod 5600 Å long and 42 Å wide. However, it is excreted in the urine largely in the form of tetramers and higher polymers of the fundamental molecule. The latter contain 5.4 per cent galactose, 2.7 per cent mannose, about 6 per cent glucosamine, 1 to 2 per cent galactosamine, 1 per cent fucose, and about 5 to 9 per cent sialic acid (NANA). Evidence from titration and enzymatic effects indicates that the sialic acid is on the terminal end of a polysaccharide with its carbonyl groups free. Dialysis studies of the isolated carbohydrate prosthetic group suggest that its molecular weight is about 12,000, thus composed of about 60 residues of monosaccharide units. If this is so, there would be about 120 prosthetic groups attached to the protein core.

T and H glycoprotein comes primarily from the renal pelvis or kidney itself, but it is too large to be filtered through the glomeruli. Human urinary casts are composed primarily of T and H glycoprotein. Other evidence also suggests that the glycoprotein is formed at the external layer of the cell membrane of the proximal convoluted tubule and is then cast off into the urine. It is further notable that the binding properties of T and H glycoprotein are claimed to be quite similar to those of cellular membranes in general. Although suspected of playing a role in renal lithiasis, the connection between kidney stones and T and H glycoprotein remains obscure. In patients with cystic fibrosis of the pancreas the glycoprotein molecules may be of abnormal size and show a tendency to aggregate.

Hormones

THYROGLOBULIN. The storage form of the thyroid hormone is a glycoprotein which contains about 10 per cent total carbohydrate divided among glucosamine (2.4 per cent), galactose (1.2 to 1.4 per cent), mannose (2.5 to 3.5 per cent), fucose (about 0.5 per cent) and sialic acid (about 1 to 2 per cent). This glycoprotein is distinctive for having two different types of carbohydrate chains according to study of glycopeptides obtained by enzymatic digestion. One type contains five residues of mannose and one of N-acetylglucosamine; the other contains all the sugar components of the original molecule and is said to resemble the carbohydrate groups of fetuin. Since the protein of thyroglobulin is presumed to be merely a carrier for the active principle, thyroxine, there is no apparent physiological significance of the carbohydrate moiety for hormonal activity.

INTERSTITIAL CELL-STIMULATING HORMONE (ICSH) AND FOLLICLE-STIMULATING HORMONE (FSH). These two pituitary gonadotropins, which have stimulating effects on the male and female gonads, contain galactose, mannose, glucosamine, galactosamine, fucose, and sialic acid. Per molecule of ICSH (also called LH, luteinizing hormone) there are four to five residues of glucosamine, two of galactosamine, four of mannose, and one each of galactose and fucose. Sialic acid occurs to a very minor extent. Porcine and human ICSH have somewhat different proportions of monosaccharide constituents, as do preparations of FSH. A significant distinction between ICSH and FSH is that the hormonal activity of ovine ICSH is not noticeably reduced by neuraminidase (cleaving off sialic acid), whereas that of ovine FSH is completely abolished. The molecular weights of various preparations of ICSH and FSH are relatively small (16 to 100×10^3).

HUMAN CHORIONIC GONADOTROPIN (HCG) AND PREGNANT MARE

Serum Gonadotropin (PMSG). Another glycoprotein hormone with a role in reproduction is HCG, which is of placental origin and excreted in urine in maximal amounts from the 60th to the 80th days of pregnancy. The constituents, galactose, mannose, and N-acetyl glucosamine, are present in slightly higher amounts than in ICSH and FSH. Galactosamine occurs to a minor extent. A more significant difference between HCG and the pituitary gonadotropins is the much higher sialic acid content in HCG. The isoelectric pH is therefore very low. The molecular weight of the glycoprotein (or glycopolypeptide) is about 30×10^3. Activity is destroyed by neuraminidase. The full structures of two types of branched oligosaccharide chains in HCG have been elucidated.

A related hormone, PMSG, is also a glycoprotein of similar molecular weight and content of carbohydrate. However, the galactose content is unusually high (13 g./100 g. dried glycoprotein) and there is some glucose present. The sialic acid content is also high.

Thyroid-Stimulating Hormone (TSH or Thyrotropin). Yet another piuitary hormone which is a glycoprotein is TSH. There is no galactose or sialic acid present, but there are 9 to 10 residues of mannose, six of glucosamine, three of galactosamine, and one of fucose per molecule. The molecular weight is similar to that of other pituitary glycoproteins.

Erythropoietin. A glycoprotein known as erythropoietin is evidently produced in the kidney, circulates in serum, stimulates erythropoiesis in the bone marrow, and fluctuates in serum concentration inversely with the hemoglobin mass. The content of hexose (galactose and mannose) is high—29 per cent by weight—and that of hexosamine (glucosamine) is also high—17.5 per cent. Sialic acid content is also considerable (13 per cent) and there is loss of biological activity when it is removed by action of neuraminidase or acid. The molecular weight is about 40×10^3.

Enyzmes

Ribonuclease. There are four ribonucleases (A, B, C, & D) secreted from bovine pancreas. Ribonuclease A is present in tenfold higher amount than ribonuclease B and contains very little carbohydrate. Ribonucleases B, C & D are glycoproteins which contain 5 to 10 per cent total reducing sugars, including glucosamine, galactosamine, mannose, glucose, galactose, fucose, and sialic acid. The carbohydrate chains appear to be linked to asparagine in the polypeptide moiety.

Ceruloplasmin. Ceruloplasmin is a copper-containing serum protein which has oxidase activity on certain polyamines, polyphenols, and ascorbic acid. It is reported to contain 2 per cent galactose, 1 per cent mannose, 0.2 per cent fucose, 1.9 per cent hexosamine, and 2 per cent sialic acid. There may be a second component of ceruloplasmin with only half as much carbohydrate. The molecular weight of the first (larger) component is about 15×10^4.

CHEMISTRY OF CARBOHYDRATES

OTHER ENZYMES. Other glycoprotein enzymes which are less well characterized, but which all contain sialic acid, are cholinesterase, γ-glutamyl transpeptidase, and rabbit serum atropinesterase. Treatment of these enzymes with neuraminidase, however, does not destroy their enzyme activity.

Gastrointestinal (GI) Tract

INTRINSIC FACTOR. A protein found in gastric mucosa has been identified as the "intrinsic factor" (IF) necessary to bind vitamin B_{12} and thus facilitate its absorption from the gastrointestinal tract. As a glycoprotein it contains 10 to 25 per cent carbohydrate (differing with the source). Besides hexoses, hexosamines and fucose, there is a significant amount of sialic acid, since neuraminidase inactivates B_{12}-free IF, though not the vitamin B_{12}-IF complex. An interesting finding is that if the released sialic acid is replaced by a strongly acid cation exchanger, the biological activity of the preparation is regained. Further evidence that the carbohydrate moiety is involved in biological activity is the fact that other glycoproteins of plasma, urine, and erythrocyte stroma have some capacity to bind vitamin B_{12}.

GLYCOPROTEINS OF GI MUCUS SECRETIONS. The secretions of the GI tract contain various heteropolysaccharides but also a large number of glycoproteins which have not been so well characterized as others with more evident biological activity. In gastric juice, intestinal juice, and rectal mucus there are found glycoproteins with a range of carbohydrate content as follows: 5 to 8 per cent hexosamine, 6 to 25 per cent hexose, 2 to 5 per cent fucose, and 1 to 7 per cent sialic acid. Mucin from the rectum contains a higher proportion of sialic acid and a lower proportion of fucose than that of the upper GI tract. Of special interest is the occurrence of a sulfated glycoprotein in sheep colonic mucin. There are also a few other examples of sulfated oligosaccharides in glycoproteins (corneal stroma, rat mammary gland) which, along with the anomalous make-up of keratan sulfate, represent heteroglycans which resemble both typical heteropolysaccharides and typical glycoproteins.

Respiratory Tract

In normal human bronchial secretion the average sugar composition of the total unprocessed glycoproteins was found to be 7.3 per cent acetylhexosamine, 5.9 per cent hexose, 4.9 per cent fucose, and 4.8 per cent sialic acid. Glycoproteins of pathological sputums are reported to contain slightly different proportions of these sugars.

Female Genital Tract

CERVICAL MUCUS. A glycoprotein of the bovine cervical mucus contains glucosamine, galactosamine, galactose, fucose, and sialic acid.

The total hexosamine content (26 to 28 per cent) and that of hexose (about 28 per cent) are higher than for most glycoproteins. The content of different sugars is virtually the same whether the mucus is obtained during estrus or during pregnancy. The molecular weight is about the same (4×10^6) in each case, but the intrinsic viscosity of the glycoprotein during estrus is about twice that during pregnancy.

Connective Tissues

The collagen from skin, tendon, sclera, cornea, lens and basement membranes typically contains multiple disaccharide units, α-D-glucopyranosyl-1,2-D-galactose, each linked O-glycosidically from galactose to a hydroxylysine residue of the peptide. To a variable extent galactose alone may be linked to some of the hydroxylysine. In another kind of oligosaccharide unit there are small amounts of mannose, fucose, hexosamines, and sialic acid in collagens, particularly those of basement membranes, such as in renal glomeruli. Other specific characteristics of glycoproteins from certain sites are as follows.

BONE, CARTILAGE, AND SYNOVIAL FLUID. From bovine long bones a glycoprotein with a relatively high content of sialic acid (15.9 per cent) and hexoses (13.4 per cent) has been isolated. Hexosamine (6.4 per cent) and fucose (2.4 per cent) are also present. Hexoses include galactose, mannose, and glucose. A glycoprotein fraction from bovine nasal cartilage contains about 10 per cent reducing sugars and 1.2 per cent sialic acid.

Characterization of glycoprotein in normal synovial fluid is lacking, but the glycoprotein from the joint fluid of patients with rheumatic arthritis was found to contain galactose, mannose, a pentose, and sialic acid. It appeared to be identical immunochemically with a glycoprotein of normal human serum, which was not orosomucoid.

ARTERIAL TISSUE. From bovine aorta a homogeneous glycoprotein containing glucosamine, galactosamine, galactose, fucose, glucose, and mannose has been identified. Normal human aorta yields a glycoprotein fraction in which there is 17.8 per cent hexosamine (ratio of glucosamine:galactosamine 2.7:1), 14.0 per cent hexose (ratio of galactose: mannose 3:1), 1.6 per cent fucose, and 8.1 per cent sialic acid.

SKIN, TENDON, AND RETICULIN. From fetal calf skin two glycoproteins have been isolated, one of which is noteworthy for containing no hexosamine. The other has a carbohydrate content of 17 per cent, which includes galactose, mannose, glucosamine, and NANA. It resembles fetuin of fetal calf serum. In bovine Achilles tendon there is a glycoprotein in which glucose, galactose, mannose, and glucosamine have been identified.

Reticulin is a fibrous protein found in most fetal tissues and in spleen, kidney, and lymphatic nodes. Its network of fibers appears to be coated with a glycoprotein, which may have some rather special structural features. The carbohydrate moieties include only glucosamine, galactose, and mannose, which occur in a ratio of 6:1:1. Furthermore, the oligosaccharide components evidently exist as double-chain bridges between two polypeptide moieties.

VITREOUS BODY AND CORNEA. From vitreous body of the anterior chamber of the eye a glycoprotein of the γ-globulin type has been prepared. Its molecular weight is about 14×10^4 and it contains 2 per cent hexosamine, 4.3 per cent hexose (ratio of galactose:mannose 4:5), only a trace of fucose, and 0.4 per cent sialic acid. In corneal stroma there is a glycoprotein containing mannose, galactose, glucosamine, and sialic acid, and it is notable for containing also an ester sulfate. This glycoprotein was present in only one tenth the amount of heteropolysaccharide in cornea. Analysis of Descemet's membrane of bovine cornea showed galactose, glucose, and mannose present in a ratio of 6:2:1. Fucose, glucosamine, and galactosamine were also found.

Miscellaneous Other Sources

In various other sources—sweat, cerebrospinal fluid, colostrum, pleural fluid—glycoproteins have been identified but are as yet only partially characterized.

Glycolipids

The chemistry of carbohydrates in glycolipids is considered together with their metabolism in Chapter 7.

REFERENCES

Egami, F., and Oshima, Y. (eds.): Biochemistry and Medicine of Mucopolysaccharides. Maruzen Co., Ltd., Tokyo, 1962.
Florkin, M., and Stotz, E. H. (eds.): Comprehensive Biochemistry, Vol. 5: Carbohydrates. Elsevier Publishing Co., Amsterdam, 1963.
Fruton, J. S., and Simmonds, S. (eds.): General Biochemistry, 2nd edition. John Wiley and Sons, Inc., New York, 1958.
Gottschalk, A. (ed.): Glycoproteins: Their Composition, Structure and Function, 2nd edition. Elsevier Publishing Co., Amsterdam, 1972.
Karlson, P.: Introduction to Modern Biochemistry, 2nd edition. Academic Press, New York, 1965.
Northcote, D. H.: Polysaccharides, Ann. Rev. Biochem. 33:51–74, 1964.
Seminars in Biophysics and Physical Chemistry of Connective Tissue, Federation Proceedings 25, No. 3, Part I, 1003–1052, 1966.
Spiro, R. G.: Glycoproteins: Their biochemistry, biology and role in human disease, New Eng. J. Med. 281:991–1001, 1043–1056, 1969.
Whelan, W. J., and Cameron, M. P. (eds.): Ciba Foundation Symposium, Control of Glycogen Metabolism. J. and A. Churchill, Ltd., London, 1964.
Wolstenholme, G. E. W., and O'Connor, M. (eds.): Ciba Foundation Symposium, Chemistry and Biology of Mucopolysaccharides. J. and A. Churchill, Ltd., London, 1958.

CHAPTER

2

DIGESTION AND ABSORPTION OF CARBOHYDRATES

HYDROLYSIS OF POLYSACCHARIDES

Dietary carbohydrates and their assimilation are important parts of the total story of carbohydrate metabolism in mammals. There is less carbohydrate in the diet of carnivores than of herbivores, and for certain herbivores, the ruminants (cow, sheep, goat), the digestion part of the story is special. Briefly, microorganisms in the reticulo-rumen (the forestomachs), and to a lesser extent the large intestine of ruminants, digest the plant polysaccharide, cellulose, to the disaccharide, cellobiose, and more extensively to the short-chain volatile fatty acids—acetic, propionic, and butyric. Starch is likewise fermented to these fatty acids in the forestomachs, where they are largely also absorbed. The short-chain fatty acids then constitute the main organic carbon source for energy fuel presented to the tissues, although the liver converts much of the propionic acid to glucose and some of the acetate and butyrate to long-chain fatty acids.

For human beings the digestion of carbohydrates is much like that for most other mammals. However, man was primarily a carnivore before he became during the Neolithic age (only 10,000 years ago) much more of a herbivore, so digestive adaptations (or failures of adaptation) may presently be significant factors. At least 50 to 60 per cent of the caloric intake of man is now commonly composed of carbohydrates, mostly plant starch in the form of bread, rice, potatoes, and other vegetables with some animal glycogen from muscle and liver tissue.

The glucose units of starch and glycogen are linked in the alpha configuration, and so are susceptible to hydrolysis by the α-amylase (ptyalin) in the secretion of the salivary glands and by the α-amylase (amyloptin) in the pancreatic secretion. As with other non-ruminants, cellulose, a polymer of β-1,4-linked glucose units, is for man indigestible "roughage." Starch is composed of two types of polysaccharide, amylose and amylopectin. Amylose is made up of very long chains of α-1,4-linked glucose units, while amylopectine, like animal glycogen, is a branched structure with occasional α-1,6-linkages providing branch points for α-1,4-linked chains. The α-amylase (α-1,4-glucan,4-glucanohydrolase) of saliva hydrolyzes amylose rapidly into the disaccharide maltose and the trisaccharide maltotriose; the latter is hydrolyzed more slowly to glucose and maltose. This enzyme also splits amylopectin into glucose, maltose, maltotriose, and a mixture of branched oligosaccharides or dextrins. The smallest of these are tetra- and pentasaccharides. Amylopectin of waxy-maize starch is found to yield about 40 per cent maltose, 30 per cent maltotriose, and most of the remainder as branched dextrins by the action of human salivary amylase.

Except in a few individuals of deliberate nature, most of the salivary digestion takes place in the stomach. This continues until the acid of the gastric secretion penetrates the food bolus and puts a stop to salivary amylase, which is inactive (and destroyed) below pH of 4.0. The hydrolysis which is begun by salivary amylase is continued, or can be initiated, by pancreatic α-amylase, which may be virtually the same enzyme deriving from another organ and secreted into the duodenal contents. Pancreatic amylase does the main part of starch digestion; indeed, some animals have no salivary amylase.

Hydrolysis of starch to disaccharides and oligosaccharides by pancreatic amylase is not a process occurring only in the luminal fluid of the intestine. There is an adsorption of this enzyme onto the intestinal mucosa where a process of "membrane-" or "contact-digestion" takes place in the pores of the striated border. In all areas of the small intestine, the membrane-digestion capacity is significantly higher (two- to tenfold) than the rate of hydrolysis by amylase free in the intestinal cavity. The distribution of membrane-digestion capacity is unequal in different parts of the gut. It is highest in the distal duodenum and proximal parts of the jejunum. The distribution gradient for membrane-digestion corresponds to the distribution gradient for "sucrase" activity (see below) and decreases in areas where the villi are shorter.

In young infants (6 months or less) the pancreatic amylase activity is not fully developed, so that the products of hydrolysis of starch emerging from the duodenum include large amounts of dextrins composed of more than 30 glucose units as well as increased amounts of maltotriose and maltotetrose in the small oligosaccharide group. Because of this incomplete amylolysis, malabsorption of starch is more common during the first year of human life than later. For infants about 8 months of age and older, the increased amylase activity of duodenal juice permits the hydrolysis of amylose and amylopectins to oligosaccharides for which the average chain length is about 3 monosaccharide units. Among these products there is little of the α-1,6-linked disaccharide isomaltose (which constitutes 3 to 4 per cent of the average starch molecule) but instead there are branched dextrins as limit products of amylase activity on amylopectin. These branched oligosaccharides are subsequently split by "isomaltase" and amylases of the intestinal mucosa. There are so far two amylases identified in human, and in rat, intestinal mucosa. One of these acts on higher oligosaccharides (perhaps it is adsorbed pancreatic amylase), while another, which is less soluble, has "maltase" activity, also.

HYDROLYSIS OF DISACCHARIDES

The disaccharides which are ultimately derived from the action of the amylases are listed in Table 2–1 together with other common and uncommon disaccharides. Until recent years, it was thought that enzymes for hydrolysis of these disaccharides were contained principally in a secretion ("succus entericus") of the small intestinal mucosa. However, the small amount found in the free luminal contents does not nearly account for the rate of hydrolysis and absorption in vivo; most of the activi-

Table 2–1. *Constituent Characteristics of Dietary Carbohydrates*

POLYSACCHARIDES	DISACCHARIDES	MONOSACCHARIDES	GLYCOSIDIC BOND
starch — amylopectin	maltose	→ glucose + glucose	α 1—4
	isomaltose	→ glucose + glucose	α 1—6
amylose	maltose	→ glucose + glucose	α 1—4
glycogen	maltose	→ glucose + glucose	α 1—4
	isomaltose	→ glucose + glucose	α 1—6
	sucrose	→ glucose + fructose	α 1—β 2
	palatinose	→ glucose + fructose	α 1—6
	trehalose	→ glucose + glucose	α 1—α 1
	lactose	→ galactose + glucose	β 1—4
cellulose	cellobiose	→ glucose + glucose	β 1—4
	gentiobiose	→ glucose + glucose	β 1—6

ties of all disaccharidases are located in small "knobs" on the brush border of the intestinal epithelial cell. The plasma membrane of this border next to the intestinal lumen is highly concentrated in these enzymes as well as in alkaline phosphatase and in Na^+-K^+ ATPase. The latter could relate also to absorption of sugars, as will be seen. When the disaccharidase enzymes are found in the luminal fluid they are associated with cellular elements which have been sloughed off the intestinal wall in the natural cycle of loss and regeneration.

Naming of digestive enzymes has proceeded largely on the basis of the substrate employed. Thus, maltase activity is largely contributed by the enzymes "sucrase" and "isomaltase," although there are altogether at least five distinct "maltase" enzymes. One nomenclature of disaccharidases (suggested by Semenza and Aurrichio according to the order of appearance on elution from Sephadex G-200) is shown in Table 2–2. All the maltases and trehalase are alpha-glycosidases, whereas the lactases are beta-glycosidases (Table 2–1). The total maltase activity is divided as follows: maltase 1 = 5 per cent, maltase 2 = 15 per cent, maltase 3 = 5 per cent, maltases 4 and 5 combined = 75 per cent. Maltase 4 has 90 per cent and maltase 3 has 10 per cent of the total sucrase activity. Lactase 1 has 90 per cent of the total cell lactase; lactase 2 is not like other disaccharidases in that it is soluble and is perhaps identical to nonspecific β-glycosidases in other cells. In normal human individuals the ratios among activities of maltase:isomaltase:sucrase:lactase:palatinase:cellobiase are approximately 6:2:2:1:0.5:0.2. Measurement of some of these ratios is important in determining specific disaccharidase deficiencies.

The constituent monosaccharides of various disaccharides and the type of glycosidic bonds linking them are shown in Table 2–1. Sucrose is the common cane and beet sugar, which in the last century or two (a moment in evolutionary time) has become so much more prominent in the diet of man. It is found also in fruits and sweet vegetables. Lactose is, of course, common milk sugar; in cow's milk there are 40 to 50 g. of lactose per liter and in human milk about 70 g. per liter. There is one group of mammals, the eared seals (e.g., walrus, California sea lion) which has no lactose in its milk nor any lactase activity in its intestinal mucosa. In fact, there are no carbohydrates in the milk of these mam-

Table 2–2. *Disaccharidases and Activities**

maltase 1
maltase 2
maltase 3 = sucrase 1
maltase 4 = sucrase 2
maltase 5 = isomaltase = palatinase
lactase 1 = cellobiase 1 = gentiobiase 1
lactase 2 = cellobiase 2 = gentiobiase 2
trehalase

*Nomenclature of Semenza and Aurrichio, Biochim. Biophys. Acta 65:172, 1962.

mals and no disaccharidases have been found in their intestine. These animals have been used as models for the study of the pathology of human disaccharidase deficiencies.

Trehalose is found in fungi (e.g., young mushrooms), seaweeds, and insects. Palatinose is a bacterial transformation product of sucrose and does not occur in a normal diet. Cellobiose, the product of bacterial action on cellulose, is present only in the colon of man and other non-ruminants. Gentiobiose is a sugar found in rare plants.

Disaccharidase activity is absent in the stomach; however, sucrose and other fructo-furanosides (e.g., the polysaccharide inulin, which is composed only of fructose and is found in Jerusalem artichoke) are slowly hydrolyzed by warm, dilute mineral acid and therefore are susceptible to the HCl in the stomach. Disaccharidase activity is generally found throughout the small intestine with somewhat lower values in the proximal duodenum and distal ileum and highest values usually in the upper jejunum. In some species there are some minor differences in distribution of maltase, sucrase, and lactase activity along the intestinal tract, but in man these activities occur in a fairly constant ratio. Unlike other disaccharidases, trehalase diminishes progressively in activity from the upper third of the small intestine to the lower third. There is almost no disaccharidase activity in the large intestinal mucosa, but bacterial digestion in the colon can include disaccharides.

No significant quantities of maltose or sucrose are allowed to pass unhydrolyzed through the intestinal wall and into the blood stream, except in certain small bowel disorders. However, a lactosuria can more readily be provoked (in adult animals) by administration of large amounts of lactose. This could be related to the relatively low activity of lactase in adult animals. If any sucrose or lactose passes into the blood, it cannot be hydrolyzed and is excreted in the urine; maltose, however, can be hydrolyzed after parenteral administration or intact absorption without hydrolysis.

Sucrase and maltase are almost fully developed early in gestation in man, but lactase continues to develop up until or shortly after the time of birth. Therefore, premature babies may show a relative intolerance to milk in the newborn period. Among animals (e.g., rat, rabbit, guinea pig) lactase activity is highest at birth, then falls off after weaning, or after the normal duration of the suckling period, to about 10 per cent of its previous level. It is not adaptive to the feeding of milk beyond the suckling period. (Whether adult cats are any exception to these findings ought to be investigated.) The alpha-glycosidases (e.g., sucrase, maltase) show a converse time pattern of activity in non-ruminant animals. Low activity at birth is followed somewhat later by a rise, so that the sucrase/lactase ratio in adult rats is somewhat higher than it is in man. In ruminants, the digestion of sucrose depends on microorganisms in both the small and large intestine, where the glucose from sucrose is converted largely to volatile fatty acids.

Man is different from animals in that all intestinal disaccharidases are fully developed at or shortly after birth and commonly remain at almost the same level throughout life. The persistence of lactase after weaning is an evolutionary change in man which corresponds to the widespread practice of drinking milk of domestic animals throughout life. Tolerance for this practice is not widely bestowed, however. In certain racial groups and geographical locations the evolutionary change has occurred only partially; large and even major segments of the population have relatively low lactase after infancy or early childhood (see below).

There is no clear evidence for substrate induction of lactase activity. Some studies have suggested a loss of symptoms of milk intolerance after prolonged intake, but it is not established that this involves enzyme adaptation. Rats fed a high sucrose diet have significantly greater sucrase activity than when the diet is carbohydrate-free or even contains equivalent amounts of glucose. Certain monosaccharides (galactose and fructose) and maltose also induce the sucrase activity. In man, also, sucrase and maltase activity of the jejunal mucosa was found to be adaptive to sucrose or fructose in the diet but not to equal amounts of several other monosaccharides or disaccharides. Lactase activity was not altered by sucrose or fructose.

The adaptiveness of sucrase could possibly have an impact on the rate of absorption of glucose or fructose in individuals habitually consuming large amounts of sucrose, although the hydrolysis of sucrose or of maltose has not appeared to be the limiting factor in the rate of absorption of the glucose contained in these disaccharides. The rates of absorption of sucrose, maltose, and trehalose are all about the same and similar to that of glucose during the first two hours after intubation in the rat. Yet, the rates of hydrolysis of the three disaccharides by mucosal extracts in vitro are markedly different. This implies that the potential capacity for hydrolysis is not utilized in vivo or that some step subsequent to hydrolysis, presumably the transport of glucose into the blood, is rate-limiting. Nevertheless, after ingestion of sucrose by humans fructose appears more rapidly in the peripheral blood than after ingestion of an equal quantity of glucose plus fructose. Sucrose-^{14}C ingested by rats is oxidized faster to $^{14}CO_2$ than equal amounts of glucose-^{14}C plus fructose-^{14}C or of maltose-^{14}C; no consistent difference in oxidation of sucrose-^{14}C vs. glucose-fructose-^{14}C is found in humans, however.

Whereas hydrolysis of disaccharides and glucose transport are distinct processes (e.g., phlorizin, a plant glycoside, does not interfere with hydrolysis while completely blocking transport), there is much reason to believe that the two functions are closely linked. If glucose oxidase is present on the luminal side of the intestinal mucosa, there is less interference with absorption of glucose formed from sucrose than with that formed from glucose-6-phosphate. Other evidence also suggests that, in Crane's phrase, there is a "kinetic advantage" for transport of glucose specifically formed from sucrose, maltose, isomaltose, and lactose over that formed from glucose-6-phosphate or trehalose and over free glucose in the luminal medium. A diagram which depicts some of these presumed relationships is shown in Figure 2-1 (after Crane). The activity of sucrase in the intestinal wall is Na^+-dependent, as is the transport of monosaccharides, which further suggests a close relationship of the two functions. The kinetics of Na^+ activation are similar and change in similar ways from one species to another.

Unlike sucrose or maltose, lactose shows a slower absorption in the adult human than an equivalent mixture of its constituent monosaccharides, glucose and galactose. Whether this is also true of the normal infant is not known. In the lower part of the intestine, where the activity

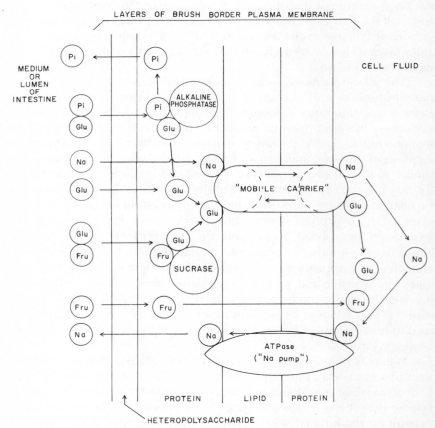

Figure 2-1. Modified from Crane, R. K. *In* Carbohydrate Metabolism and Its Disorders, Vol. I, p. 37, Academic Press, 1968.

of trehalase falls off, this enzyme appears to be the rate-limiting factor for the absorption of trehalose.

ABSORPTION AND TRANSPORT OF MONOSACCHARIDES

Free monosaccharides are not commonly present in the diet in significant quantities. There is some free glucose and fructose in honey and certain fruits. The absorption of free glucose and galactose occurs by a transport mechanism which may also be shared with these same monosaccharides as they are formed from sucrose, maltose, isomaltose, and lactose (Fig. 2-1). Transport of glucose, galactose, and some other monosaccharides through the membrane of the brush border is an active process, i.e., it occurs against a concentration gradient, consumes energy, and is metabolically supported. This transport system exhibits competition for the active site among different sugars and between sugars and typical inhibitors, e.g., phlorizin. The latter plant glycoside is a complex of glucose with phloretin, a polycyclic aromatic hydrocarbon.

The presence of Na^+ is vital to the process of glucose absorption. In the absence of Na^+ there is very slow penetration into the intestinal epithelium, and a concentration of glucose against a gradient is not possible. Crane has postulated that the "mobile carrier" for glucose has another binding site for Na^+ (Fig. 2-1). In the absence of Na^+ the carrier is less efficient in transferring substrate. Studies of inhibition of transport by phlorizin suggest that the latter displaces Na^+ at its binding site. Various inhibitory cations (Ca^{++}, K^+, Li^+, and NH_4^+) may also exert their effect by this means. Sulfhydryl-reacting compounds interfere with binding of glucose to isolated brush border preparations, which may be another clue to the interaction of glucose with the "carrier."

Probably the transport of sodium ion is the link with the metabolic system which provides the driving force for absorption of glucose against a concentration gradient. The "Na^+ pump," which in many kinds of cells is an activity associated with the enzyme ATPase, operates at some portion(s) of the epithelial cell wall to eject sodium out of the cell against a concentration gradient. Sodium entering the cell thus flows "downhill" with a gradient and at the same time via the mobile carrier serves to carry glucose "uphill" against a gradient. This mechanism is particularly operative at low concentrations of glucose, since high concentrations of sodium in the external fluid increase the affinity of glucose for the carrier. Conversely, the low concentration of sodium within the cell would have the effect of facilitating dissociation of the carrier-glucose complex. In fact, if the system actually does work to translocate glucose on the basis of differential affinities of carrier for glucose on two sides of a cellular membrane, there is no logical necessity to invoke a Na^+-carrier-sugar complex to explain the effects of Na^+ on glucose transport.

The location of a Na^+ pump in the intestinal epithelial cell wall is not yet established. Although the brush border has a high concentration of ATPase (two separate kinds), this area of the cell, the microvillus, is far removed from the mitochondrial source of most of the ATP. It has been suggested that the site of the "pump" is not the mucosal but the serosal surface, since ouabain at the serosal surface is more effective in inhibiting Na^+ transport than is the presence of this substance at the mucosal surface. Ouabain, which is a "cardiac" glycoside composed of rhamnose and the steroid-like compound strophanthidin (Chap. 3), also inhibits intestinal transport of glucose. There could be some point to the question of whether there is a Na^+ pump at the brush border; if so, it ought to interact more directly with the Na^+ concentration within the intestinal lumen, which in turn should be somewhat dependent on the Na^+ intake at the time of glucose absorption. The dietary NaCl should affect the glucose carrier, in any case.

Not only is sugar transport dependent on Na^+, but there is a stimulatory effect of glucose on Na^+ transport. Moreover, the effect of the presence of glucose on sodium transport is not, as once thought, due to the energy-yielding metabolism of that glucose, but is connected with the transport of glucose. Thus, 3-*O*-methylglucose, an artificial sugar, which is actively transported but not metabolized, can, like glucose, stimulate Na^+ transport.

Nobody has yet isolated a mobile carrier or an intact, identifiable Na^+ pump, so the above concepts are still rather speculative. More complex ideas have also been advanced. Differences between glucose and galactose in uptake vs. appearance on the serosal side of the intestinal mucosa wall have suggested a two-stage process of entry by "facilitated diffusion" (Chap. 3) and then, distal to a zone of metabolism, a concentrating mechanism. Besides sugars the small intestinal wall actively transports amino acids, pyrimidines, and bile salts. In all cases, the rate and extent of transport are dependent on the Na^+ gradient. Although many have supposed that there are different mobile carriers for each class or subclass of compounds, mutual interactions and cross-inhibitions have given rise to the concept of a common polyfunctional carrier, in which there is a series of separate binding sites in a sort of mosaic. Cross-inhibitions between classes of compounds may occur because of allosteric interactions between the associated binding sites.

An old theory held that glucose enters the intestinal mucosa via a process of phosphorylation and that the high concentration of alkaline phosphatase in the brush border is responsible for dephosphorylation. More likely alkaline phosphatase is useful for hydrolysis of sugar-phosphate esters in the diet. Various evidence contradicts the phosphorylation hypothesis. Sugars which have no hydroxyl function at C-1 or C-6, and hence are not phosphorylated by the known kinases, can still be actively transported. Glucose transported from the mucosal border to the serosal fluid does not mix actively with the metabolic glucose derivatives, e.g., glucose-6-phosphate (Gl-6-P). This was demonstrated with galactose-^{14}C, which was converted more extensively to Gl-6-P than to glucose in transport.

Characteristics of monosaccharides which seem to be most essential for assuring active transport include (1) OH in glucose configuration at C-2, (2) a pyranose ring, and (3) a methyl or substituted methyl group at C-5 of the ring. This is illustrated in Table 2–3, which lists relative rates of absorption of a large number of monosaccharides by an isolated rat intestinal segment in vivo. As in other species, glucose and galactose are absorbed most rapidly. Various departures from the glucose configuration are accompanied by lesser rates of absorption. Epimerization at C-2 (D-mannose) markedly lowers the affinity of substrate for carrier. Epimerization at C-3 (D-allose) or C-4 (D-galactose) has little or no effect, but epimerization at both C-3 and C-4 (D-gulose) strongly reduces the transportability. Although the absorption of most of the sugars listed in Table 2–3 has been assumed to occur by a process of "simple diffusion," if this were true, then rates for all such sugars should be more nearly identical than they are. This observation coupled with recent information that at very low concentrations some of these sugars are indeed transported against a gradient suggests that active processes (of a low order of specificity) are involved in the transport of many sugars. The absorption of fructose is clearly different from that of glucose or galactose, since it cannot take place against a gradient and appears to be independent of Na^+ transport.

Different regions of the small intestine have varying capacities for absorption of monosaccharides. For glucose or galactose (in the rat and

Table 2–3. *Absorption Rates of Some Carbohydrates**

CARBOHYDRATE	MEAN RELATIVE ABSORPTION RATE
D-Glucose	100
D-Galactose	99
D-Allose	67
L-Xylose	46
D-Ribose	45
D-Lyxose	38
D-Xylose	37
N-Acetyl-D-Glucosamine	34
D-Altrose	24
D-Talose	22
D-Glucosamine	22
D-Idose	20
L-Arabinose	17
D-Arabinose	16
D-Mannose	8
D-Gulose	6

*From Kohn, P., Dawes, E. D., and Duke, J. W., Biochim. Biophys. Acta *107*:358, 1965.

hamster), the maximum capacity occurs in about the middle of the small intestine (jejunum). Little absorption of glucose or other monosaccharides occurs in the colon, and this section of the intestine has no capacity for concentrating glucose.

During the absorption of glucose relatively little (10 per cent or less) is converted to lactic acid or other products. Almost all absorbed glucose is delivered promptly to the portal blood which leads to the liver. Galactose also seems to pass mainly unaltered into the portal blood stream. Fructose, on the other hand, is converted by the intestinal epithelium in substantial amounts to glucose, lactic acid, and other products with variations among species (see Chap. 4).

DEFICIENCIES OF DIGESTION AND ABSORPTION

There are no common conditions or diseases in which the capacity of salivary and pancreatic amylase is impaired or exceeded relative to usual starch intakes in the diet. A rather rare disease, cystic fibrosis, involves a deficiency of pancreatic secretory enzymes, but the failure of digestion of fat is more evident than that of carbohydrates.

In rats, with aging there is a gradual flattening of the villous projections of the intestinal mucosa. Concomitantly, there is a marked increase of cavitary digestion and a gradual decrease in "membrane-digestion" and overall hydrolytic activity in vivo. It has been postulated that cases of starch malabsorption in children or adults, secondary to gastrointestinal disorder, may be associated with pathological atrophy of the intestinal villi and loss of membrane-digestion capacity.

Deficiencies of disaccharidase activity are well recognized and of multiple types. They may be of either congenital or acquired origin (Table 2–4). Any extensive disaccharidase deficiency produces typical symptoms of watery diarrhea, abdominal cramps, and bloating, which are the consequence of the osmotic influx of fluid into the lower intestinal tract containing the large excess of low-molecular weight compounds — not only the particular disaccharide but monosaccharides, lactic acid, and volatile fatty acids formed by bacterial fermentation. Owing to the organic acids the pH of the stool is typically low — less than 5.5. Irritation of the mucosa by the fermentative products may contribute to the symptoms.

Congenital lactase deficiency in infants shows the above symptoms plus failure to thrive on milk and sometimes severe dehydration and electrolyte disturbances. Because of its higher lactose content human milk can be even more deleterious than cow's milk. As such children grow older, they may be able to tolerate moderate amounts of milk, probably because of the larger total absorbing surface of the mucosa.

Lactase deficiency in infants is rare; there is a type of genetic lactase deficiency in adults which is much more common and is expressed varia-

Table 2-4. *Classification of Malabsorption of Disaccharides and Monosaccharides**

A. Disaccharidase Deficiency
  1. Inherited (congenital) forms:
     a. Lactase deficiency.
     b. Sucrase-isomaltase deficiency: at least two different enzymes are missing.
     c. Sucrase-isomaltase-maltase deficiency: at least four different enzymes are missing (may not occur as purely inherited form).
  2. Acquired forms:
     a. General disaccharidase deficiency: all activities are low, but lactase usually lowest and most likely to give clinical symptoms.
     b. Acquired specific lactase deficiency: lactase activity is very low, while other disaccharidase activities and mucosal morphology are normal.
B. Monosaccharide Malabsorption
  1. Inherited (congenital) form:
     a. Glucose-galactose malabsorption: these two monosaccharides are absorbed very slowly because of a defect in the carrier; fructose is absorbed normally.
  2. Acquired form:
     a. Malabsorption of monosaccharides secondary to different kinds of small-intestinal disease (e.g., infection): glucose, galactose, and fructose are all absorbed slowly.

*From Dahlqvist, A., Scand. J. Clin. and Lab. Invest. *19*, Supp. 100:96, 1967.

bly in different racial groups. One study showed a 70 per cent incidence in American Negroes and 10 per cent in Caucasians, as determined by the common criterion of a flat blood glucose curve after ingestion of 1 g. lactose/kg. body weight. Such lactose non-absorbers show normal elevations of blood glucose after ingestion of equivalent amounts of galactose and glucose. There are other indications of racial differences. In East Africa, a Bantu Negro tribe has a very high incidence of lactase deficiency, while another adjacent tribe of predominantly Hamitic origin, although living under essentially the same conditions, has a low incidence. In some Mediterranean areas (e.g., Greece) and in India, milk intolerance is common and possibly due to congenital lactase deficiency. Since the inheritance of this condition is recessive, the number of heterozygotes is much larger than the number of homozygotes, who would have more obvious clinical symptoms. Heterozygotes may not show milk intolerance except to excessive amounts.

A modification of the lactose tolerance test involves measurement of galactose in the blood instead of glucose, when the test is preceded by an oral dose of ethanol. The latter compound blocks the utilization of galactose, probably because a step in galactose utilization is sensitive to the NAD^+:$NADH$ ratio, which is decreased by ethanol (Chaps. 4 and 8). With this kind of test galactose rises markedly in the blood of normal subjects but not of those with lactase deficiency. Another kind of lactose tolerance test is the measurement of $^{14}CO_2$ in the breath after oral administration of lactose labeled with ^{14}C. This test shows promise of good recognition of lactose intolerance in a relatively simple clinical procedure. The test (like others of this type, Chap. 8) could also be done with the stable, non-radioactive isotope, ^{13}C. This kind of test could well be applied to states of malabsorption of other sugars and of starch.

Another kind of diagnosis of deficiency of intestinal lactase (or any other particular disaccharidase) is measurement of hydrolytic activity for the appropriate disaccharide in a jejunal mucosal biopsy taken through oral intubation. This technique is easier than one might suppose, and is becoming fairly common. Although this is a more specific diagnosis of disaccharidase deficiency, particularly when ratios of two activities (e.g., sucrase and lactase) are measured, the small biopsy may not always be representative of the entire small intestinal activity, so this method should be accompanied by a lactose tolerance test.

There is a special kind of lactose intolerance in infants, which is characterized by lactosuria, but which also includes aminoaciduria, vomiting, and renal damage. This condition (the Durand syndrome) is not always improved by removal of lactose from the diet, whereas true lactase deficiency is invariably improved. A defect in lactase is not clearly present in the Durand syndrome.

In some gastrointestinal diseases, e.g., celiac disease, tropical sprue, kwashiorkor, ulcerative colitis, regional enteritis, and infectious diarrhea, there is a relatively high incidence of lactase deficiency, which is generally secondary and somehow acquired in connection with the disease. In fact, there is often a generalized deficiency, but that of lactase is most prominent. Possibly this may be due to the lesser normal activity and lesser margin of reserve of lactase. Also, other disaccharidases seem to have a better recovery capacity than lactase, which remains low for long after correction of a gastrointestinal disturbance. This has been particularly shown for the severe state of malnutrition in children, kwashiorkor. Extensive intestinal resection or gastric resection may be accompanied by relative lactase deficiency and symptoms of milk intolerance. Some skin diseases are for some reason also associated with lactase deficiency.

Some authorities claim there is an acquired lactase deficiency (Table 2-4) which is different from the congenital type but which is not accompanied by any other disaccharidase deficiency or any other mucosal abnormality and which can be manifest as milk intolerance even without any superimposition of bowel disease or loss of bowel surface. Since the incidence of congenital lactase deficiency increases progressively with age, it is difficult to see the distinction between congenital and acquired lactase deficiency without overt cause. It has been suggested that primary lactase deficiency, whatever its origin, may, in view of its very common incidence and because its effects impose an abnormal strain on the bowel mucosa, be partly responsible for susceptibility to some gastrointestinal diseases, thus helping to explain the high correlation of these with lactase deficiency.

Deficiencies of sucrase and isomaltase occur as invariably associated phenomena in a recessively inherited condition which is uncommon but which when present is manifest very early in infants with the symptoms already described. Genetically, it is quite interesting because it is rare that such an "inborn error of metabolism" affects more than one enzyme. Mutation of a regulator gene controlling several enzymes or of a polypeptide component common to several enzymes might be an explanation for the result. In contrast to lactase deficiency in which enzyme levels are low but never absent, individuals with sucrase-isomaltase malabsorption show a total loss of sucrase activity, although isomaltase activity is not totally lost. Low values of sucrase in some parents and siblings of affected individuals suggest heterozygous carriers of a recessive trait. The incidence of the homozygotic deficiency state in the U.S. population is estimated at about 0.2 per cent, while the frequency of heterozygous individuals may be almost 10 per cent. In infancy and early childhood, sucrose must be excluded from the diet of homozygotes. In older children and adults, some tolerance seems to develop (gastric HCl may hydrolyze some sucrose) but watery stools are common. Commercial sucrase can be provided in the diet occasionally (in a stroke of medical humanitarianism) to allow occasional indulgence in sweets by children at parties, a practice not recommended to excess, however (see Chap. 8).

In sucrase-isomaltase deficiency there is about a 25 per cent retention of maltase activity in maltases 1 and 2 (Table 2-2), so absorption of starch is not usually a major problem. Since there is not much isomaltose relative to maltose, the isomaltose malabsorption is a minor consideration. There could be some merit in using those kinds of starch (corn and rice) which are low in amylopectin and therefore potential isomaltose, rather than those starches (wheat and potato) which are more highly branched.

Another uncommon cause of severe diarrhea in human infants has been recognized in recent years to be a disorder of absorption of glucose and galactose and other sugars which are specifically transported by the "mobile carrier" in the luminal brush border of the intestinal epithelial cell. Glucose-galactose malabsorption (GGM) can be distinguished from disaccharidase deficiency and other causes of voluminous, watery, acid stools in infants by tolerance tests with the appropriate carbohydrates. Fructose, of course, is tolerated because it is absorbed passively without the mediation of the active carrier system and is the carbohydrate food of choice. Sucrose, because of its fructose content, causes less diarrhea than glucose, galactose, or lactose. It is not yet clear whether the syndrome of GGM is independent of any decrease in disaccharidase activity. In patients with GGM the renal threshold for glucose may be lowered with the occurrence of intermittent glycosuria. This suggests a dysfunction of a common transport system in the renal tubule and in the intestinal mucosa. Some adults have been noted to have GGM, so the prognosis may be relatively good if the patient survives the neonatal period. Besides fructose another carbohydrate dietary substitute in older children is the polysaccharide inulin, the polymer of fructose, which is hydrolyzed by gastric HCl.

Some extensive infectious diarrheas are accompanied by poor absorption of any monosaccharides as well as disaccharides. This state can be distinguished from GGM by the relative clinical effects of glucose and fructose in the diet, or, as stated earlier, by carrying out laboratory tests of absorptive capacity, such as the relative rates of oxidation to expired CO_2 of the monosaccharides labeled with ^{14}C or ^{13}C.

REFERENCES

General

Crane, R. K.: *In*:Carbohydrate Metabolism and Its Disorders, ed. by Dickens, Randle and Whelan, Vol. I, pp. 25–49, Academic Press, New York, 1968.
Haemmerli, U. P., and Kistler, H.: Disaccharide malabsorption. Disease-A-Month, July, 1–51, 1966.
Newey, H.: Absorption of carbohydrates, Brit. Med. Bull. *23*:236–240, 1967.
Prader, A., and Aurrichio, S.: Defects of intestinal disaccharide absorption. Ann. Rev. Med. *16*:345–358, 1965.

For Hydrolysis of Polysaccharides

Aurrichio, S., Della Pietra, D., and Vegnente, A.: Studies on intestinal digestion of starch in man. II. Intestinal hydrolysis of amylopectin in infants and children. Pediatrics, *39*:853–862, 1967.
Jesuitova, N. N., De Laey, P., and Ugolev, A. M.: Digestion of starch *in vivo* and *in vitro* in a rat intestine. Biochim. Biophys. Acta, *86*:205–210, 1964.

For Hydrolysis of Disaccharides

Aurrichio, S., Rubino, A., Prader, A., Rey, J., Jos, J., Fregal, J., and Davidson, M.: Intestinal glycosidase activities in congenital malabsorption of disaccharides. J. Pediatrics *66*:555–564, 1965.
Dahlqvist, A., and Thompson, D. L.: The digestion and absorption of maltose and trehalose by the intact rat. Acta. Physiol. Scand. *59*:111–125, 1963.
Isselbacher, K. J., and Senior, J. R.: The intestinal absorption of carbohydrate and fat. Gastroenterology *46*:287–298, 1964.
Mansford, K. R. L.: Absorption of nutrients from the intestine. Recent studies on carbohydrate absorption. Proc. Nutr. Soc., *26*:27–34, 1967.
Peternel, W. W.: Disaccharidase deficiency. Med. Clin. of N. Am., *52*:1355–1365, 1968.
Rosensweig, N. A., and Herman, H. R.: Control of jejunal sucrase and maltase activity by dietary sucrose or fructose in man. A model for the study of enzyme regulation in man. J. Clin. Invest., *47*:2253–2262, 1968.
Sasaki, Y., Iio, M., Kameda, H., Ueda, H., Aoyagi, T., Christopher, N. L., Bayless, T. M., and Wagner, H. N., Jr.: Measurement of ^{14}C-lactose absorption in the diagnosis of lactase deficiency. J. Lab. and Clin. Med. *76*:824–835, 1970.
Townley, R. R. W.: Disaccharidase deficiency in infants and childhood. Pediatrics *38*:127–141, 1966.

For Absorption of Monosaccharides

Alvarado, F., Transport of sugars and amino acids in the intestine: evidence for a common carrier. Science *151*:1010–1013, 1966.
Faust, R. G., Leadbetter, M. G., Plenge, R. K., and McCaslin, A. J.: Active sugar transport by the small intestine. J. Gen. Physiol. *52*:482–494, 1968.
Jeffery, J. R.: Glucose-galactose malabsorption. McGill Med. J. *37*:125–130, 1968.
Kohn, P., Dawes, E. D., and Duke, J. W.: Absorption of carbohydrates from the intestine of the rat. Biochim. Biophys. Acta *107*:358–362, 1965.

CHAPTER 3

CELLULAR TRANSLOCATION OF CARBOHYDRATES

Cellular and intracellular membranes are typically layers of lipid and lipoprotein micelles (see Physiological Chemistry of Lipids in Mammals in this series), through which the movement of intensely hydrophilic sugar molecules can occur only to a limited extent by free diffusion and for which there is therefore a special facilitating mechanism. Although transport is a commonly used word for this occurrence, translocation is a broader and better term for a process which in some cases may involve movement across a membrane but possibly also along or between membranes, or even less definable but spatially organized movement. The words "transport" and "carrier" will be used often here, because they are short and colloquial, but should not necessarily be taken to imply vehicular movement.

Electron microscopy has revealed that animal and plant cells generally exhibit a profusion of membrane systems in the cell interior as well as at the surface. The outer cell membrane is often found to enfold tortuously into the cell interior, in which state it may be contiguous with or identical to the endoplasmic reticulum (ER), a fine network of a paired membrane system in the cytoplasm. The ER may also be in intimate contact with the membranes of the various organelles—nucleus, mitochondria, Golgi apparatus, lysosomes, and so forth. Evidently changes occurring at the cell surface, including movement of sugars, might rapidly "telegraph" specific messages to deeply interior cell components. The network of intracellular membranes provides for many discrete compartments and structural frameworks, which indeed are organized enzyme systems, including some of those for carbohydrate metabolism.

Thus, the distinction between penetration of any metabolizable organic substance, including sugars, into the cell and intracellular movement of substances vs. enzymatic reactions becomes blurred. It should be recognized that many enzymatic reactions, indeed many chemical reactions in general, are microscopically vectorial processes. If the system is highly anisotropic, the microscopic vectors can display macroscopic vector chemical or transport characteristics. The phenomena of translocation can have characteristics similar to enzyme reactions simply because chemically they can be very similar processes and in fact identical in some instances.

The constant chemical reactions in living systems assure the occurrence of multiple electrical micropotentials and microcurrents within cells. The coupling of various chemical reactions determines a pattern of material transfer through membranes, which results also in mass circulation currents. Within the cytoplasm eddy currents may also be set up, which are one possible source of the motion known to microscopists as protoplasmic streaming.

Although little is yet known about intracellular translocation of carbohydrates, the initial entry into the cell of glucose and some other monosaccharides has been much studied in various tissues. As mentioned in Chapter 4, it seems possible to view this process as separate from that of phosphorylation, which is considered to be the first intracellular enzymatic reaction of glucose. The initial "transport" of glucose across the outer cell membrane has a number of characteristics, including those resembling enzymes, which have led to the concept of a membrane "carrier."

CONCEPT OF A MONOSACCHARIDE CARRIER

In Chapter 2 were described some characteristics of the active transport system for intestinal absorption of glucose. The mechanism for renal tubular reabsorption of glucose against a concentration gradient is very similar. For other cells there is a kind of non-active transport system, which does not move glucose against a gradient but facilitates the "downhill" movement of glucose and other monosaccharides by a process called "facilitated diffusion." This process resembles active transport in several respects, e.g., stereospecificity, competition among sugars, saturability, and susceptibility to similar inhibitors. In these ways it also resembles an enzymatic reaction.

There is a characteristic of facilitated diffusion which has particularly led to the structural idea of a "mobile carrier" (Fig. 3-1) or, without necessarily supposing mobility, a "reorienting pore," as indicated in Figure 3-2. This characteristic is called countertransport or the "trans effect." If in Figure 3-1 glucose molecules are represented by dots and

CELLULAR TRANSLOCATION OF CARBOHYDRATES

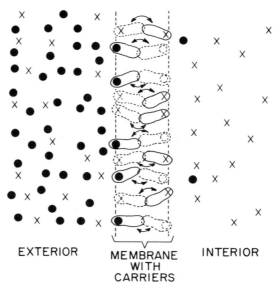

Figure 3-1. "Mobile carrier" concept of monosaccharide transport and counter transport of another (see text). (From Park, C. R., et al.: J. Gen. Physiol. 52:296, 1968.)

molecules of another, usually non-metabolizable, sugar, e.g., L-arabinose, by crosses, then an increase in concentration of glucose at the outer side of the membrane will result in a faster efflux of L-arabinose from the inner or trans side of the membrane. With the increase in glucose concentration a greater number of carriers are complexed, which somehow causes a faster rate of movement of carriers back and forth across the membrane. The greater rate of carrier movement from outside to inside allows also a greater rate in the opposite direction carrying L-arabinose out of the cell. Since L-arabinose does not compete very effectively with D-glucose for the carrier, an uphill trans-

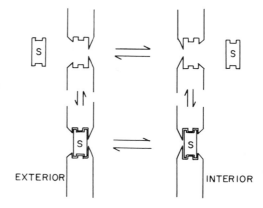

Figure 3-2. "Reorienting pore" concept of monosaccharide transport. (From Stein, W. D.: Brit. Med. Bull. 24:146, 1968.)

port of L-arabinose against a gradient can actually be caused by this mechanism. Calculations suggest that the mobility of the loaded carrier can be several times greater than that of the unloaded carrier. Thus, not only does the carrier facilitate the diffusion of the substrate, but in addition the substrate facilitates the movement of the carrier. One wonders whether there may be comparable behavior in enzymatic reactions, particularly in sequential array.

Non-active transport seems to be in some way different from active transport in that the non-active type cannot use metabolic energy to move glucose against a concentration gradient. That is, it seems not to be coupled positively to a process like the "sodium pump," which utilizes the energy of ATP. Facilitated diffusion is not dependent on the presence of Na^+, as is active transport. However, recently the effect of insulin on glucose oxidation and presumably transport (since it is rate-limiting) in the fat cell has been shown to be greatly facilitated by the presence of Na^+. Possibly, therefore, facilitated diffusion of glucose through adipose tissue membranes involves the simultaneous movement of Na^+ and glucose, just as does active transport of glucose. Moreover, sugar transport in the rat hemidiaphragm is increased by agents which inhibit the sodium pump and decreased by factors which activate the pump. This appears to be opposite to the effects expected with active transport (Chap. 2), but nevertheless argues for some kind of fundamental ionic mechanism involved in both types of transport. Apropos to the situation is the biological adage that electrolytes may be translocated through membranes independently of non-electrolytes but the converse is not true.

Models have been constructed of carriers in two mobile forms (Fig. 3–3) which can account for either equilibrating or concentrative transport, depending on whether the dissociation constants (K and K′) of free carrier (C) and carrier-substrate complex (AC) for the two forms are

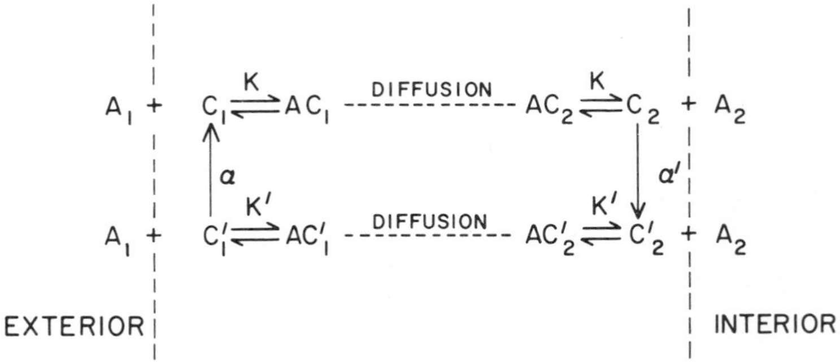

$$A_1 + C_1 \underset{}{\overset{K}{\rightleftharpoons}} AC_1 \text{-----DIFFUSION-----} AC_2 \underset{}{\overset{K}{\rightleftharpoons}} C_2 + A_2$$

$$\alpha \updownarrow \qquad\qquad\qquad\qquad\qquad \alpha' \updownarrow$$

$$A_1 + C_1' \underset{}{\overset{K'}{\rightleftharpoons}} AC_1' \text{-----DIFFUSION-----} AC_2' \underset{}{\overset{K'}{\rightleftharpoons}} C_2' + A_2$$

EXTERIOR　　　　　　　　　　　　　　　　　　　　INTERIOR

Figure 3–3. Concept of two mobile forms of monosaccharide carrier existing in equal or dissimilar proportion on two sides of a membrane (see text). (From Silverman, M., and Goresky, C. A.: Biophys. J. 5:487, 1965.)

very nearly equal to each other (equilibrating) or very dissimilar (concentrative). The two forms of carrier (C and C′) are assumed to exist in different proportions on the two sides of the membrane, depending on different rate constants, α and α′. In this model there is a sliding scale of degree of concentrative ability related to the degree of similarity of the rate constants α and α′ and of K and K′. Although there is no proof for such a model, it could help account for the differing net rates of transport of different carbohydrates as well as systems which show only facilitated diffusion at high concentrations but concentrative ability at low concentrations.

STEREOSPECIFICITY OF MONOSACCHARIDE TRANSLOCATION

A high degree of specificity for steric configuration is an indication of the particularity of association between glucose or other monosaccharides and a certain membrane component. As studied in great detail in the human erythrocyte by LeFevre, the three-dimensional configuration of the sugar is the most important factor determining the affinity. The "chair form" of D-glucose called the C-1 type (Chap. 1) is presumably favored. The affinity of other hexoses and pentoses is correlated with the relative extent of the stabilities of their C-1 conformations in water. L-glucose, which cannot exist in the C-1 shape at all, but rather the mirror-image 1-C, has a very poor affinity for the transport site.

Since D-glucose and most sugars with a high affinity for the carrier have their hydroxyl groups equatorially situated, this may be a clue to an important structural factor involved in affinity. Since none of the hydroxyls on carbons 1,2,3,4 and 6 appears to be essential, the binding between sugar and carrier may involve association through several hydrogen bonds. Presence of hydroxyl groups at C-2 and C-6 appears to be particularly important.

Transfer of sugars by non-active transport involves two fundamental parameters: the affinity factor and the capacity factor. Different sugars have different affinities for the transport system, but the order may vary as a function not only of cell and species but of the sugar concentration at which measurements are made, the presence of other sugars, and other environmental factors such as temperature. For glucose transfer into erythrocytes various methods suggest a half-saturation concentration (K_m)* of 7 to 10 mM. For the penetration under one set of conditions of some other sugars into red blood cells (RBC) the values of K_m and the maximal transfer rate, V_{max}, are given in Table 3–1. Evidently the affinity varies widely but the maximal transfer

---

*K_m, in enzyme kinetics, refers to the Michaelis-Menten constant, which is the concentration of substrate required to half-saturate the enzyme.

Table 3-1. *Parameters of Sugar Transfer in Human Red Cells**

SUGAR	HALF-SATURATION, K_m (mM)	MAXIMAL TRANSFER RATE, V_{max} (isotonic units/min.)
D-mannose	13	2.1
D-galactose	36–57	2.2–2.5
D-xylose	48–69	2.0–2.2
L-arabinose	214–247	2.0–2.6
D-ribose	1870–2710	1.7–1.8
D-arabinose	5120	2.0

*From Widdas, W. F., *in* Carbohydrate Metabolism and Its Disorders, Vol. I, pp. 1–23, 1968.

rate, expressing the diffusivity of the sugar-carrier complex, is about the same for all sugars.

There is a theory that mutarotase, an enzyme which catalyzes the interconversion of the anomers (alpha and beta forms) of sugars which possess a structure similar to glucose in carbons 1, 2, and 3, may be involved in sugar transport. Although some have considered that mutarotation is an essential part of transport, presently it seems that the binding site for mutarotation either coincides with or is located close to that for sugar transport. The anomer form during transport is not certainly known.

MOLECULAR NATURE OF THE MEMBRANE BINDING SITE

Because transport does not occur once the cell structure has been disrupted, it has been difficult to define the actual membrane components which are involved. Studies of specific combination of glucose and other sugars with isolated cell components have implicated phospholipids, but certain evidence suggests that proteins, perhaps together with lipids, are the types of components of the carrier. Proteins, rather than lipids, could provide the high degree of stereospecificity which is indicated by the large range of affinities for different sugars. The heteroglycans in glycolipids and glycoproteins could also confer high stereospecificity. Proteins are more or less intimately involved in the carrier, because the latter is susceptible to non-competitive inhibition by sulfhydryl-binding agents such as *N*-ethyl-maleimide and organic mercurial compounds. This suggests importance of the thiol groups of cysteine residues. Some sulfhydryl agents which inhibit intracellular enzymes have no effect on glucose transport, which suggests that only particular S-H groups are relevant. Also, combination with only 1 to 3 per cent of the total sulfhydryl groups on the membrane (of erythrocytes) is sufficient to realize the full effect of sulfhydryl inhibition. Dinitrofluorobenzene, which combines with protein, is a strong inhibitor of glucose transport. Some evidence identifies a linkage between glucose and the terminal amine group of a lysine moiety in a protein. Other evidence has implicated tryptophan.

Considerable insight, particularly for physiological purposes, has been gained by study of competitive inhibition of sugar transport by the glycoside phlorizin (Chap. 2). The diphenolic component, phloretin (called the aglycone), is also inhibitory. There is differing effectiveness of phlorizin and phloretin, depending on the particular tissue and species. Alvarado suggests two adjacent binding sites, one (g in Fig. 3-4) for free glucose and the glucose moiety of phlorizin and another (p in Fig. 3-4) a phenol-binding site for Ring B of phloretin. If the binding sites are closer together in some kinds of cells, then phloretin might have inhibitory capacity approaching (or exceeding) that of phlorizin. Besides the hydroxyl group at C-2 of glucose, those at C-3, C-4, or C-6 or those of Ring A of phloretin could also be involved in binding. Because phlorizin binds to both sites its affinity for the carrier can be (in the hamster intestine) 3 to 4 orders of magnitude higher than that of the physiological substrate, glucose. The hypothesis of these two binding sites may be relevant to interaction of transport of glucose and of amino acids (see below and Chap. 2).

Typically phlorizin is a much stronger inhibitor than phloretin in the "uphill" type of active transport found in kidney and intestinal mucosa. From its early use in physiology the effect of phlorizin to produce glycosuria in vivo became known as "phlorizin diabetes." On the other hand, for the transport of glucose by the erythrocyte, where the movement of glucose is one of "facilitated diffusion," phloretin is more inhibitory than phlorizin. This may be possible because phloretin by its lack of attachment to glucose is less restricted than phlorizin in its

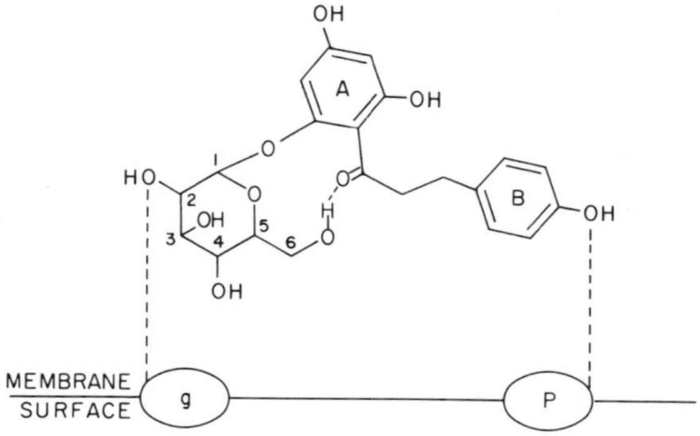

Figure 3-4. Interaction of phlorizin with cell membrane (glucose carbons numbered) (After Alvarado, P.: Biochim. Biophys. Acta. *135*:483, 1967.)

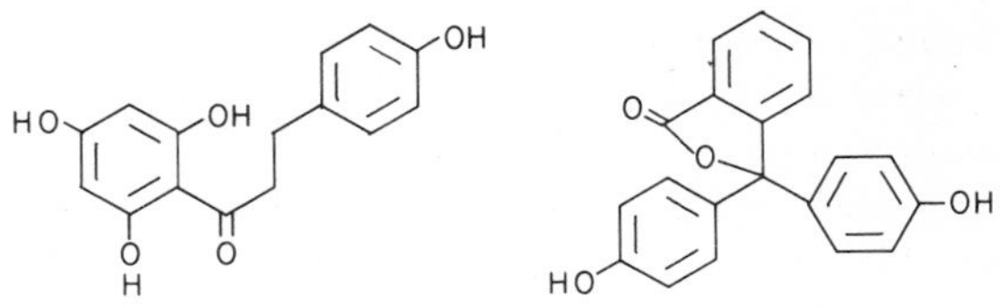

PHLORETIN   PHENOLPHTHALEIN

Figure 3–5.

possible attachment to the phenol-binding site, which could involve any of its four phenolic oxygens. In fact, the hydroxyl group at C-4 of Ring B, which is mandatory for inhibition by phlorizin, is not required in phloretin.

The phenol-binding site may bind other phenols. Phenolphthalein can strongly inhibit intestinal transport of glucose (as well as that of sodium), which may be the basis of its action as a cathartic. Phenolphthalein also competitively inhibits glucose transfer in red cells. Its structural resemblance to phloretin is evident in Figure 3–5. Search for the active binding sites of phloretin led LeFevre to the finding that the synthetic estrogen diethylstilbestrol is likewise a strong inhibitor of glucose transport in RBC, indeed more so than phloretin. Conversely, phloretin has definite estrogenic activity. Estradiol increases glucose uptake by its target organs, which could be due to interaction of the natural phenolic steroid with a phenol-binding site in such a way as somehow to facilitate rather than interfere with the binding of glucose to its carrier site on the cell membrane. The structures of diethylstilbestrol and estradiol (Fig. 3–6) indicate their similarity to each other and to phloretin. It may be that other steroids active on carbohydrate metabolism, e.g., adrenal glucocorticoids, though not phenolic, are sufficiently analogous in other respects (spacing of oxygen functions) to be active at the membrane site. Such effects are indeed known. Deoxycorticosterone glucoside, for example, is about half as effective as phlorizin in inhibiting tubular reabsorption of glucose. It should be noted that the natural steroids commonly form glucuronides through the hydroxyl on carbon 3 of Ring A, in which form they would be structurally more related to phlorizin.

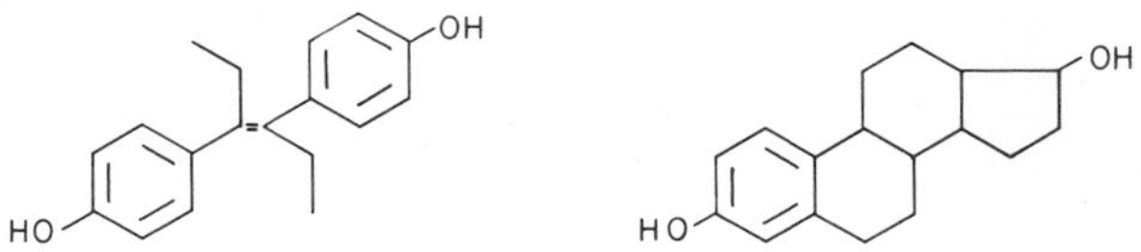

DIETHYLSTILBESTROL   ESTRADIOL

Figure 3–6.

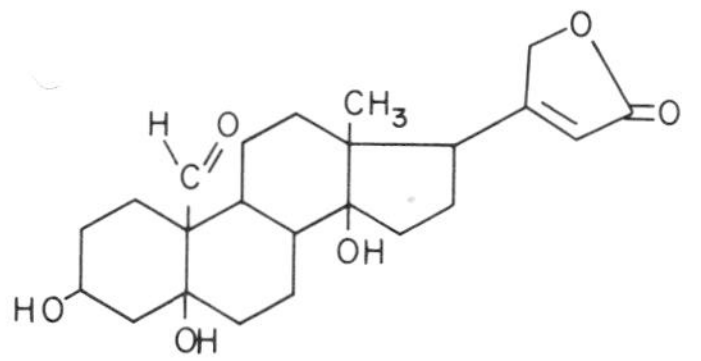

CELLULAR TRANSLOCATION OF CARBOHYDRATES

STROPHANTHIDIN

Figure 3-7.

Other vital compounds may interact with the sugar transport site. The uptake of pyrimidine nucleosides is inhibited by phlorizin, galactose, and 3-O-methylglucose (a non-metabolizable analogue), which has suggested that the sugar carrier could be involved in the transport of these nucleosides. The utilization (and perhaps transport) of glucose in brain and muscle is dependent on provision by the liver of some circulating factor(s), which have been tentatively identified as the pyrimidine nucleosides, cytidine and uridine. A class of pharmacological compounds which is membrane-active and has effects on transport of carbohydrate as well as Na^+ and K^+ is that of the cardiac glycosides, so known because of their clinically useful effects on contractility of heart muscle. One of these, ouabain, is available in crystalline form and has therefore been most studied. As mentioned in Chapter 2, it inhibits intestinal absorption of both sodium and glucose, but in muscle and adipose tissue it has an opposite effect on transport of cations and monosaccharides. The aglycone part of ouabain, strophanthidin (Fig. 3-7), has a steroid-like nucleus to which is attached an unsaturated lactone ring at C-17. The oral antidiabetic drug phenethylbiguanide is a compound with one phenol ring and a basic side chain. This compound also impairs intestinal absorption of glucose, but improves the uptake of glucose by muscles in diabetics, which has been related to its effect on increasing the muscle "clearance" of plasma insulin. Perhaps the binding of insulin to the membrane, which would be at or near the carrier site, is facilitated. The effects of tryptophan and its metabolites, including 5-hydroxytryptamine (serotonin), on glucose metabolism (Chaps. 5 and 8) should also be considered in the light of their chemical structural resemblance to steroids and like compounds and therefore their possible activity at membrane sites of glucose transport.

HORMONAL EFFECTS ON GLUCOSE TRANSPORT

The glucose transport system is particularly important and interesting because it has proved to be a major focus of physiological regulation by hormones. Although the major facts of such hormonal control may be known, we are only beginning to learn the details, and the role for various hormones in a number of organs is less clear than for others.

Insulin

The hormone most obviously important in controlling translocation of glucose through the cell membrane is insulin from the pancreas. Conversely, the most clearly demonstrated action of insulin on the cells of major organs is that of promoting glucose transport. Levine's experiments on distribution of D-galactose in the eviscerated nephrectomized dog showed that insulin increased appreciably the body space occupied by the galactose, essentially unmetabolized by the remaining tissues. Levine's group defined certain sugars (D-galactose, L-arabinose, D-xylose) as responsive to insulin and presumably sharing a transport system with glucose, whereas other sugars (D-fructose, L-sorbose, D-sorbitol, D-arabinose, L-rhamnose) were unresponsive. They postulated the requirement of a certain configuration of hydrogens and hydroxyl (like that of D-glucose) about carbon atoms 1, 2, and 3, but exceptions to this rule have since been noted. Park further demonstrated that in heart muscle under certain conditions insulin increases the intracellular concentration of glucose, which suggested its action on transport rather than metabolism.

Many, perhaps most, of the myriad of other effects of insulin on carbohydrate metabolism could be rationally explained as secondary to a primary action on transport of glucose (or amino acids or electrolytes). That there is only very tenuous evidence for any kind of action of insulin on broken-cell preparations is another argument for the primacy of action on transport. Furthermore, the action of insulin on glucose transport is not dependent on synthesis of RNA or protein, as may be the case with other hormones.

Probably the action of insulin on glucose transport is only one part of a general effect of this hormone to stimulate the penetration into the cell of various types of small molecules, including electrolytes and amino acids as well as sugars. Various sugars, e.g., glucose, galactose, and fructose, inhibit the absorption of amino acids into rat kidney cortex slices and through intestinal mucosa, which has suggested that sugars and amino acids compete for a common carrier. Insulin is well known to cause a lowering of concentration of several amino acids in plasma. Acceleration of the influx of K^+ into muscle cells, probably with concomitant egress of Na^+, is an established effect of insulin, which incidentally occurs at much lower insulin concentrations than those required to show an effect on glucose transport. There seems to be no stoichiometric relationship between movements of K^+ and glucose as affected by insulin, which further suggests that, at least in muscle, there is no intimate molecular association of electrolyte and non-electrolyte on a common carrier. Nevertheless, a certain difference between exchange and non-exchange passage of glucose* could obscure a stoichiometric pattern.

Attempts to demonstrate that insulin changes particularly the K_m or the V_{max} of movement of glucose have produced diverse findings in different studies and different tissues. Insulin may exert an effect on the ratio of mobilities of carrier and carrier-substrate complex or on the ratio of "active" and "inactive" forms of the carrier (Fig. 3-3). It could modify the binding sites to allow firmer binding of glucose; by lowering the K_m insulin could allow greater net transfer at lower concentrations of glucose, since transport, like enzymatic reactions, is more efficient at concentrations near the K_m.

Whatever the way in which insulin acts to stimulate transport, its action is blocked by sulfhydryl-binding agents. There is other evidence that insulin binds to the cell membrane through formation of disulfide bonds. With isolated fat cells it was shown that insulin must be continually present to exert its effect on glucose transport. Upon being washed off, the insulin effect is immediately lost and the cell becomes responsive to another aliquot of the hormone. The insulin which becomes bound to the fat cell membrane is metabolized or altered to products which are still unknown, but which no longer react as insulin by immunologic or biologic assay.

Steroids

Effects of steroids on glucose transport may be positive or negative, depending on the particular tissue and type of steroid, as mentioned above in the case of estrogens. Whereas glucocorticoids inhibit the entry of glucose into adipose, muscle, and thymus cells, they probably stimulate uptake by the intestinal mucosa (another example of differing effects on facilitated diffusion and active transport). Hydrocortisone increases the uptake of the non-metabolizable amino acid, amino-isobutyrate, into the isolated perfused rat liver, but whether this has any bearing on glucose transport in liver is not clear. In any case, the major physiological effect of glucocorticoids on carbohydrate metabolism is thought not to occur on glucose transport but rather on gluconeogenesis (Chap. 5).

Within 1 to 2 hours after injection in vivo of estradiol into ovariectomized rats the transport of the non-metabolized 3-O-methylglucose (3-O-MG) into the rat uterus in vitro is increased, but addition of es-

---

*Studies with ^{14}C-labeled glucose show that exchange between the glucose in the cell and that in the medium is several times faster than the maximal net rate of exit of sugars into a sugar-free medium. This suggests that the glucose-carrier complex makes several oscillations between the two sides of the cell membrane with exchanges of glucose molecules for each time there is a more complete dissociation.

tradiol in vitro has no effect. The V_{max} of a transport system obeying Michaelis-Menten kinetics is doubled, but there is no change in K_m. There are conflicting reports as to whether the effect of estradiol to stimulate transport of 3-O-MG into the intracellular space of uterine muscle is dependent on RNA or protein synthesis. The transport effects of other steroids, e.g., that of aldosterone on Na^+, are now thought to be preceded by an action on RNA and protein synthesis. Thus, it becomes necessary to consider whether the steroid itself must be translocated through the cell membrane to an intracellular site as a part of its action on carbohydrate transport. Perhaps relevant to this question is the observation that phloretin and diethylstilbestrol are both highly concentrated in erythrocytes by uptake from the medium. The glucose carrier which is affected by these agents and presumably by steroids may be related to a carrier for the agents and steroids, also.

Pituitary Hormones and Hypothalamus

Some of the trophic hormones of the pituitary, i.e., adrenocorticotrophic hormone (ACTH) and thyroid-stimulating hormone (TSH), seem to exert their effects on their target organs via stimulation of initial phases of glucose metabolism, but whether this may occur by an effect on glucose transport is not yet known. In rat heart the sensitivity to insulin of galactose transport shows a seasonal variation, since it is lower during the winter months. Likewise, rabbits kept in natural lighting conditions at controlled temperature are more sensitive to the hypoglycemic effect of insulin in summer than in winter. The photoperiodicity of this phenomenon suggests hypothalamic control of elaboration of an insulin antagonist. Possibly this antagonist, by selectively decreasing sensitivity to insulin of the glucose transport in muscle but not adipose tissue, favors lipogenesis from glucose in adipose tissue, thus fat storage during winter months.

MONOSACCHARIDE TRANSPORT IN DIFFERENT ORGANS AND SPECIES

Muscle

Transport of glucose in muscle is particularly significant because under most circumstances for both skeletal and cardiac muscle it is rate-limiting for over-all utilization of glucose. After stimulation by insulin or other influences phosphorylation may become rate-limiting. Besides the susceptibility to control by insulin and other hormones there are some "built-in" factors operating on glucose transport to maintain the integrity and augment the function of muscle. Exercise has long been known to have an insulin-like effect and to diminish the requirements of diabetics for insulin. A humoral factor released from exercised muscle can activate the glucose transport system in a cross-transfused dog. The same kind of sugar stereospecificity is exhibited as for insulin. This exercise or muscle activity factor, which is very labile, can be precipitated by ammonium sulfate from the incubation medium of isolated exercised skeletal muscle and then shown to increase uptake of 3-O-MG by another muscle specimen. Likewise in cardiac muscle, glucose uptake becomes progressively accelerated as the ventricle (of a perfused rat heart) develops more pressure. In the working rat heart there is stimulation up to five-fold in the rate of glucose uptake over that of the resting heart. Moreover, muscular work (at least in the heart) increases the sensitivity of the transport system to insulin.

The early experiments on the exercise factor used electrical stimulation as the muscle stimulant. Natural exercise (swimming, running) in rats increases the space of D-xylose distribution in muscles, and previous training increases the response of the transport system to natural exercise as well as to electrically stimulated contraction when the stimulation is submaximal. When the stimulation is maximal, there is no greater capacity for transport in trained muscles. Skeletal muscle in frogs (to depart from mammals) becomes more permeable to sugar when contractions are elicited not only by electrical stimulation but also by caffeine, which produces contractures without depolarizing the membrane. The increase in sugar permeability seems not to be related directly to depolarization of the membrane but more definitely to an increase of Ca^{++} concentration in the "myoplasm."

Anoxia is a strong stimulant to glucose uptake in skeletal and cardiac muscle. In the rat heart, anoxia increases glucose uptake five- to ten-fold, which is similar to the percentage change effected by insulin. Furthermore, under anaerobic conditions the heart muscle becomes 10 times more sensitive to insulin than the well-oxygenated heart. The action of work to increase glucose transport does not depend upon the development of anoxia, but maintenance of oxidative metabolism (e.g., by supply of pyruvate) can counteract the enhancing effect on sugar permeability of a substance which gradually appears in the perfusion fluid of the beating heart. This substance may be related to the humoral exercise factor of contracting skeletal muscle. One theory holds that the presence of ATP at the membrane site somehow restrains sugar transport and that anoxia (or dinitrophenol) acts by decreasing the available ATP. This theory suggests the action of insulin is rather secondary in that its function depends on promotion of phosphorylation and the resultant use of ATP.

There is a difference, however, in the way that insulin and anoxia stimulate glucose transport in skeletal muscle. Insulin lowers the apparent K_m of glucose uptake, i.e., increases the affinity, whereas anoxia increases the V_{max}, i.e., increases the capacity of the transport system. The effect of insulin remains evident under anaerobic conditions, which further suggests that the two influences do not share a common intermediate. One possible view is that anoxia opens up more transport sites, whereas insulin facilitates the binding of glucose to the available sites. In the perfused rat heart, insulin seems to increase both the K_m and the V_{max} of transport.

Glucose uptake by diaphragm increases in proportion to the concentration of Na^+, but the presence of Na^+ is not essential for glucose (or galactose) uptake; substitution of Na^+ by Li^+ even increases the uptake. Lithium, even in low concentrations of 1 to 5mM, exerts an insulin-like effect on glucose uptake and glycogenesis, though the incremental effect is much less than that of insulin. The effect of insulin is to increase K^+ content of muscle but augment release of lactate, whereas Li^+ has the opposite effect on both these constituents.

The effects of Li^+ are not unlike those of ouabain, the cardiac glycoside, which also stimulates glycogenesis in diaphragm muscle and inhibits production of lactic acid. These effects are directly proportional to the concentration of Na^+ in the medium. The omission of K^+ from the medium has consequences similar to that of ouabain. Ouabain inhibits the (Na^+-K^+) ATPase which is the "sodium pump" that is conceived as acting in concert with active glucose transport in intestinal mucosa kidney. Whether ouabain actually increases glucose transport in muscle in conjunction with its glycogenic effect is not clear. In any case, these considerations lie close to the mechanism of action of insulin on glucose transport and metabolism. An important effect of insulin (in various organs) is to cause a decrease in 3′,5′-cyclic AMP (cAMP), which is formed by adenyl cyclase from ATP in another kind of ATPase occurring at the cell membrane. Insulin reduces cAMP by inhibiting adenyl cyclase (Fig. 3-8).

Epinephrine stimulates glucose uptake by muscle, but this effect may be secondary to stimulation of glycolysis, as in adipose tissue (see below). Long-chain fatty acids inhibit glucose transport in muscle, particularly under conditions of work. This seems to depend upon oxidation of the fatty acids. The decrease in sensitivity of glucose transport to insulin, which is seen as an effect of pituitary growth or somatotrophic hormone (STH) and adrenal glucocorticoids, may be related to the increase in concentration of fatty acids which is a consequence of action of these hormones. Hyperosmolarity of the external medium, induced by various means, also stimulates glucose uptake by diaphragm muscle and adipose tissue.

The question of glucose transport in smooth muscle remains largely unexplored except for the studies with uterus (see above). There is much reason to suspect an abnormality in sodium transport in the smooth muscle of arteriolar walls as a component of the pathogenesis of hyper-

CELLULAR TRANSLOCATION OF CARBOHYDRATES 67

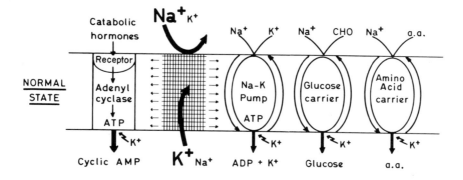

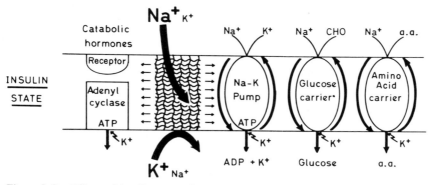

Figure 3-8. Effects of insulin on conditions and events at the cell membrane. (According to M. Rodbell, by permission, from Metabolism Lecture, FASEB Mtg., 1969.)

tension. It would be logical also to suppose that the function of ATPase, which is intimately associated with the contractile elements of actomyosin, may be disturbed in hypertension. Evidence now connects hypertension with carbohydrate intolerance and abnormalities of plasma insulin. Interesting disclosures can be expected from further investigation of the glucose transport system in normal and pathological smooth muscle.

Adipose Tissue

In the fat cell, like muscle, glucose transport is a limiting factor in its utilization, and therefore physiological control mechanisms may be expected at the transport site. As in the case of muscle, insulin promotes the facilitated diffusion of glucose across the adipose cell membrane and at lower concentrations of insulin than are required for effects on glucose transport in muscle. Again suggesting relatedness of glucose and electrolyte transport, there is a decrease in the stimulation by insulin of oxidation of glucose-^{14}C to $^{14}CO_2$ in cells from epididymal fat pads when

Na$^+$ is replaced by K$^+$ or other cations in the medium. Ouabain has insulin-like effects on fat cells just as in muscle—stimulation of glucose uptake and glycogenesis as well as antilipolytic effects—and, like insulin, ouabain reduces adenyl cyclase activity. These effects can be simulated by complete omission of K$^+$ from the medium, which further relates the carbohydrate actions of both insulin and ouabain fundamentally to electrolytes.

Some insight into the mechanism of action of insulin on the fat cell membrane has come from the use of phospholipases and proteases applied to cell membranes of fat cells isolated by collagenase; such application strips away the outer "basement" membrane. At concentrations below those which rupture the remaining plasma membrane, which still contains the carrier mechanism, these lytic enzymes have insulin-like effects on glucose uptake and lipolysis. Electron microscopy suggests a change in configuration of lipoproteins of the membrane whereby laminar components are changed to a micellar state.

There is conflicting evidence as to whether the lipolytic hormones, e.g., epinephrine, glucagon, ACTH, affect glucose uptake. All these hormones increase the intracellular level of free fatty acids (FFA). The latter, when presented externally to fat cells, increase glucose uptake, which is in contrast to their effect on muscle. This is an important difference because the effect of several hormones which act commonly to increase intracellular FFA may be a diversion of glucose uptake from muscle to adipose tissue. The above lipolytic hormones all cause an increase in adenyl cyclase activity in fat cell membranes. The effect of epinephrine to increase uptake of the non-metabolized sugar, L-arabinose, by a mechanism not dependent on its lipolytic effect has been explained as a function of the stimulatory action of cAMP on a key enzyme in glycolysis, phosphofructokinase. With increase in glycolysis there is a decrease in glucose-6-phosphate, which could be a feedback inhibitor of transport of glucose (or L-arabinose), as it is for hexokinase (Chapter 4). This theory would seem to require that the effect of insulin on glucose transport is not mediated by a decrease of cAMP but that changes in glucose transport and cAMP are both consequences of an effect of insulin on the membrane.

Rodbell has advanced the notion that the actions of insulin on glucose transport, amino acid transport, adenyl cyclase, and the (Na$^+$-K$^+$) ATPase are all secondary to the effect of insulin to decrease the affinity or permeability of the membrane for potassium relative to sodium. This is pictured as occurring coincident with a change in the state of configuration of the lipoproteins of the membrane (Fig. 3-8). Such a change could thereby explain intracellular retention of potassium, change in resting potential of the membrane, decreased activity or sensitivity of adenyl cyclase to catabolic hormones, and increased activity of the carrier systems for glucose and amino acids.

Rodbell noted that fatty acids, when confined at high concentrations within fat cells, may themselves cause defects in the plasma membrane as well as release of enzymes. He suggested that FFA may thereby have transport effects parallel to those of insulin and further that the prostaglandins, a special class of circulating fatty acids which have strong antilipolytic action (see Physiological Chemistry of Lipids in Mammals), could also be membrane active.

As with other tissues, sulfhydryl groups on the membrane surface of fat cells interact with insulin. One of the sulfhydryl-blocking agents, N-ethylmaleimide, has biphasic effects on glucose transport with an insulin-like action under some circumstances but an inhibitory effect on insulin stimulation upon more prolonged exposure.

Some other influences may operate to promote glucose transport by fat cells. It could be asked whether the exercise factor from muscle has an effect on adipose tissue which is in proportion to that of insulin on the two tissues. Certain divalent metal ions, Co^{++} and Ni^{++}, enhance the uptake of glucose by the epididymal fat pads and particularly increase the oxidation of C-1 of glucose to CO_2. Nickel increases incorporation of glucose into glycogen. Zinc stimulates glucose uptake by adipose tissue from zinc-deficient rats; these have high levels of plasma FFA and rapidly lose body fat. Zinc readily forms complexes with insulin and could be involved in its physiological action. Lithium increases uptake of glucose by rat epididymal fat pads, as it does for muscle. Some polyene antibiotics have insulin-like effects on glucose transport in fat cells, possibly by combining with cholesterol-containing lipids in the membrane.

Various species differ markedly in the activity of uptake of glucose into adipose tissue. Slices of epididymal or perirenal fat of the rat show a rate of glucose uptake 10 times that of corresponding tissues of rabbit or hamster with guinea pig intermediate. Responses to insulin are proportionate. Human adipose tissue is also slow in uptake relative to that of the rat. The body sites of adipose tissue may also vary in activity—in the human being omental fat utilizes glucose more rapidly than does subcutaneous fat.

Brain

The brain normally utilizes only glucose as a nutrient and at rest is the organ which consumes more of the blood glucose than any other. Therefore, glucose transport in the brain is of vital importance (relevant, one might say, to social as well as physiological problems). In general, water-soluble substances do not pass readily into the brain, which has prompted the concept of a blood-brain barrier (b-b-b), but glucose penetrates the brain so rapidly that a special mechanism, presumably a "carrier" as in cell membranes of other organs, has been postulated. Krogh first proposed that the b-b-b behaves more like a typical cell plasma membrane than like the usual capillary wall. Where the b-b-b is located is still a matter of speculation, but some evidence suggests the endothelial cells which line the brain capillaries in an overlapping manner. The cerebral capillaries contain ATPase in the basement membrane and in the endothelial cell, which further indicates a regulatory function on transport of both glucose and ions.

The brain shows stereospecificity and substrate competition among monosaccharides, which are evidence for a specific carrier mechanism. Saturation kinetics are exhibited for D-glucose and D-mannose, both of which have high affinity for the carrier while D-galactose and D-xylose have low affinities and do not show saturation at high concentrations by evidence in the mouse. Curiously, brain levels of glucose in the rat are much lower than those in the mouse. As tested with perfusion of glucose-^{14}C into the carotid circulation of dogs, the carrier system for glucose is saturated when the concentration of glucose exceeds 70 mg./dl.* (Both common and clinical experience suggest that at least some parts of the human brain are better nourished at higher concentrations.) About 25 per cent of the glucose-^{14}C in the dog is extracted in one passage through the brain circulation (at a concentration of 80 to 100 mg./dl.) but the net extraction at this level is about 10 to 15 per cent, so there must be a passage of glucose in both directions across the b-b-b. Glucose-^{14}C passes the b-b-b 6 to 7 times as fast as fructose-^{14}C, which further indicates specificity and also explains why fructose is relatively ineffective in alleviating the symptoms of hypoglycemia.

Slices of guinea pig cerebral cortex show saturation kinetics for 2-deoxy-D-glucose, D-glucose, D-arabinose, and D-xylose in order of diminishing affinity, while D-ribose, D-galactose, D-mannose, L-sorbose, and D-fructose enter by simple diffusion or have low affinities for the carrier. Thus, guinea pig cortex in vitro differs from rat and mouse brain in vivo in its affinity for D-mannose. In rat brain there is saturable translocation of 3-O-MG, which is inhibited competitively by D-galactose, D-glucose, 2-deoxy-D-glucose, D-xylose, and D-mannose and which exhibits countertransport in response to infusion of mannose or glucose.

Until recently no studies indicated that insulin increases glucose transport or utilization in the brain, but some evidence with human beings has shown that insulin may lower the threshold for glucose uptake by the brain and that the infusion of insulin together with glucose increases the uptake by 50 per cent over the value with glucose alone. Low uptakes by patients with cerebrovascular disease or diabetes could be restored nearly to normal by insulin. In regard to particular brain areas, the uptake of gold thioglucose (which parallels that of glucose) by the satiety center of the hypothalamus is subnormal in diabetic mice and improved by insulin. Insulin increases glucose uptake and utilization by

---

*dl. is more succinct than the commonly used equivalent, 100 ml.

the anterior pituitary. The effect of a hepatic extract, or of cytidine plus uridine, to maintain glucose uptake by the perfused cat brain has been mentioned. Other natural regulators or influences on glucose transport of brain are not presently known.

Ouabain at high concentrations inhibits uptake of arabinose by cat brain, which suggests a connection between sugar and ion transport. Furthermore, this implies the existence of a concentrating type of transport as found in the intestine and kidney, where ouabain is also inhibiting. In general, transport of sugars against a gradient across the b-b-b has not been observed. Phlorizin is not inhibitory to uptake of arabinose by cat brain. The effect of Li^+ on glucose transport in brain may become of interest in view of its recent use in treatment of manic episodes in psychiatric patients at serum levels (0.5 to 5.0 mM) which are sufficient to stimulate deposition of glycogen in muscle.

Red Blood Cells

Translocation of monosaccharides in red cells has served as a model for detailed studies of stereospecificity, substrate competition, and kinetic aspects of the membrane carrier. Relative specificity (affinity) for various sugars has been listed in Table 3-1. Both aldoses and ketoses seem to share a common carrier mechanism in RBC. Although fructose has a very high K_m for entry into red cells, a trans effect is exerted by glucose at high intracellular concentration on the movement of fructose-^{14}C into the cell, which suggests a common carrier.

Remarkable differences exist among and even within species in the rate of glucose transport into red cells. Cells from most laboratory animals are almost impermeable to glucose compared with the high rate in humans and other primates. The maximum rate in rabbit erythrocytes is only about 0.4 per cent of that of human red cells. Another difference is that human cells are very sensitive to sulfhydryl-blocking agents, whereas rabbit cells are not. There are similar stereospecificities in human and rabbit cells, however. Curiously, fetal erythrocytes among non-primates have a high rate of transport like that in human cells. The relatively slow rate of penetration for animal red cells is probably of the order of the rate of glucose utilization by these cells, while for human cells transport is many times faster than utilization. According to the saturation kinetics there are at most 600,000 binding sites per human red cell and each site can transport about 150 glucose molecules per second (at 25° C). Thus, about 25 mg. of glucose can enter 1 ml. of packed erythrocytes in one minute.

An intraspecies difference in fructose entrance into erythrocytes has been demonstrated with cattle. Some individuals have RBC which are 5 to 10 times faster in fructose accumulation than those of other bovines in a herd. No corollary changes are yet known which would help explain or identify this interesting phenomenon. There are other species differences in fructose transport; e.g., glucose is a potent inhibitor of fructose transport in human RBC but not in those of the rabbit.

Although phlorizin and other phenolic compounds inhibit glucose transport in red cells (see before) and sulfhydryl groups are typically involved with the carrier site, insulin exerts no accelerating effect on glucose transport in red cells.

Monosaccharide transport into white blood cells appears to be little studied. Characteristics may not be similar to those for red cells, since the cell types have different embryonic origins.

Kidney

In renal tubular cells, as in the intestine, there is a transport from lumen to cell of glucose against a concentration gradient. Both D-glucose and D-galactose are transported by an active Na^+-dependent mechanism, which is strongly inhibited by phlorizin. There is competition between these two sugars; D-xylose and α-methyl-D-glucoside also share the common carrier system. If the Na^+ concentration of fluid perfused through a rat kidney is decreased below 50 mM (about one-third of plasma concentration) there is a gradual decline of glucose reabsorption, although about 25 per cent of tubular glucose is reabsorbed even in Na^+-free medium. Ouabain and other cardiac glycosides decrease reabsorption of Na^+ and glucose concomitantly; it may be assumed that, as with the intestine, a (Na^+-K^+) ATPase is involved in the energetic mechanism for active glucose reabsorption. The latter occurs progressively all along the proximal renal tubule, and the tubular fluid/plasma ratio for glucose concentration declines steadily along the proximal tubule. By contrast, although both water and Na^+ are also reabsorbed, the proximal tubular fluid remains equimolar with plasma in Na^+ concentration.

Although a concentrative, ouabain-inhibited mechanism of transport appears to exist at the luminal brush border of the tubular epithelial cell, the exit of glucose at the basal cell membrane probably occurs by facilitated diffusion, as inferred from studies with galactose. The net transfer of glucose from lumen to plasma will be a composite result not only of translocation across the two opposite borders of the tubular cell but of glucose utilization by various pathways and also of gluconeogenesis by the cell.

Since the carrier mechanism for glucose transport into the cell exhibits saturation kinetics, there is a limit to the reabsorptive capacity which for the whole kidney is known as the "tubular maximum" for glucose, or Tm_G. Generally, if the concentration of glucose in the plasma rises above 180 to 200 mg./dl. and the glomerular filtration rate is in the normal range (in man about 120 ml./min.), then the Tm_G will be exceeded and glucose will begin to appear in the urine (glucosuria). This has long been a gross index of the existence of diabetes mellitus and was responsible for the name of the disease. The presence of abnormally high concentration of glucose in the tubular fluid interferes osmotically with reabsorption of water, which then results in excessive urinary volume. Diabetes in Greek means "to run through a siphon" and mellitus refers to "sweet." At normal fasting blood glucose concentrations the kidneys of man reabsorb about 100 to 150 mg. glucose/min. The Tm_G which is approached asymptotically at concentrations above 150 mg./100 ml., is about 250 mg./min. By study of the quantitative aspects of the depression of Tm_G as a function of the number of binding sites occupied by ^{14}C-labeled phlorizin in the dog kidney, the maximum number of operating carriers per cell has appeared to be about 10^7 — about 20 times the maximum number for a red cell. On this basis the average minimal turnover rate for the glucose-carrier complex is 1400 glucose molecules/carrier site/min. Thus the carriers in kidney cells appear to be much more numerous but much slower than those of red cells.

As indicated by studies with 3-O-MG and L-glucose, the sugar transport carrier in kidney tubules is bidirectional in nature and exhibits countertransport effects. An interaction of amino acid and glucose transport in kidney has previously been mentioned. Some amino acids (lysine, glycine, and alanine) but not others (aspartic acid and leucine) cause a significant depression of glucose reabsorption by the dog kidney tubule. However, the effects are not necessarily competitive but may depend on general alterations of membrane permeability or intracellular osmotic pressure. Similar considerations pertain to the "competition" seen between glucose and phosphate for renal tubular reabsorption, which may or may not represent a common carrier phenomenon. There is a positive effect of parathyroid hormone (PTH) on tubular reabsorption of glucose. It has been hypothesized that inhibition of phosphate reabsorption by PTH permits greater availability of the carrier for transport of glucose.

In the disease of true renal glucosuria, which is a benign and often familial condition, glucose is not adequately reabsorbed even in the absence of hyperglycemia. The nature of the transport defect is not known; possibilities for either low capacity or low affinity of the carrier for glucose have been considered. Renal glucosuria need not be accompanied by abnormalities in intestinal glucose transport, which might be expected as a common genetic defect because of the similar type of carrier. The phenotypic variability in familial renal glucosuria suggests that several different mutations are involved. Glucose reabsorption, like that of sodium, decreases with some other pathological conditions, e.g., uremia due to progressive renal disease. Diabetic patients sometimes have very low or very high thresholds for glucosuria; for this there is no adequate explanation. Low thresholds in some diabetics may be related to the condition of pseudorenal glucosuria, which is more common than true renal glucosuria, requires plasma glucose concentrations somewhat

1. 2 glucose + 2 ATP $\longrightarrow$ 2 glucose-6-P + 2 ADP
2. glucose-6-P $\longrightarrow$ glucose-1-P
3. UTP + glucose-1-P $\longrightarrow$ UDPG + PP$_i$
4. UDPG + glucose-6-P $\longrightarrow$ trehalose-6-P + UDP
5. trehalose-6-P $\longrightarrow$ trehalose + P$_i$
6. trehalose $\longrightarrow$ 2 glucose

Sum: 2 glucose + 2 ATP + UTP $\longrightarrow$ 2 glucose + 2 ADP + UDP + PP$_i$

Figure 3-9. A proposed mechanism for reabsorption of glucose in the kidney. (From Sacktor, B.: Proc. Nat. Acad. Sci. 60:1007, 1968.)

above 100 mg./dl. to produce glucosuria, and may precede the appearance of diabetes. Insulin has not been observed to act directly on tubular reabsorption of glucose. Pregnancy can be accompanied by renal glucosuria due to increased glomerular filtration rate.

Recently trehalase, as well as enzymes which perform the biosynthesis of the disaccharide trehalose (Chap. 2), has been identified in the renal corticoid tubules of 10 mammalian species, including man. Sacktor has suggested that these enzymes could function as a mechanism for reabsorption of glucose in the kidney, and possibly intestine, through the set of reactions in Figure 3-9. Trehalose-6-P phosphatase has a pH optimum of 9.3; high concentration of alkaline phosphatase at the luminal border of kidney and intestinal epithelial cells has long been recognized. There is a wide range of activities of tubular trehalase among different mammalian species, with primates having intermediate levels. In man the activity of trehalase and other enzymes requisite for the reaction of Figure 3-9 is sufficient to account for glucose reabsorption, but a physiological role remains to be proved. No labeled trehalose could be demonstrated in extracts of kidneys incubated with glucose-1-^{14}C.

Liver

The liver cell has long seemed to be freely permeable to glucose, since concentrations of glucose in intracellular water are usually similar to those in plasma water and equilibration takes place rapidly, though not so fast as mixing of glucose in plasma and interstitial water of liver sinusoids. In the postabsorptive state, when blood glucose and insulin levels are low, the liver is a producer of blood glucose and intracellular levels are higher than extracellular, whereas during hyperglycemia the liver normally has a balance of net utilization of glucose and the concentration gradient is inward.

Glucose moves both into and out of liver cells at a rate of about 1 mg./g. of liver/min., which is about 10 times the rate of glucose production. Glucose is exchanged between the intra- and extracellular phases

CELLULAR TRANSLOCATION OF CARBOHYDRATES

with a half-time of only about 1.5 minutes. However, glucose does not enter (or probably leave) the liver cell simply by free diffusion but by an extremely fast transport system. This was shown by comparing the rates of simultaneous equilibration of ^{3}H-labeled D-glucose and ^{14}C-labeled L-glucose. Also, phlorizin lowers the rate of penetration of monosaccharides into the liver.

In contrast to muscle, fat cells, and some other tissues, insulin does not seem to promote facilitated transport of glucose into the liver. L-glucose and 3-O-MG distribute into about 85 per cent of total liver tissue water and this is not increased by insulin (but see Chap. 8). Insulin in the diabetic dog can promptly increase the uptake of glucose by the liver or permit it to occur at elevated blood levels, but this is supposedly an action on the enzymes, glucokinase or glycogen synthetase. However, the possibility of a membrane transport effect should not be excluded, particularly in view of the intimacy of glycogen synthetase with the cell membrane. Insulin promptly aids the perfused liver to reincorporate K^+ into the cell and inhibits the release of glucose into the perfusate—which, again, could be an enzymatic inhibition of glycogenolysis or gluconeogenesis (Chaps. 5 and 6) rather than a transport effect.

Other Organs and Tissues

Translocation of glucose across the placenta shows saturation at high concentrations and competition with galactose and fructose. There is a facilitated diffusion type of transfer with a gradient from maternal to fetal blood. Changes in maternal blood glucose levels are reflected by similar changes in the fetal blood glucose. As the concentration of blood glucose is raised on the maternal side of the isolated perfused human placenta, the transplacental gradient increases but the apparent net transfer rate may actually decrease. To explain this a carrier mechanism is postulated in which the transfer rate is proportional to the difference in number of saturated carriers at the two sides of the membrane. This factor could operate to minimize changes in the fetal blood glucose and protect it from large fluctuations of concentration in the maternal blood. (Nevertheless, there is evidence that the fetus is affected by abnormally high concentrations of blood glucose in diabetic mothers.)

In the choroid plexus, which is a network of special capillaries between the brain ventricles, there occurs a concentrative type of transport system for glucose, galactose, and other sugars such as 3-O-MG, which is inhibited by anoxia, dinitrophenol, phlorizin, lack of sodium, and ouabain. There is a brush border type of microscopic structure like that of intestine and kidney. Active transport at the choroid plexus may operate to maintain the glucose concentration of cerebrospinal fluid below that of plasma. In the ciliary body of the eye there is a similar concentrative transport, which may govern the glucose concentration of the aqueous humor.

Characteristics of uptake of 3-O-MG by isolated bone cells indicate that concentrative transport of monosaccharide does not occur, nor do hormones (PTH, thyrocalcitonin) otherwise active on these cells have any apparent effect on glucose uptake. Some mucosal structures other than the intestine are permeable to glucose. The urinary bladder is able to absorb D-glucose from the mucosal side, as indicated by the rate of conversion of glucose-1-^{14}C to $^{14}CO_2$ in vivo after intravesical injection in Swiss mice. The rate was almost as fast as after intraperitoneal injection. Absorption of ^{14}C-labeled glucose from the oropharyngeal cavity occurs with a curious concentration through the face and brain which indicates a passageway more direct than through the general circulation.

The penetration of glucose into the beta cells of the pancreas could be important because of the major stimulatory effect of hyperglycemia on insulin production and release. Over a wide range of external glucose levels (5 to 100 mM) the intracellular glucose concentration of pancreatic islets is nearly equal to that of plasma, which suggests ready permeability similar to that of liver. The fact that ouabain produces a rise in circulating insulin levels with hypoglycemia in the intact dog is some suggestion of a membrane effect. The sensitivity of insulin release to agents which regulate the level of adenyl cyclase is further circumstantial evidence of significant events occurring at the membrane. The release of insulin from the secretory granule depends upon some alteration of the conformation of the membrane, which could be related to transport of glucose or other substances.

INTRACELLULAR TRANSLOCATION OF CARBOHYDRATES

Compared with the knowledge of transport through the outer cell membrane, that which concerns intracellular translocation is at present minimal. Various analyses with non-metabolized sugars suggest that they distribute into part but not all of the intracellular aqueous space and that insulin in some instances can increase the fraction of intracellular space available to glucose or other sugars. The changing rate of efflux of galactose from the kidney cell suggests two intracellular compartments for galactose.

Mitochondria

There is growing information about the translocation through the mitochondrial membrane of carbohydrates of the tricarboxylic acid (TCA) cycle and related compounds (see Chap. 4). An ATPase is a component of the (inner) mitochondrial membrane. While its phosphorylative function is linked to that of electron transfer, there may be associated conformational changes that could affect permeability of the mitochondrial membrane to carbohydrates. The penetration of the dicarboxylic acids, malate and oxaloacetate, into isolated mitochondria is energy-facilitated, being stimulated by ATP and inhibited by cyanide. The kinetic data, high activation energies and stereospecificities indicate special permeability mechanisms for di- and tricarboxylic acids rather than free diffusion. Some dicarboxylic acids (D- and L-malate, succinate, mesotartrate) but not others (fumarate, maleate, D- and L-tartrate) enter by a specific mechanism. Butylmalonate inhibits transport but not enzymatic reaction of L-malate. The presence of L-malate (or an analogue, e.g., malonate) somehow activates the influx or efflux of other organic anions, e.g., 2-oxyglutarate and the tricarboxylic acids, citrate, cis-aconitate, and isocitrate. There are separate transporting systems (located at the inner, enfolded mitochondrial membrane) for the dicarboxylic acids, for the tricarboxylic acids, and for oxyglutarate. (Presumably another alpha-keto acid, pyruvate, shares the latter.) Thus, the transfer of oxyglutarate is very active in heart mitochondria, but not that of citrate; in liver and kidney both systems are active. Rates of entry of malate and oxaloacetate into isolated mitochondria of rat liver, kidney, and heart have been measured. The K_m values for penetration into liver mitochondria were 130 μM and 40 μM, respectively, for malate and oxaloacetate. The concentration of cytoplasmic malate is usually well above this measured K_m, but that of oxaloacetate is lower.

Many metabolic functions of cells, e.g., oxidation of substrates, lipogenesis, and gluconeogenesis, are activities coordinated between enzyme components within and outside the mitochondrion and requiring passage of such compounds as malate, citrate, pyruvate, and others through the mitochondrial membrane (Fig. 3-10). Although the cytoplasm of the liver cell in some species contains enzymes which can be coupled to form glucose from pyruvate, the process of gluconeogenesis in other species seems to require movement of pyruvate into the mitochondrion, carboxylation there to oxaloacetate, then conversion of oxaloacetate to malate and translocation of the latter through the membrane to the extramitochondrial space (Fig. 3-10) (see Chap. 5). Since there is no appreciable transfer through the mitochondrial membrane of the coenzymes nicotinamide-adenine dinucleotide (NAD) and NAD-phosphate (NADP), which shuttle hydrogen and electrons (reducing equivalents) in processes of oxidation and biosynthesis, the movement of these reducing equivalents across the mitochondrial membrane must occur via substrates, such as malate and glycerol-phosphate. In the diagram of the composite, idealized cell of Figure 3-10 it can be seen that malate can shuttle reducing equivalents from either NADH or NADPH into the mitochondrion after their generation in the cytoplasm. Details of the indicated reactions are found in other chapters. On the other hand, in other metabolic circumstances or particular tissues the movement of malate out of the mitochondrion may provide transfer of reduc-

78   PHYSIOLOGICAL CHEMISTRY OF CARBOHYDRATES IN MAMMALS

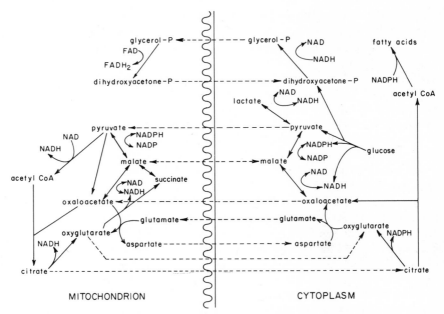

Figure 3–10.   Flow of various substrates across the mitochondrial membrane in the coordinated activity of metabolic processes in mitochondrion and cytoplasm.

ing equivalents for synthesis of fatty acids, glucose, or other cell constituents. Within the mitochondrion of the adrenal cortical cell the reducing equivalents of malate may be used for hydroxylation of steroid intermediates via NADPH. Malate is formed inside the mitochondrion mainly from succinate via the TCA cycle. Isotopic studies suggest that succinate is a good source of reducing equivalents for reductive processes in the cytoplasm.

Oxaloacetate does not move readily out of the mitochondrion but is linked through various reactions to malate, to oxyglutarate, to citrate, and to the amino acids, glutamate and aspartate, all of which can move through the mitochondrial membrane (Fig. 3-10) to provide further avenues for carbon transfer. These multiple hydrogen and carbon transfers may result in different concentrations of reduced and oxidized substrates on the two sides of the membrane in the dynamic living state, thus setting up differences in redox potential between cytoplasmic and mitochondrial compartments, an effect which has been experimentally observed. Citrate moves from intra- to extramitochondrial location as a means of transferring acetyl CoA for biosynthesis of fatty acids (Fig. 3-10). Enzymes for formation and oxidation of glycerol phosphate in cytoplasm and mitochondrion are coupled with the movement of glycerol phosphate into the mitochondrion, which allows for another means for transfer of reducing equivalents. This pathway has implications for the metabolic efficiency of the cell because of its lower P/O ratio than for other oxidative pathways.

In liver cells the redox pairs of substrates such as malate/oxaloacetate, glycerol phosphate/dihydroxyacetone phosphate, and lactate/pyruvate seem to be coupled through the common coenzyme pair NADH/NAD$^+$, since the ratios for all these redox pairs fluctuate in the same direction together. In diabetes these ratios are higher than normal, and insulin administered in vivo promptly (within 1 hour) lowers them, an effect which although not yet explained could suggest a locus of action at the mitochondrial site. The effect of thyroid hormone also is toward maintenance of lower levels. There are other indications that the level of the redox pairs is sensitively poised in response to the energy state of the mitochondria. In muscle, unlike liver, the redox pairs are not strongly coupled, which may indicate compartmentation. Also, in muscle mitochondria, only pyruvate and glycerol phosphate are oxidized rapidly and transport systems for other compounds shown in Figure 3-10 may be relatively inactive. This serves partly to explain the accumulation by working muscle of lactate, which is not oxidized readily by muscle, but instead when formed in excess leaves the muscle cell to be delivered by the circulation to other tissues (Chap. 8).

REFERENCES

For General Discussion of Monosaccharide Transport

Csaky, T. Z.: Transport through biological membranes. Ann. Rev. Physiol. 27:415–450, 1965.
Heinz, E.: Transport through biological membranes. Ann. Rev. Physiol. 29:21–58, 1967.
LeFevre, P. G.: Transport of carbohydrates in animal cells. *In:* Metabolic Pathways, Vol. 6, ed. by L. E. Hokin, Academic Press, New York, 1972, pp. 385–454.
Park, C. R., Crofford, O. B., and Kono, T.: Mediated (nonactive) transport of glucose in mammalian cells and its regulation. J. Gen. Physiol. 52:296–318, 1968.
Stein, W. D.: The transport of sugars. British Med. Bull. 24:146–149, 1968.
Widdas, W. F.: Membrane transport of sugars. *In:* Carbohydrate Metabolism and Its Disorders, ed. by Dickens, Randle, and Whelen, Academic Press, New York, 1:1–23, 1968.

For Molecular Nature of the Binding Site

Alvarado, P.: Hypothesis for the interaction of phlorizin and phloretin with membrane carriers for sugars. Biochim. Biophys. Acta 135:483, 1967.

For Hormonal Effects on Monosaccharide Transport

Levine, R., and Goldstein, M. S.: Action of insulin on tissues. New York State J. Med. 62:1236–1245, 1962.
Roskoski, R., Jr., and Steiner, D. F.: The effect of estrogen on sugar transport in the rat uterus. Biochim. Biophys. Acta 135:717–726, 1967.

For Monosaccharide Transport in Muscle

Bihler, I., Cavert, H. M., and Fisher, R. B.: The uptake of pentoses by the perfused isolated rabbit heart. J. Physiol. 180:157–167, 1965.

Chaudry, I. H., and Gould, M. K.: Kinetics of glucose uptake in isolated soleus muscle. Biochim. Biophys. Acta *177*:527–536, 1969.

Gould, M. K., and Rawlinson, W. A.: Effect of electrical stimulation and training on muscle pentose transport. Am. J. Physiol. *211*:141–146, 1966.

For Monosaccharide Transport in Adipose Tissue

Ho, R. J., and Jeanrenaud, B.: Insulin-like action of ouabain. I. Effect on carbohydrate metabolism. Biochim. Biophys. Acta *144*:61–73, 1967.

Rodbell, M., Jones, A. B., Chiappe de Cingolani, G. E., and Birnbaumer, L.: The action of insulin and catabolic hormones on the plasma membrane of the fat cells. Recent Progr. in Hormone Research *24*:215–254, 1968.

For Monosaccharide Transport in Brain

Crone, C.: Facilitated transfer of glucose from blood into brain tissue. J. Physiol. *181*:103–113, 1965.

Le Fevre, P. G., and Peters, A. A.: Evidence of mediated transfer of monosaccharides from blood to brain in rodents. J. Neurochem. *13*:35–46, 1966.

For Monosaccharide Transport in Kidney

Diedrich, D. F.: Glucose transport carrier in dog kidney: its concentration and turnover number. Amer. J. Physiol. *211*:581–587, 1966.

Sacktor, B.: Trehalase and the transport of glucose in the mammalian kidney and intestine. Proc. Nat. Acad. Sci. *60*:1007–1014, 1968.

For Monosaccharide Transport in Liver

Hetenyi, G., Jr., Norwich, K. H., Studney, D. R., and Hall, J. D.: The exchange of glucose across the liver cell membrane. Canad. J. Physiol. and Pharmacol. *47*:361–367, 1969.

For Monosaccharide Transport in Placenta

Howard, J. M., and Krantz, K. E.: Transfer and use of glucose in the human placenta during in vitro perfusion and the associated effects of oxytocin and papaverine. Am. J. Obst. and Gyn. *98*:445–458, 1967.

For Transport of Carbohydrates in Mitochondria

Chappell, J. B.: Systems used for the transport of substrates into mitochondria. Brit. Med. Bull. *24*:150–157, 1968.

Hohorst, H. J., Arese, P., Bartel, H., Stratmann, D., and Talke, H.: L(+)-lactic acid and the steady state of cellular Red/Ox systems. Ann. N.Y. Acad. Sci. *119*:974–994, 1965.

CHAPTER

4

INTRACELLULAR UTILIZATION OF MONOSACCHARIDES

Since glucose is the principal circulating carbohydrate source of energy, the enzymatic apparatus of most mammalian tissues is well geared for the catabolism of this monosaccharide. However, significant differences exist among different tissues and organs, and among different species, as will be discussed. Important in mammalian metabolism also is the utilization of fructose, galactose, and mannose. The utilization of some of the pentoses and more special carbohydrates is discussed in other chapters. The oxidation of intermediary carbohydrates, which is common to the metabolism of various monosaccharides as well as amino acids, is considered in the present chapter.

Controversy has existed concerning whether the initial metabolic utilization of glucose, which for the most part in mammals consists of its phosphorylation at position C-6, is sequential to or inseparably linked to the transport of glucose through the cell membrane. Studies of R. Levine and of C. R. Park and others have shown that transport is not necessarily linked to phosphorylation. For instance, at temperatures much lower than body temperature phosphorylation is almost completely abolished, but translocation of glucose into muscle cells can be observed. Translocation of glucose and other monosaccharides from extra- to intracellular phase is described in Chapter 3. Nevertheless, there is still much to be learned about the relationship between transport and enzymatic utilization of carbohydrates. Certain phosphorylated sugars, e.g., glucose-6-phosphate, are rapidly dephosphorylated after introduction into the circulation and are utilized by certain tissues in vitro (e.g., liver slices and intact diaphragm). Some studies with the isolated perfused rat heart suggest that certain glycolytic enzymes (phosphohexose isomerase, aldolase, triosephosphate isomerase) are attached to the superficial cell membrane and may exert their actions on substrates external as well as internal to the cell membrane.

PHOSPHORYLATION OF GLUCOSE

Any mammalian tissue which utilizes glucose must have the capacity for its phosphorylation. The value to the cellular energy exchange of phosphorylation of substrates in general lies in the universal dependence of cells (not only mammalian) on carbon-phosphate, nitrogen-phosphate, and phosphate-phosphate bonds for storage and transfer of energy for purposes of biosynthesis, osmotic work, and so forth. The mechanisms of this energy exchange process are considered in detail in another monograph of this series.

The reaction of glucose phosphorylation is one which has received much attention because (1) the reaction has often appeared to be rate-limiting in the over-all utilization of glucose, (2) various research has pointed to the possibility that hormones affecting carbohydrate metabolism (insulin, adrenal glucocorticoids, pituitary somatotrophic hormone) exert controlling action on this metabolic step, (3) certain enzymes for phosphorylation in some species are highly adaptive to the amount of glucose presented to the interior of the cells, and (4) the reaction for hexokinase is essentially irreversible with a thermodynamic equilibrium very far to the direction of formation of glucose-6-phosphate (Gl-6-P). (This last characteristic usually makes a biochemical reaction more vital and interesting, because it affords a better possibility of a control point in a sequence of reactions.)

Reactions

Various mechanisms by which glucose, and some other hexoses, may become phosphorylated and enter into further metabolic reactions are indicated in Figure 4-1. Knowledge about glucose phosphorylation has been growing rapidly in complexity in recent years. The general reaction of ATP-glucose phosphotransferase to form Gl-6-P was considered at one time to be the property of a single enzyme, called hexokinase, since it was found to act also on other hexoses. Now at least four distinct types (isozymes) of hexokinase (Types I, II, III, and IV) are recognized, with variable mixtures of these occurring in different tissues. Type IV hexokinase ("glucokinase" of liver) has been found by starch gel electrophoresis to consist of two components, "fast" and "slow."

All hexokinases utilize ATP as the preferred phosphoryl donor.

Other triphosphonucleotides (e.g., ITP) react only at extremely low rates. Typically, Mg^{++} ion is involved in the reaction—by complexing with two adjacent phosphates of the nucleotide. Possibly Mn^{++} may substitute for Mg^{++}. Different hexokinases differ sharply in their affinities for the substrate, glucose. Type I hexokinase of liver has a high affinity with a K_m of $10^{-5}M$. Type IV hexokinase (glucokinase) in the same organ has a relatively low affinity (K_m of $1-2 \times 10^{-2}M$). Since this K_m is higher than normal concentrations of blood glucose, the rate of phosphorylation by this enzyme should be more dependent on fluctuating blood glucose levels than is phosphorylation by other hexokinases.

For some hexokinases inhibition of the reaction is caused by accumulation of either of the reaction products, ADP or Gl-6-P. This inhibition may be competitive (with the substrate, glucose) or non-competitive ("allosteric") in nature. Where the inhibition is produced by very low concentrations of Gl-6-P, it is generally non-competitive. With Type IV hexokinase, inhibition by Gl-6-P is virtually absent, occurring only at very high concentrations (65 mM). Inhibition by Gl-6-P may be relieved by inorganic orthophosphate (Pi), which, however, does not relieve inhibition by the other product of the reaction, ADP.

The activity of some hexokinases (Types II and IV) is depressed by starvation or a carbohydrate-free diet or in alloxan diabetes and is restored by carbohydrate or, in the case of diabetes, by insulin. This adaptivity of the enzyme is dependent upon changes in de novo synthesis of the enzyme. However, some evidence suggests that insulin may in some way protect the enzyme from inactivation. Glucokinase is also depressed in biotin-deficient rats.

Another type of phosphotransferase, which forms glucose-6-phosphate, has been identified in liver and kidney as an enzymic entity which may be the same as glucose-6-phosphatase but which can act in the reverse direction by the employment of certain phosphoryl donors. Besides pyrophosphate (PPi) and cytidine diphosphate (CDP) (Fig. 4-1) other possible phosphoryl donors are mannose-6-P and several other nucleotides in addition to CDP. This phosphotransferase activity (perhaps a "family" of enzymes) is found in the microsomal fraction of the cell. Under optimal conditions measured rates of activity for normal rat liver are 40 μmoles/g. wet liver/hr. for PPi-glucose phosphotransferase and 24 μmoles/g./hr. for CTP-glucose phosphotransferase—values which are lower than those for Type IV hexokinase but higher than for Type I hexokinase (at physiological glucose concentrations). In diabetic livers the activities may increase several-fold—as is typical for glucose-6-phosphatase. Drawbacks to the postulation of a physiological role for this group of phosphotransferases are (1) the K_m for glucose for all types of phosphoryl donor is high (90 mM), (2) the K_m for the phosphoryl donors is also somewhat high, and (3) the pH range for activity is rather low (4 to 6). The system using CDP, however, is active even at pH 6.4.

84 PHYSIOLOGICAL CHEMISTRY OF CARBOHYDRATES IN MAMMALS

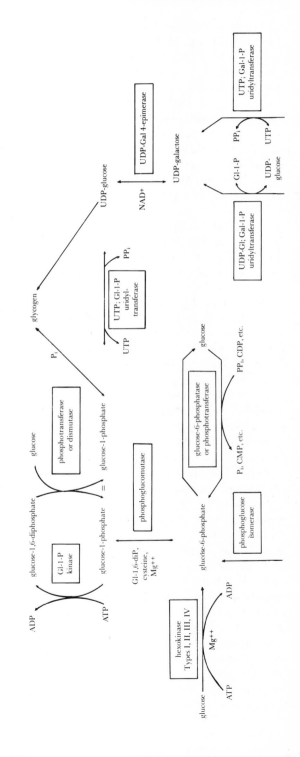

INTRACELLULAR UTILIZATION OF MONOSACCHARIDES 85

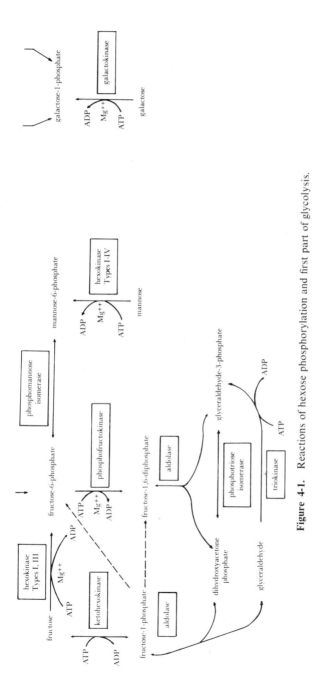

Figure 4-1. Reactions of hexose phosphorylation and first part of glycolysis.

Further investigation is needed to determine the true physiological significance of this type of phosphotransferase.

No hexokinase has been found which transphosphorylates from ATP to the C-1 position of glucose. However, there is in muscle a phosphotransferase or "dismutase" which forms from glucose and glucose-1,6-diphosphate (Gl-1,6-diP) two molecules of glucose-1-phosphate (Gl-l-P), or vice versa. According to one suggestion (Smith, Taylor, and Whelan) this enzymatic activity could be combined with that of α-Gl-1-P kinase, which phosphorylates α-Gl-1-P at the C-6 position with ATP as phosphoryl donor, to form a cycle, the net effect of which is the formation of α-Gl-1-P from glucose. This scheme requires initially a catalytic amount of Gl-1-P, which could be formed from Gl-6-P by phosphoglucomutase or from the breakdown of glycogen by phosphorylase (Fig. 4–1). While this enzymic system is known only for muscle, it could possibly explain findings in both muscle and liver which have suggested that glycogen can be formed from glucose by a series of reactions not requiring Gl-6-P as an intermediate (see below). It has been suggested that a form of glycogen storage disease (excessive accumulation of glycogen) characterized by a deficiency of phosphoglucomutase is also better explained by this mechanism than by a formation of glycogen from glucose only through the phosphoglucomutase reaction.

Activities and Functions in Various Organs and Species

BRAIN. The highest activities of hexokinase, measured in vitro, have been found in brain. This is consistent with the vital dependence of brain upon glucose for metabolic fuel. Whereas most of the enzymes of glycolysis are in the soluble portion of the cytoplasm, brain hexokinase is largely associated with the mitochondria. This is not true for hexokinase in most other tissues, although it is so for the photoreceptor cells of the retina, embryonic liver, and some tumor tissue. The product, Gl-6-P, can effect reversible release of hexokinase from the mitochondria of brain (also observed for heart muscle and ascites cells), which may be significant as a control mechanism. Inhibition (allosteric) by Gl-6-P occurs at a relatively low concentration (90 per cent at 1 mM).

Brain has long appeared to be one of those tissues not sensitive to insulin. One theory of insulin action considers the possibility that in insulin-sensitive tissues this hormone facilitates attachment of extramitochondrial hexokinase to the respiratory apparatus of the mitochondria with consequent utilization of mitochondrial ATP for phosphorylation of glucose. If brain hexokinase were invariably attached to mitochondria, it would therefore not be sensitive to insulin. However, some recent evidence that insulin may, after all, in some circumstances increase glucose uptake by the brain in vivo could be related to the effect of Gl-6-P on the attachment of the enzyme to the mitochondria. Brain does not contain either the Type II or Type IV hexokinases which have been found in other tissues to be responsive to various physiological and hormonal changes.

MUSCLE. Studies of glucose transport and phosphorylation in skeletal and cardiac muscle suggest that under usual circumstances the transport rather than phosphorylation is the limiting factor in glucose utilization. However, the presence of Type II hexokinase suggests the occurrence of physiological variations. Hexokinase activity in hearts of diabetic animals is low, but relationships of insulin to muscle hexokinase have not been clearly established. As indicated above, the equilibrium between soluble and particle-bound forms of hexokinase in cardiac muscle is affected by Gl-6-P and depends also upon pH and ionic strength. The depressed activity of hexokinase in diabetic heart muscle is associated with increased levels of Gl-6-P and depends upon the presence of cortisol and growth hormone.

As studied in enzymic extracts of rat tissues, hexokinase of heart has about one-half the potential activity of that of brain, while skeletal muscle hexokinase has about one-fifth.

The mechanism described above for net production of glucose-l-phosphate from glucose by operation of the muscle enzymes, Gl-l-P kinase and Gl-1,6-diP-glucose phosphotransferase, may help to explain certain findings (with rat diaphragm incubated with Gl-6-P-^{14}C or glucose-^{14}C) which have suggested that glucose is converted more extensively to glycogen and less extensively to lactate and CO_2 via glycolysis than is glucose-6-phosphate. Examination of Figure 4-1 indicates why this could be so. It further could explain why the conversion of glucose to glycogen is more responsive to insulin than is the conversion of Gl-6-P. The enzyme system (glycogen synthetase) which converts Gl-l-P to glycogen is particularly increased by insulin (Chap. 6).

However, another explanation for these differences between metabolism of glucose and that of Gl-6-P has been the suggestion that two "pools" of Gl-6-P are present in the cell, to one of which glucose but not extracellular Gl-6-P has access. Landau and Sims suggested that only a certain area of the cell membrane (the sarcotubule) might contain this insulin-responsive pathway to glycogen. The idea of separation of pools gains some inferential support from the increasing observation of multiplicity of isozymes of hexokinase.

LIVER. Glucose absorbed from the intestine travels in the circulation first to the liver, where a significant amount can be metabolized to glycogen and other products. In this metabolism several hexokinases may take part. Most unique in liver is the Type IV hexokinase, or glucokinase, which has such a uniquely high K_m for glucose but which, is more specific for glucose than are other hexokinases. As discussed in Chapter 3, the liver parenchymal cell is rapidly permeable to glucose, so that intracellular concentrations can be similar to those of blood. Because of the ready permeability phosphorylation is more likely to be the initial limiting factor in utilization.

In the livers of normal fed rats at a concentration of glucose in circulating blood of 90 mg/dl. about half of the total phosphorylation of glucose occurs by the action of glucokinase and about half by the other hexokinases. Since in prolonged starvation (48 hours in a rat) the glucokinase activity may drop to 0 to 10 per cent of the normal rate, the rate of total phosphorylation can drop to about half of normal. The same can be said also of the severely diabetic rat.

The phosphorylation of glucose by glucokinase presents an example of a biochemical property which can have important differences among species. In particular it demonstrates the hazard of extrapolating from the laboratory rat, and some other animals, to man. In omnivorous mammals (such as rat, mouse, pig), in non-ruminant herbivorous mammals (guinea pig, rabbit), and in the dog the activities of glucokinase and the rates of glucose incorporation into glycogen are high. In ruminants (sheep, cattle), in ruminant-like marsupials, in the cat, and in man there are low or absent activities of glucokinase and the rates of conversion of glucose to glycogen are relatively low. Also, in some species where the enzyme is found it has not appeared to be sensitive to starvation or diabetes as it is in the laboratory rat. A definite role for glucokinase (Type IV hexokinase) in carbohydrate metabolism, particularly in man, cannot yet be fully evaluated.

Since ruminants largely convert ingested carbohydrates to short-chain fatty acid by bacterial action in the rumen, there is little call for the livers of these animals to remove glucose rapidly from the blood. Likewise, strict carnivores, such as the felines in their natural state, would have little need to dispose of any large amount of carbohydrate in the diet. It is worth reflecting that man is like these species in the low level of hepatic glucokinase perhaps because he was primarily a carnivore in his biological evolution. Only since Neolithic times has carbohydrate been a major constituent of his diet—and even more recently, refined and concentrated amounts of the disaccharide sucrose. The common, and rising, occurrence of diabetes in the human population may be only one example of the strain of his adaptation to it.

There may be present in liver either one or both of the two alternative mechanisms (Fig. 4-1) for phosphorylation of glucose otherwise than by action of hexokinase. The occurrence in liver of the two-enzyme cycle forming glucose-l-phosphate from glucose can only be postulated, since the enzymes have not yet been found there. However, as in the case of muscle, the operation of this system could help explain data from studies with liver slices and homogenates which indicate that glucose and Gl-6-P are metabolized to glycogen and CO_2 in different proportions. The reversible action of glucose-6-phosphatase via a high-energy phosphoryl donor, while demonstrable for liver in vitro, cannot yet be assigned a definite physiological role in view of drawbacks already discussed.

ADIPOSE TISSUE. Under ordinary circumstances the uptake of glucose by adipose tissue may be dependent on transport phenomena. Only at very high rates of uptake, in the presence of insulin and high glucose concentration in the medium, does the rate of phosphorylation appear to become limiting. As with other tissues the hexokinase of adipose tissue can be shown to be inhibited by both of its products, Gl-6-P and ADP. Type II hexokinase has been identified and shown to be variable in response to physiological and hormonal changes. This may prove to be significant in helping to explain the marked changes in formation, storage, and accumulation of fat which obviously occur under different conditions of carbohydrate nutrition.

OTHER ORGANS AND TISSUES. Hexokinase is present in intestinal mucosa with a fairly high activity, but various evidence suggests that it is not a direct intermediary in the active transport of glucose from the lumen through the epithelial cell (see Chap. 2). It could be involved in an energy-yielding process related to transport.

Erythrocytes depend strongly on hexokinase, and succeeding glycolytic reactions, for their only source of ATP. The activity of hexokinase in red cells is under control by Gl-6-P as an inhibitor, but to what extent this is a major control mechanism in different species is still in doubt. The inhibition by Gl-6-P is relieved by Pi.

Hexokinase has also been identified in uterus, spleen, pancreas, lung, and kidney, and probably occurs in various other tissues. There is evidence that the "reverse glucose-6-phosphatase" type of phosphotransferase may be operative in reabsorption of glucose from the tubular lumen of the nephron.

GLYCOLYSIS

The process of anaerobic breakdown of glycogen to lactic acid in mammalian muscle was termed glycolysis by its early investigator, Meyerhof. Subsequent elucidation by several pioneer biochemists during the period 1900 to 1950 led to an understanding of a sequence of about a dozen enzymatic steps resulting in the formation of 2 molecules of lactic acid and 2 of ATP from one glycogen-glucosyl residue, 2 molecules of ADP, and 2 of inorganic phosphate. This is now known as the Embden-Meyerhof-Parnas (E-M-P) glycolytic pathway, which consists of reactions and enzymes quite analogous to those used for fermentation by yeast. The first part of this pathway is indicated in Figure 4-1 and the second part in Figure 4-2. The first reaction in this sequence, formation of glucose-l-phosphate (Gl-l-P) from Pi and glycogen, will be further considered in Chapter 6.

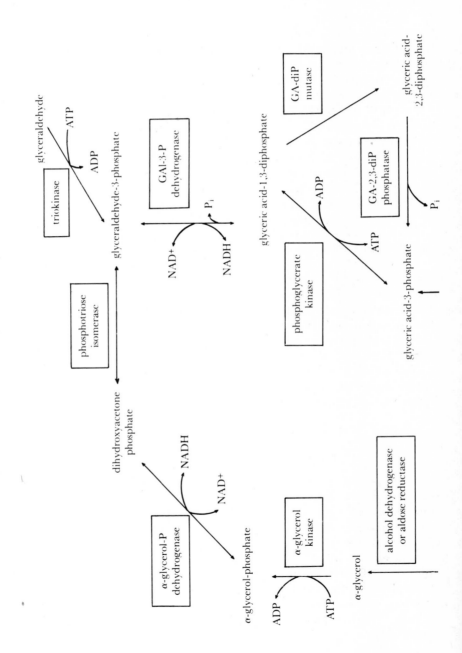

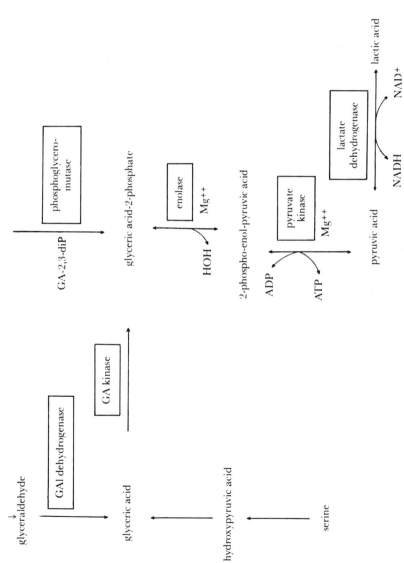

Figure 4-2. Reactions of glycerol and second part of glycolysis.

Reactions

The transformation of Gl-1-P to Gl-6-P is carried out by the enzyme phosphoglucomutase, which has been crystallized from extracts of muscle. The equilibrium of the isolated reaction lies far (95 per cent) in the direction of formation of Gl-6-P. Of course, under in vivo conditions the reaction could be readily reversed by continual withdrawal of product Gl-1-P to form UDP-glucose, glycogen, and other compounds.

Cysteine and Mg^{++} appears to be essential for activity of phosphoglucomutase, but the crude enzyme is also activated by manganese, chromium, or cobalt ions. The mechanism of the reaction involves the intermediate, glucose-1,6-diphosphate (Gl-1,6-diP), since Gl-1-P is successively phosphorylated at C-6 and dephosphorylated at C-1, or vice versa. Gl-1, 6-diP is also involved in a closely related enzymic cycle, as discussed in a preceding section (Fig. 4-1).

According to Figure 4-1 the phosphorolytic breakdown of glycogen and the phosphorylation of free glucose result in the same product, Gl-6-P, which would presumably contribute to the same substrate pool for further reactions. This view, however, may be an oversimplification, as suggested in the preceding section on phosphorylation.

After formation of Gl-6-P from glucose or glycogen the next step in glycolysis is an internal rearrangement of the hexose to form fructose-6-phosphate (Fr-6-P). This reaction is catalyzed by phosphoglucose isomerase and involves no significant energy exchange. Nevertheless, this reaction can be a possibly rate-limiting step in glycolysis of skeletal muscle, both in the resting and in the stimulated state, according to study of the concentration of phosphorylated hexose intermediates in gastrocnemius muscle of cats in situ.

The succeeding reaction of transfer of high-energy phosphate from ATP to Fr-6-P to form fructose-1,6-diphosphate (Fr-1,6-diP or FDP) is a kinase reaction similar to that of the intitial phosphorylation of glucose. As with the hexokinases the equilibrium of the reaction catalyzed by phosphofructokinase (PFK) is far toward formation of the doubly phosphorylated hexose. The main reason for this is that the energy content of the phosphate group in the primary hydroxyl (at C-1 or C-6) is much less than in ATP. However, in succeeding glycolytic reactions the two phosphates of Fr-1,6-diP are again raised to the energy level of ATP.

A regulatory role for PFK has been demonstrated not only for liver and skeletal muscle but also for brain, heart and kidney cortex, as well as tumor tissues. Phosphofructokinase is subject to a number of stimulatory and inhibitory influences. Notably the enzyme is inhibited by high concentrations of one of its reactants, ATP, and by citrate. Since both compounds are products of the aerobic metabolism of glycolytic end products, there is obviously a possible mechanism of feedback control of the glycolytic rate through these inhibitory effects. The inhibition of PFK by

ATP is released by Pi, various nucleotides (e.g., 3', 5'-cyclic AMP, which is notably involved in hormonal expression), and NH_4^+. The product of the reaction, FDP, is itself a potent activator of PFK. It is apparent that phosphofructokinase, like hexokinase, can be readily influenced by numerous intracellular constituents.*

The next step in glycolysis (Fig. 4-1) cleaves the nearly symmetrical Fr-1,6-diP to form two 3-carbon sugars, dihydroxyacetone phosphate (DHAP) and glyceraldehyde-3-phosphate (GAl-3-P) by the action of the enzyme aldolase. This reaction, like some others preceding and following in glycolysis, does not involve any significant energy change, is freely reversible, and is evidently employed in the biosynthesis of hexose from 3-carbon precursors (Chap. 5). No coenzymes or activators are required by animal aldolase. Some inhibitory effects may be exerted by adenine nucleotides (see section on phosphorylation of fructose).

Subsequent glycolytic reactions consist of further metabolism of GAl-3-P, but the other half of the hexose is also available for the same fate because of the enzyme phosphotriose isomerase, which converts DHAP to GAl-3-P, and vice versa. This enzyme is so active that there is relatively rapid equilibration between the two 3-carbon forms. Isotopic studies indicate that under fasting conditions in vivo the equilibration between the two trioses is almost complete, while in the fed state it is less so, owing to changes in flux rates of related metabolic processes.

As can be seen in Figure 4-2, there are alternative possibilities for metabolism of the triose phosphates. On the one hand they may continue to follow the glycolytic pathway to pyruvate and lactate, or there may be a reduction of DHAP to glycerol phosphate by the enzyme α-glycerol phosphate dehydrogenase. Various hormonal and dietary changes appear to exert appreciable influence on the relative activities of these two routes of disposal of triose phosphate. For instance, as determined by study with glucose-1-^{14}C and glucose-6-^{14}C, the amount of utilized glucose which is converted to glycerol, relative to that which is converted to fatty acids (deriving from pyruvate), is increased by epinephrine and pituitary growth hormone and decreased by insulin, when these hormones are added to isolated adipose tissue.

The reaction of oxidation of the aldehyde of GAl-3-P to the carboxylic acid level, like other reactions of this type, is one which releases large amounts of potentially useful energy. Some of this energy is captured in a form usable by the cell through a mechanism for converting ADP and inorganic phosphate (Pi) to ATP. The enzyme glyceraldehyde-3-P dehydrogenase contains side chains with cysteine residues. The aldehyde first condenses with the sulfhydryl groups of these cysteine

---

*Control of activity of PFK is probably the mechanism for much of the "Pasteur effect," i.e., an increase in rate of glycolysis secondary to a decrease of the oxidative metabolism.

moieties. The condensation product is then dehydrogenated (with reduction of NAD^+ to NADH) to form an acyl mercaptide or thioester, which has a high energy level like that of a high-energy carboxyl-phosphate bond. The latter is formed by phosphorolysis with Pi to give glyceric acid-1,3-diphosphate (GA-1,3-diP) and again restore the enzyme to the reduced form. Glyceraldehyde-3-P dehydrogenase differs from most NAD^+-coupled dehydrogenases in that its coenzyme, NAD^+, is rather firmly bound.

The high-energy phosphate at C-1 of the glyceric acid is then transferred by the enzyme phosphoglycerate kinase to ADP to form ATP. In the case of muscle there may be a direct transfer via creatine kinase to form phosphocreatine.

The further glycolytic reactions (Fig. 4–2) effectively transform the low-energy phosphate in glyceric acid-3-phosphate (GA-3-P) to a high-energy state preparatory to additional formation of ATP. The phosphate is transferred from C-3 to C-2 of glyceric acid by the enzyme phosphoglyceromutase. It is probable that glyceric acid-2,3-diphosphate is an intermediate of this reaction. Then a dehydration of glyceric acid-2-P by the enzyme, enolase, provides 2-phosphoenolpyruvate (PEP), in which a redistribution of energy within the molecule leaves the phosphate in a high-energy state. Enolase requires Mg^{++} and is inhibited by fluoride in the presence of Pi. (In order to prevent glycolysis in vitro in blood specimens prior to their analysis for glucose concentrations, fluoride is used as an additive to such specimens.)

The last enzyme in the main glycolytic sequence is pyruvate kinase, which transfers the phosphate of PEP to ADP, after which enol-pyruvic acid can readily rearrange to form pyruvic acid. Pyruvate kinase, unlike phosphoglycerate kinase, can also utilize IDP, CDP, and UDP. The mechanism of the pyruvate kinase reaction involves $MgADP^-$ as the coenzyme complex, while K^+ functions to activate the enzyme. There are two types of pyruvate kinase according to electrophoretic and immunologic procedures. The M type is the usual muscle enzyme. In liver both types M and L are present and it is type L that fluctuates in activity under various hormonal and dietary conditions. Type M is distributed in muscle, brain, heart, liver, and kidney, while type L is contained only in liver and kidney. Type L pyruvate kinase may have more significance as a controlling factor in gluconeogenesis than in glycolysis (Chap. 5). Like some other "control points" in metabolism, the transfer of phosphate is not commonly reversed because the reaction is so exergonic in the direction of ATP formation.

The phosphoglycerate kinase and the pyruvate kinase reactions are both instances of nucleotide phosphorylation "at the substrate level." This is in contrast to the mechanism of formation of the bulk of ATP, which occurs as a result of operation of the electron carrier system into which feed the "oxidative dehydrogenations" of the Krebs or TCA cycle (see later section). In the glycolytic reactions two moles of ATP are invested per mole of glucose (one mole of ATP if starting from glycogen) in order to obtain finally by means of the substrate-linked phosphorylation four moles of ATP (thus a net gain of two moles, or three if counting from glycogen). While this ATP is only a small fraction (about 10 to 15 per cent) of the total ATP which is finally produced from the complete oxidation of glucose to carbon dioxide, it is an important fraction, since it can be realized if necessary under anaerobic conditions. Therefore, it can for a limited time sustain the intense activity of muscle contraction, which utilizes ATP, although the accumulation of lactic acid will eventually interfere with various cell processes.

Anaerobically glycolysis gives lactic acid as a main end product through reduction of pyruvic acid by lactate dehydrogenase (LDH). The latter enzyme is found very commonly in tissues, is located in the "soluble" (cytoplasmic) phase, and notably is found in the form of multiple isozymes, i.e., a family of structurally similar proteins with the same enzymatic function. The physiological significance of multiple isozymes of lactate dehydrogenase is not clear. There may be distinctive elevations of particular isozymes in the serum in certain diseases of the heart or liver, since liver and skeletal muscle contain isozymes different from those of heart muscle. In a form of muscular dystrophy there is a deficiency of LDH_3-isozyme.

The reduction of pyruvic acid with the reducing equivalents (electrons and hydrogen) transferred to NAD^+ by phosphoglycerate dehydrogenase permits the regeneration of NAD^+ and the continuation of the glycolytic process. However, besides the substrate, pyruvate, and the enzyme, lactate dehydrogenase, there are other substrates and extramitochondrial enzymes which can alternatively accept the reducing equivalents. For instance, DHAP (also formed in glycolysis) can be reduced to α-glycerol phosphate by α-glycerol phosphate dehydrogenase. Also acetoacetate can be reduced to β-hydroxy-butyrate by the appropriate dehydrogenase. In fact, it seems likely that under aerobic conditions it is one of these other reductions, and particularly that of DHAP, which maintains the glycolytic process. This is because a corresponding enzyme (phosphoglycerol oxidase or another β-hydroxybutyrate dehydrogenase) within the mitochondria further transfers the hydrogen to the main electron transport system of the cell, which is itself intramitochondrial. (It is this electron transport system which finally combines the electrons and hydrogen with oxygen to form water.) The mitochondrial membrane is readily permeable to glycerol or malate or β-hydroxybutyrate, so these substrates serve as carriers of the reducing equivalents from the site of glycolysis to the site of further electron transport (Chap. 3). An interesting aspect of these "shuttle" mechanisms is that they are evidently defective in most malignant tissues, which lack both glycerophosphate dehydrogenase and β-hydroxybutyrate dehydrogenase. This lack is one possible reason that lactic acid is formed in especially large amounts in malignant cells even under aerobic conditions.

Activities and Functions in Various Organs and Species

MUSCLE. By far the largest "glycolytic factory" in the body exists in skeletal muscle. In sheer mass it comprises 40 per cent of the body weight and moreover has a relatively high activity of the critical rate-limiting enzyme phosphofructokinase. Virtually all the glucose catabolized by skeletal muscle proceeds via glycolysis—activity of the alternative pentose cycle, for example, is very low.

The value of glycolysis in muscle is connected with the sporadically intense activity characteristic of this tissue. Full oxidative metabolism is not suited to such rapid and extensive changes in rate requirements, partly because the circulation and the characteristics of dissociation of oxyhemoglobin are not designed for coping with large changes in the necessary supply of oxygen. Coincident with muscle contraction, glycolysis is markedly enhanced. In resting muscle, aerobically the rate may be only a few μmoles glucose/g. wet wgt./hr. but may increase manyfold during intense activity. One hypothesis for the mechanism of this increase concerns the effect of Ca^{++} ion, the movement of which may be triggered by membrane depolarization. Through a mechanism similar to that for the activation of actomyosin ATPase (which is in the contractile muscle element itself) certain key glycolytic enzymes, e.g., phosphorylase and phosphofructokinase, could be thus critically activated, perhaps, as suggested by Helmreich and Cori, by promoting access of reactants (substrates) to catalyst (enzyme).

In various animal tissues the capacity for glycolysis appears to be correlated with the amount of phosphagen (phosphocreatine and phosphoarginine) contained in the tissue. Average striated skeletal muscle contains five times as much as the more slowly acting smooth muscle, and cardiac muscle also contains relatively little. Between types of skeletal muscle the faster "white" muscle shows higher concentrations of most glycolytic enzymes than the slower "red" muscle. The latter has, however, a higher rate of aerobic oxidation of glucose and lactate to CO_2, which is consistent with higher mitochondrial concentration.

The rate of anaerobic glycolysis in exercising skeletal muscle may be influenced by the rate at which lactate can be removed from the circulation, which may further depend on the rate of lactate disposal from the circulation at sites elsewhere than muscle (see Chap. 8). However, the importance of this factor relative to the capacity for hydrogen disposal in muscle through mitochondrial mechanisms has not yet been evaluated.

In heart muscle certain glycolytic enzymes (phosphohexoseisomerase, aldolase, and triosephosphate isomerase) have been found attached to superficial cell membranes, where they may exert their action on substrates external as well as internal to the cell membrane. In skeletal muscle two glycolytic systems have been proposed, one which is presumably associated with the cell membrane (intermediates of this system exchange with extracellular phosphorylated analogues) and another which is more internal in the cell (intermediates do not exchange with phosphorylated analogues). The latter system requires bicarbonate for activity of PFK, and is responsive to insulin.

BRAIN AND RETINA. Glycolysis is especially important for the central nervous system because of the strict dependence upon glucose for fuel (except on long fasting). The rate of aerobic glucose consumption may be estimated at about 20 μmoles/g./hr., which is higher than for most tissues. Anaerobically the lactate production suggests rates up to 10 times as high; i.e., a strong "Pasteur effect" is evident. Different structures of the brain may have somewhat differing glycolytic rates. The enzymes of the E-M-P pathway are more active in "gray matter" (nerve cell bodies) than in "white matter" (myelinated nerve fibers). Those of the pentose cycle are much lower in either type of brain tissue than the glycolytic enzymes, but are relatively higher in white than in gray matter. The activities of some glycolytic enzymes in vitro suggest a potential maximal capacity for human brain which could be at least twofold higher than the maximal glucose consumption estimated in vivo during short periods of high activity (convulsions). Even higher enzymatic activities are found in lower mammals, a species difference which could be ego-disturbing.

The retina, which is derived embryologically from the forebrain, has very high rates of both aerobic and anaerobic glycolysis. Anaerobically the rate is three to four times higher than that of brain. As in the case of red cells, in retina and adjacent ciliary body glycolysis has been shown to be geared to the transport of ions (Na^+ and K^+).

BONE MARROW AND BLOOD CELLS. Bone marrow is among the tissues with high glycolytic rates; anaerobically the rate is similar to that of brain. White blood cells also have very active glycolysis. For polymorphonuclear (PMN) leukocytes the aerobic uptake of glucose is about 50 μmoles/g./hr., of which about 85 per cent is metabolized to lactic acid. Although both pentose cycle and TCA cycle activities are present, they are very minimal compared with glycolysis. Lymphocytes show a somewhat lesser (one-half to one-seventh) activity of glycolysis compared with PMN leukocytes. Small Pasteur and Crabtree* effects are said to occur with lymphocytes. With resting, intact human PMN leukocytes about a three-fold increase in glycolytic rate occurs upon incubating cells in serum anaerobically instead of aerobically. However, if the leukocytes are "damaged" in preparation or if they are from exudative sites, the

---

*The Crabtree effect is a kind of reverse Pasteur effect, i.e., an inhibition of oxidative phosphorylation by excessive catabolism of glucose through the glycolytic pathway.

aerobic glycolysis is higher and the Pasteur effect is lower. As with brain, the rates of glycolytic enzymes measured individually in leukocyte homogenates are several times higher than the observed maximal anaerobic production of lactate by intact leukocytes—five-fold higher even for the least active enzyme, hexokinase. Factors other than enzyme potential, therefore, appear to limit the rate of glycolysis in vivo. Since the pentose cycle enzymes, including transketolase and transaldolase, are relatively active in PMN leukocytes, pentoses and their precursors can be significant sources of lactic acid production in addition to glucose and glycogen.

The glycolytic rate of human platelets is about one-half to two-thirds that of PMN leukocytes and the extent of Pasteur effect is similar.

In red cells glycolysis is of vital importance; it is the only significant source of formation of ATP in the mature red cell, which has no functioning TCA cycle, although the reticulocyte has. Nevertheless, the rate of glycolysis is considerably lower than in WBC—only about 2μmoles glucose/g./hr. for human RBC and only one-tenth of that for RBC of the pig. Glycolysis is involved in the active transport of cations by red cells through its provision of substrate ATP for the activity of ATPase, an enzyme which is a participant in the active transport of sodium and potassium. It is estimated that 30 per cent of the energy from glycolysis is probably required for the active cation fluxes.

In red cells rate-limiting steps of glycolysis have been defined at various steps: those catalyzed by hexokinase, phosphofructokinase, glyceraldehyde phosphate dehydrogenase, and pyruvate kinase. Different steps may become limiting under different conditions, but coordinated increases or decreases of several enzymes generally occur. Asynchronous glycolysis, i.e., an uncoordinated change in the ATP-consuming and ATP-yielding reactions of glycolysis with intermittent accumulation of glycolytic intermediates, is not thought to be common normally, although a lactate deficit and accumulation of triose phosphates can be produced by high phosphate concentrations. There are accumulations of certain intermediates in specific deficiencies of enzymes of glycolysis in RBC. These deficiencies are usually genetically determined and have been identified with certain clinical hematologic abnormalities.

One of the intermediates which may accumulate is 2,3-diphosphoglycerate. Whereas this compound is a catalytic intermediate in the general glycolytic operation of phosphoglyceromutase to produce GA-2-P from GA-3-P, in RBC there is a special "shunt" reaction converting GA-1,3-diP to GA-3-P with GA-2,3-diP as an intermediate. Moreover, in this shunt, as seen in Figure 4-2, there is a loss of high-energy phosphate to form Pi, whereas in the parallel phosphoglycerokinase reaction the high-energy transformation serves to form ATP. This flexibility may be useful to the cell in regulating the supply of ATP independent of the rate of glycolysis, but it may also be a source of derangement of metabolism.

Some studies of glycolysis in red cells have indicated that an increase of pH increases glycolysis, as does an increase in Pi concentration. On the other hand, interference (by addition of ouabain) with ATPase activity (and with active cation transport) slows down glycolysis in red cells, possibly by an increase of the ATP/ADP ratio, which has an inhibitory effect on PFK. A crucial level of ADP is needed, also, to accept high-energy phosphate from GA-1,3-diP. In red cells the sulfate ion as well as Pi may stimulate the PFK reaction.

Another special function of glycolysis in RBC lies in provision of reducing equivalents to form NADH, which is utilized to maintain methemoglobin in the reduced state. Possibly the other reduced pyridine nucleotide, NADPH, aids in this function.

Some study has been done of the cellular location of glycolytic enzymes in RBC. Fractionation of beef RBC yields "membranous entities" which perform the entire glycolytic sequence. The over-all glycolytic activity is concentrated there 24 times relative to protein quantity. There is a wide divergence in firmness of association of individual glycolytic enzymes to the membrane. Some, like triosephosphate dehydrogenase, are completely retained in the membrane fraction. Others, especially the "non-rate-limiting" isomerases, are recovered on the membrane in very low yield. These membranous entities may be attached to the interior surface of the erythrocyte stroma.

ADIPOSE TISSUE. In this tissue glycolysis, at least the first part of it, appears to be very important as a means of providing glycerol phosphate, which is needed for continual esterification of fatty acids and storage of triglycerides. To a variable extent, fatty acids are also formed from glucose in adipose tissue; this metabolic fate of glucose, while high in some rodents, is very low in humans. Rates of glucose uptake are not high in comparison with some other tissues—as measured in isolated epididymal adipose tissue of rats, guinea pig, rabbit, and hamster the uptake at glucose concentrations of about 25 mg./dl. amounted to about 0.5 to 1.0 μmoles/g./hr., although at a glucose concentration ten times higher the uptakes increased about five times. When human adipose tissue, subcutaneous and omental, is incubated with glucose at concentrations of about 200 mg./dl. the uptake, into neutral lipids, is only 0.2 to 0.5 μmoles/g./hr.,—with omental tissue about twice as active as subcutaneous. In adipose tissue the specific glucose uptake is low. Yet owing to the large bulk of this tissue in the total body, there is a considerable amount of total glucose metabolized. Probably this occurs mainly through the glycolytic pathway to glycerol. Furthermore, it is quite variable and highly subject to the influences of glucose concentration and hormonal factors.

KIDNEY. In considering glycolysis in kidney tissue a clear distinction must be made between two main areas of the kidney, the cortex and the medulla. The cortex, consisting largely of the glomeruli and proximal and distal convoluted tubules, shows a primarily aerobic type of metabolism in which glucose oxidation accounts for only about 10 per cent of the total oxygen consumption. Even though large quantities of glucose are reabsorbed through the proximal convoluted tubule, little of this available substrate is utilized for energy purposes via glycolysis and the TCA cycle. Measurements in rats and rabbits indicate aerobic lactate production equivalent to about 1 to 2 μmoles glucose/g./hr. Anaerobically the yields of lactate were two to three times higher. Studies of oxidation of fatty acids by kidney cortex more directly indicate that fatty acids rather than glucose are the main energy source for this part of the kidney.

By contrast, kidney medulla (loop of Henle and collecting ducts) has a much lower (one-sixth) oxygen consumption compared with cortex, but an anaerobic glycolysis which is about four times higher. Since sodium transport occurs in the ascending limb of the loop of Henle, it appears that anaerobic glycolysis is the main energy source for this process and perhaps also for other ionic changes performed by the collecting ducts, such as sodium reabsorption, hydrogen ion and potassium secretion, and ammonia production.

LIVER. The enzymes of glycolysis have been relatively well studied in liver, but the function of glycolysis in liver seems to be of minor significance in comparison with other facets of carbohydrate metabolism—gluconeogenesis, glycogen storage and metabolism, the TCA cycle, the glucuronic acid cycle—occurring in that organ. Glycolysis is not required for the provision of glycerol phosphate, since glycerol kinase in liver can utilize glycerol transported to the liver from other tissues. However, glucose may provide an auxiliary source of glycerol phosphate and also a precursor of fatty acid synthesis. The rate of anaerobic glycolysis in rat liver has been found to be about 20 μmoles/g. wet wgt./hr.

MISCELLANEOUS TISSUES AND ORGANS. Other tissues which have been noted to have moderately high rates of anaerobic glycolysis (exceeded only by brain, retina, bone marrow, and embryonic organs in general) are placenta, spleen, and spermatozoa. Lung, studied in the dog and rat, appears to have an exceedingly low rate of glucose utilization. Glycolysis can be judged to be a relatively minor pathway also for pancreatic islet cells, in which activities of PFKase and aldolase have been found to be quite low.

PENTOSE CYCLE

A route of oxidative metabolism of glucose-6-phosphate operates partly in parallel to the E-M-P glycolytic pathway, interconnects with it, and occurs to a varying extent in different tissues. It appears to serve special but widely necessary functions, which supplement those provided by glycolysis and the TCA cycle. Because oxidations occur directly on the hexose carbon chain and at earlier stages than in the E-M-P pathway, it has been called the direct oxidative pathway for hexose monophosphate (HMP). Other names are the HMP "shunt" pathway and the Warburg-Dickens-Horecker (W-D-H) pathway. Pentoses (or, rather, their 5'-phosphates) are important intermediates, so it has been called the pentose-phosphate pathway or cycle. Potentially it forms a cycle by regeneration of Gl-6-P, and isotopic studies indicate that it does operate in a cyclic manner. Some reactions are essentially irreversible; therefore, in part it is a unidirectional pathway.

Reactions

As seen in Figure 4-3 the first set of reactions of the pentose cycle are primarily oxidations on the carbon chain of Gl-6-P, first at position 1 to form 6-phosphogluconate and then at position 3 to form 3-keto-6-phosphogluconate. After the first dehydrogenation 6-P-gluconolactone* is an intermediate which upon hydrolysis (catalyzed by a hydrolase) is converted to 6-P-gluconate. The enzyme, 6-phosphogluconate dehydrogenase, which catalyzes the second dehydrogenation at C-3, is ordinarily slower and so more rate-limiting.

A very important characteristic of the two dehydrogenases is that they operate far better with the coenzyme $NADP^+$ than with NAD^+ (one or two orders of magnitude difference in mammalian tissues). Since NADPH is usually highly favored over NADH as a coenzyme for biosynthetic reductions (e.g., in synthesis of fatty acids or steroids), this part of the pentose cycle is evidently linked to biosynthetic processes. Both of the dehydrogenations are catalyzed by divalent cations (Fig. 4-3), which may act by lowering the K_m for the substrate or else for the coenzyme. The oxidative enzymes of the pentose cycle appear to fluctuate considerably according to dietary and hormonal conditions. They are to be classed among those enzymes which are highly inducible by substrate availability. There is also evidence that glucose-6-phosphate dehydrogenase (G6PDH) is inhibited and activated by long-chain acyl CoA compounds, which is a possible mechanism of metabolic control.

---

*A variety of mammalian livers (cat, dog, sheep, ox, rat) and also mammalian brain contain an enzyme, glucose dehydrogenase, which converts glucose to gluconolactone without need for prior phosphorylation. Gluconokinase, which phosphorylates gluconolactone with ATP, has been noted in rat liver and kidney and hog kidney. The dehydrogenase enzyme can utilize NAD^+ or $NADP^+$ as coenzyme. This route of metabolism has not been thought to be very active in cellular utilization of glucose because of the low catalytic activity of the enzyme, the high K_m, and a strong inhibition by Gl-6-P. However, the true role of this set of enzymes in vivo has yet to be evaluated in mammalian species.

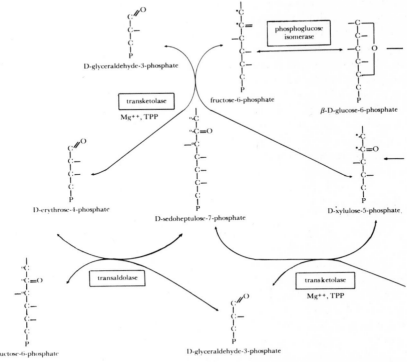

Figure 4-3. Reactions of the pentose cycle. (Unattached valence lines signify hydroxyl groups; unentered valence lines signify hydrogen atoms.)

The compound 3-keto-6-phosphogluconate is a fleeting intermediate which readily decarboxylates to the ketopentose ribulose-5-phosphate. This reaction is not reversible to any significant extent under physiological conditions. Ribulose-5-P can then undergo two reactions, both of which are involved in the cycle. The enzyme ribose-5-phosphate ketoisomerase isomerizes the keto sugar to the aldose form. A significant result of the operation of the pentose cycle (PC) pathway up to this point is the provision of the pentose ribose, which can be further utilized for formation of ribonucleotides. Another enzyme, ribulose-5-phosphate epimerase, alters the configuration about carbon-3 to form xylulose-5-phosphate from another molecule of ribulose-5-P.

If two molecules of xylulose-5-P and one of ribose-5-P are available, then the cycle is completed by a series of reactions shown in Figure 4-3. Xylulose-5-P has the configuration of the first three carbon atoms, which makes it suitable as a substrate donor of a 2-carbon unit known as "active glycolaldehyde." The enzyme transketolase transfers this unit to ribose-5-P to form the 7-carbon sugar, D-sedoheptulose-7-P, and glyceraldehyde-3-P. This reaction utilizes Mg^{++} ion and also thiamine pyrophosphate (TPP). The latter forms a complex with the active glycolaldehyde

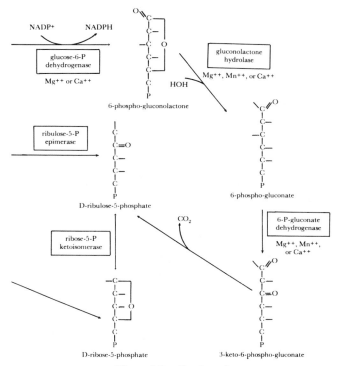

Figure 4-3 *Continued.*

much as does "active acetaldehyde" in the oxidation of pyruvate. A method of evaluation of thiamine (vitamin B_1) deficiency involves measurement of the activity of the transketolase in erythrocytes in man and animals and its response to added TPP.

A following reaction catalyzed by transaldolase next transfers the first three carbons of sedoheptulose-7-P to the aldose, glyceraldehyde-3-P, to form fructose-6-P and D-erythrose-4-P. Another reaction, also catalyzed by transketolase, can transfer the active glycolaldehyde from another molecule of xylulose-5-P to erythrose-4-P to form another molecule of Fr-6-P and one of glyceraldehyde-3-P. The cycle is completed by the action of hexose isomerase to convert Fr-6-P to Gl-6-P. However, for each two such hexose molecules re-formed one has been converted to three molecules of CO_2 and one of glyceraldehyde-3-P.

If some other enzymes of the E-M-P pathway (triose isomerase, aldolase) plus the enzyme fructose-1,6-diphosphatase are available to work in a direction opposite to that for glycolysis, then GAl-3-P can be reconverted to hexose and continued operation of this more extensive cycle can serve for the complete oxidation of glucose to CO_2. In some circumstances it might operate for this purpose. More likely the cycle is used for the supply of pentoses and reduced nicotinamide-adenine dinucleotide phosphate (NADPH). The evaluation of its importance as an energy-providing (i.e., ATP-forming) mechanism hinges largely on the existence and extent of activity in various tissues of a mechanism for transhydrogenation from NADPH to NADH, since only the latter pyridine nucleotide has the possibility of transferring electrons via the ATP-generating electron transport system. Furthermore, since the enzymes of the PC pathway are located in the soluble, extramitochondrial cell phase, another necessity would be the transfer of electrons to the mitochondrial phase via an appropriate substrate. Such mechanisms have been observed more fully in recent years (see section on glycolysis and Chapter 3), but the extent of their operation is not yet known.

Activities and Functions in Various Organs and Species

As indicated above, the PC pathway appears to serve two main functions of providing (1) ribose for nucleotide synthesis and (2) the form of reduced coenzyme (NADPH) most useful to the cell for reductive biosynthesis. In each case there are alternative metabolic mechanisms for synthesizing either ribose or NADPH, and the cruciality of the intact cycle for these functions in different tissues is still not clear. Ribose can also be synthesized by reversal of the non-oxidative reactions of the cycle, which do operate reversibly by isotopic evidence (Chap. 5). However, most evidence indicates that the cycle works primarily in the oxidative direction and the calculated through-put of glucose into the cycle probably exceeds considerably the amount of ribose needed by the cell for growth or nucleic acid turnover. Nevertheless, in growing tissues (fetal rat heart, early placenta) the activity of the PC pathway is higher than in the corresponding mature structures.

In contrast to the probably comfortable margin for ribose synthesis, calculations indicate that adequate supply of NADPH (at least for fatty acid synthesis in adipose tissue) would be barely maintained by the PC pathway in the fasting state; under the stimulus of insulin or other factors, which promote increased triglyceride synthesis and turnover, there would be a need for other sources of NADPH. Therefore, the PC pathway may be more critical, at least in some tissues, for supply of NADPH than for ribose.

In the biosynthesis of fatty acids NADPH is used in reduction of ketoacyl to hydroxyacyl units and in subsequent reduction of double bonds to form the saturated fatty acid chain. At three steps in the synthesis of steroids—ring closure, hydroxylation, and reductive methylation—NADPH is utilized. Other important functions of NADPH are reduction of ribonucleotides to deoxyribonucleotides and hydroxylation in the case of aromatic amino acids (e.g., tyrosine). Certain sugar-like polyols (sorbitol, galactitol, xylitol) are formed from corresponding aldoses (glucose, galactose, xylose) through the intermediation of polyol dehydrogenase and NADPH. Glutathione reductase utilizes NADPH; therefore sulfhydryl groups in general are maintained in the reduced state.*

There have been extensive investigations to determine the activity of the PC pathway in various tissues and in particular in relation to that of the E-M-P glycolytic pathway. Much of this work has made use of ^{14}C-labeled glucose with the label in specific carbon positions, e.g., C-1,C-6, or C-2. No carbon dioxide is liberated in the course of the E-M-P pathway, whereas, as seen in Figure 4-3, CO_2 is evolved from C-1 of glucose in the early phase of the pentose cycle. In the E-M-P pathway carbons 1 and 6 of glucose become mixed at the triose stage and both identified as C-3 of the end products, pyruvate and lactate. Subsequent oxidation of pyruvate in the TCA cycle liberates carbon dioxide equally, thereby, from original C-1 or C-6 of glucose. Also through these and subsequent pathways labeled fatty acids are formed equally from C-1 and C-6. However, equivalence of C-1 and C-6 for formation of either carbon dioxide or fatty acids depends on the condition that the two triose phosphates, DHAP and GAl-3-P, are equilibrated. As discussed in a previous section, the equilibration is usually not complete.

The greater the activity of the pentose cycle relative to glycolysis plus oxidation in the TCA cycle, the higher will be the ratio of C-1/C-6 (referring to the type of labeled glucose) in formation of $^{14}CO_2$ and the lower the ratio of C-1/C-6 in fatty acids. Equations for calculation from such data of percentages of glucose utilized by the various pathways have been developed, but their validity depends on several assumptions, including that of triose-P equilibration. In some cases these assumptions are reasonably justified on the basis of evidence, but in others they are not. Factors such as differential dilution of intermediate products in the separate pathways, draining-off of ^{14}C in non-equilibrated side-reactions, and recycling of products of the pentose cycle can introduce errors in the estimation. Specific yields of $^{14}CO_2$ will generally allow a fair estimation of the activity of the pentose cycle only if the yields of CO_2 from C-1 of glucose are twice or more those from C-6 and synthesis of glycerol-P (from DHAP) does not exceed 20 per cent of the utilized glucose.

Combined evidence from various analyses can minimize the needed assumptions and errors arising from them. For instance, combining data on formation of both $^{14}CO_2$ and fatty acid-^{14}C can eliminate the necessity for assumption of triose-P equilibration, as can analysis of both glycerol

---

*Production of cataracts in the lens may involve oxidation of protein sulfhydryl groups to disulfides. When intracellular levels of aldoses are high (as in galactosemia or diabetes, in both of which cataracts may occur) reduction of these aldoses to the polyols by NADPH is promoted, thus possibly making the coenzyme unavailable for the essential function of sulfhydryl maintenance.

and fatty acid moieties of triglycerides. A more recent method, which avoids this and some other assumptions, is based on the fact that ^{14}C originally in the C-2 position of glucose is rearranged into the 1 and 3 positions by the action of the pentose cycle. Measurement of the relative activities of ^{14}C in the 1,2, and 3 positions of glycogen-glucose provides information from which the percentage of utilized glucose channeled through the pentose cycle can be calculated. Measurement of ^{14}C in the other carbons (4,5, and 6) of glucose, and in such other products as glycerol and lactate, can yield further information on equilibration of various reactions and other fates of utilized glucose. From a thorough study of this type in isolated adipose tissue virtually a complete carbon balance on utilized glucose has been achieved.

Relative activities of the PC pathway in various tissues have been assessed by radioisotopic methods such as described above and by measurement of enzymatic activities in preparations of cell fractions.

ADRENAL GLAND. Adrenal cortex has very high activities of the oxidative enzymes of the cycle (as measured in glands from rat, rabbit, cow, and sheep). This is consonant with the supposition that these reactions are a main source of supply of NADPH for reduction of intermediates in the synthesis of cholesterol and the hormonally active steroids. Also the ratio of $^{14}CO_2$ yield from C-1 and C-6 of labeled glucose (in the cow adrenal) indicates a major role for the pentose cycle in glucose oxidation. Whereas the oxidative phase of this pathway is highly active in adrenal cortex, the utilization of ribose-5-P is of the same order as for other tissues (liver, kidney, duodenum) in which the oxidative enzymes are much less active than in adrenal cortex. This seems to imply the occurrence of, or at least the potential for, ample net synthesis of ribose. The adrenal medulla shows somewhat lower levels of the oxidative enzymes than the cortex.

MAMMARY GLAND. During lactation the mammary gland of the rat shows remarkable increases of activity of the oxidative enzymes to levels which are sixtyfold for G6PDH and twentyfold for 6PGDH above those in the non-lactating state. The levels reached are in the same range or higher than those normal for the adrenal gland. Perfusion of the cow's udder with certain ^{14}C-labeled compounds followed by analysis of distribution of ^{14}C in the galactose of milk has indicated that normally about 25 to 40 per cent of glucose metabolism proceeds through the pentose cycle, much of the remainder of the glucose is converted into lactose, and only about 10 per cent traverses the E-M-P pathway. There is a close temporal relationship of the high activity of the PC pathway in the lactating mammary gland and a high rate of synthesis of fatty acids from acetate.

BRAIN. There is somewhat variable evidence for the activity of the pentose cycle in this tissue. Levels of non-stimulatory dehydrogenase activity, as observed in cow, sheep, and rat, seem to be quite moderate. For slices of mature rat and guinea pig cerebral cortex, and for the circulation through human brain, the evidence from use of variously labeled glucose suggests that virtually all glucose catabolism proceeds via glycolysis with little activity of the pentose cycle. Activity of the cycle, however, may be quite enhanced by articificial electron acceptors coupled for oxidation of NADPH in slices of mature guinea pig brain cortex. Electrical stimulation of these slices does not increase pentose cycle activity but does increase glycolysis and TCA cycle oxidation. There is normally some utilization of pentose cycle in newborn dog brain cortex, unlike the mature dog brain or that of other species. Very substantial activity has been indicated by $^{14}CO_2$ yields from glucose-1-^{14}C and glucose-6-^{14}C perfused alternately through arterially isolated calf brain. Randomization of ^{14}C in the first three carbons of the glycosyl residue of rat brain glycogen after metabolism of glucose-2-^{14}C in vivo has suggested very little cycling in the pentose pathway. Differences among the above findings remain to be resolved possibly on the bases of species, age, or particular area of brain under study (myelinated areas show more activity of pentose cycle enzymes than unmyelinated). There is a possibility that the pentose cycle may play a role in providing NADPH for the synthesis of cerebral lipids, particularly in immature brain.

LIVER. Although liver is a notable factory of lipid synthesis, pentose cycle dehydrogenases in isolated liver tissue have only moderate activity similar to that found in duodenum, salivary gland, kidney, and brain. Nonetheless, these activities can increase markedly in response to glucose administration, refeeding, and provision of insulin. Livers from female adult rats show higher activity of dehydrogenases than those of male adult rats. Rat liver has four times the dehydrogenase activity of ruminant liver, which may indicate that rat liver is more dependent on glucose than is liver of ruminants, for which glucose is a less important circulating fuel. Estimates of cycle activity from $^{14}CO_2$ yields from labeled glucose have been high for liver of mouse and rat (possibly overestimated in early studies), but low for cow and sheep. Later studies with glucose-2-^{14}C in vivo indicate low cycle activity in rat liver. In mouse liver in vivo there is very little specific transfer of glucose-1-^{3}H to fatty acids compared with other tritium-labeled carbohydrates (e.g., malate, succinate or lactate), whereas much more specificity is shown in isolated adipose tissue and extrahepatic tissues in vivo.

ADIPOSE TISSUE. Extensive ^{14}C studies in this tissue (mostly isolated rat epididymal fat pad) have suggested that 10 to 20 per cent of utilized glucose follows the pentose cycle. There is evidence for even higher activity in vivo. As with the liver there is a stimulation of this pathway in adipose tissue by insulin, which increases the proportion of glucose utilized by the pentose cycle of the fat pad to 20 to 30 per cent. Refeeding of rats after fasting strongly stimulates the PC dehydrogenases. On the other hand, epinephrine, growth hormone, or thyrotropic hormone decreases the contribution of the cycle to less than 10 per cent of total utilized glucose. Measurement of G6PDH in human omental and subcutaneous adipose tissue indicates higher activity of this enzyme in the omental source, which is consonant with other evidence of a higher rate of lipogenesis at this site.

WHITE BLOOD CELLS. Polymorphonuclear leukocytes have been shown to have a very high ratio of production of $^{14}CO_2$ from glucose-1-^{14}C vs. glucose-6-^{14}C (about 20:1), which indicates major utilization through the PC pathway. Furthermore, the large increase in oxidation of glucose during phagocytosis is accountable mainly as metabolism through this pathway. In contrast to the high activity in PMN leukocytes, 6 per cent or less of glucose metabolism was found to proceed via the pentose cycle in plasma cells (from a patient with multiple myeloma) and in lymphocytes.

RED BLOOD CELLS. Estimates (simplified by the lack of occurrence of a TCA cycle in RBC) of the pentose cycle according to the yield of $^{14}CO_2$ in comparison with the amount of utilized glucose have suggested that only 5 to 10 per cent of the latter followed this pathway. This was true for both porcine and human erythrocytes, although the latter utilize glucose ten times as fast. Measurement of the dehydrogenases in the RBC of the rat also indicates relatively low activity.

Although the pentose cycle may be quite minor relative to the E-M-P pathway, nevertheless genetic deficiency of Gl-6-P dehydrogenase in humans is readily expressed in the occurrence of hemolytic anemias, which can be precipitated by various drugs (e.g., primaquine, sulfonamides, salicylates), infections, and other agents (e.g., fava bean, methylene blue). Deficiency of G6PDH is one of the most prevalent of genetic abnormalities with potentiality for clinical expression; more than 100 million people may be affected by this enzymatic deficiency. Negroes, some Mediterranean peoples, and various Oriental populations are often affected. The deficiency is of a different nature and less severe in black races than in white. Those persons with the deficiency in the erythrocyte have also been found to have a low level of the enzyme in the lens, platelets, liver, skin, and vasculature. It has been suggested that the mechanism of the tendency to hemolysis in individuals who lack G6PDH in their red cells is susceptibility to hydrogen peroxide, which is detoxified by reduced glutathione. The latter is kept in the reduced state by NADPH.

KIDNEY. Studies of kidneys from cows and sheep suggest, both from $^{14}CO_2$ yields and from activities of hexose-P dehydrogenase via NADP, that this organ may have a more active pentose cycle in ruminants than liver, mammary gland, duodenum, or brain. Studies with ^{14}C of rat kidney slices, however, have indicated lower activity than in liver slices. The G6PDH activity of various parts of the human nephron measured after microdissection have shown activity of the order: glomerulus > distal tubule > proximal tubule.

VASCULAR STRUCTURES. Comparisons of $^{14}CO_2$ yield from 1- and 6-labeled glucose incubated with isolated segments of aorta, artery, and vein of various species (dog, cow, guinea pig, rat) have shown widely differing results. Highest ratios were found with guinea pig aorta, lowest with that of the rat. This suggests some relationship between the high resistance of the rat to atherosclerosis and reduced lipogenesis due to less NADPH. Arterioles of the kidney were found to have G6PDH activity similar to that of the glomerulus. Activity of G6PDH has also been observed in human aorta, coronary artery, and pulmonary artery. It is of interest that enzymatic activity in both aorta and coronary artery was reported to be lower for women than for men, which may bear upon differences between the sexes in lipid content of the vascular structures. In this same regard coronary artery disease is less frequent among Negroes with the deficiency type of G6PDH than among those with the normal type.

MUSCLE. Several studies of metabolism of ^{14}C-labeled glucose in isolated skeletal muscle and cardiac muscle indicate that no more than 2 to 5 per cent of utilized glucose follows the PC pathway in these tissues. Measurements of G6PDH and 6PGDH activities also indicate relatively low levels of these enzymes. It has been shown, however, with rat skeletal muscle, that the oxidative phase of the pentose cycle can be stimulated by the addition of pyruvate, which increases threefold the amount of $^{14}CO_2$ produced from glucose-1-^{14}C anaerobically (probably by acting as an electron acceptor from NADPH with formation of malate: see below). However, even with stimulation by pyruvate the amount of $^{14}CO_2$ produced anaerobically is only one-fifth of that produced aerobically. (A similar effect of pyruvate on the pentose cycle has been shown for slices of brown adipose tissue of rats and for bovine corneal epithelium).

OTHER TISSUES AND ORGANS. The pentose cycle may be the predominant aerobic route for glucose in corneal epithelium because the rate of oxygen uptake is similar to that of liver, but there is relatively little TCA cycle activity. Pancreatic islet-cell tumors (composed largely of β-cells) from human subjects show a relatively high ratio of $^{14}CO_2$ from 1- and 6-labeled glucose (6:1), suggesting that the pentose cycle pathway is perhaps a major one in this tissue. This finding may be pertinent to the mechanism of action of glucose in stimulation of production and release of insulin. The salivary glands of cow and sheep show activities of the PC dehydrogenases which are similar to those for kidney and higher than for most other organs. Spleen, thymus, and lymph nodes (of rats) contain levels of G6PDH and 6PGDH which are moderately active, i.e., similar to adrenal medulla, liver, and kidney. A variety of organs with fairly low enzymatic activities are thyroid, ovary, testis, prostate, seminal vesicles, placenta, and pituitary. Nevertheless, rather high C-1/C-6 ratios for $^{14}CO_2$ are reported for ovary, testis, anterior pituitary, posterior pituitary, and thyroid, as well as for the parathyroid gland. The placenta shows greater utilization of the pentose cycle in the early weeks of gestation (i.e., in the time of greater placental growth) than at term, when it is virtually non-functional.

WHOLE BODY. When glucose-1-^{14}C and glucose-6-^{14}C are on separate occasions injected intravenously or given orally to normal human beings there is a slightly higher (15 to 20 per cent) and earlier yield of $^{14}CO_2$ in the breath from glucose-1-^{14}C. The ratio is higher for women during the menstrual years than for men and declines in women in the postmenopausal period. There is some evidence that other hormonal variations, e.g., diabetes or thyroid disease, may be accompanied by changes in the C-1/C-6 ratio of $^{14}CO_2$ after whole-body administration of labeled glucose.

UTILIZATION OF PYRUVATE

As indicated in Figure 4-4, pyruvic acid occupies a central position in the metabolism of carbohydrates. It is the end product of aerobic glycolysis. Depending on the state of oxygenation of the tisssue, pyruvic acid generally has two primary fates: (1) oxidation to acetyl CoA or (2) reduction to lactic acid, the chief end product in anaerobic glycolysis. Under some circumstances and in certain organs there may be other significant and major fates of pyruvic acid (Fig. 4-4). For instance, in the liver carboxylation to dicarboxylic acids (oxaloacetate and malate) takes place extensively in the course of gluconeogenesis (Chapter 5). In some tissues (e.g., muscle) the formation of phosphoenolpyruvate occurs more directly via pyruvate kinase. The metabolism of pyruvate is linked to that of protein via its relationships with various amino acids, e.g., the transaminative formation of alanine from pyruvate (and vice versa), dehydration and deamination of serine to pyruvate, and conversion of threonine to pyruvate through methylglyoxal. Transamination occurs in several tissue sites, whereas the latter two enzymatic activities have been observed particularly in the liver.

The main route for utilization of pyruvate in tissues under respiring conditions is oxidative decarboxylation to acetyl CoA. In the isolated perfused rat heart approximately two-thirds of the utilized pyruvate is metabolized via this route. As seen in Figure 4-4 the over-all process involves several coenzymes, among which is a notable array of vitamins: thiamine, lipoic (thioctic) acid, flavin (in FAD), nicotinamide (in NAD^+), and pantothenic acid (in coenzyme A). Also recent evidence suggests that vitamin E, as well as selenium, may somehow be required for proper oxidation of pyruvate. Pyruvate initially reacts with enzyme-bound thiamine pyrophosphate (TPP), whereupon the pyruvate is decarboxylated leaving the acetaldehyde attached at the carbonyl group to TPP. Lipoic acid, an 8-carbon fatty acid containing two sulfhydryl groups on

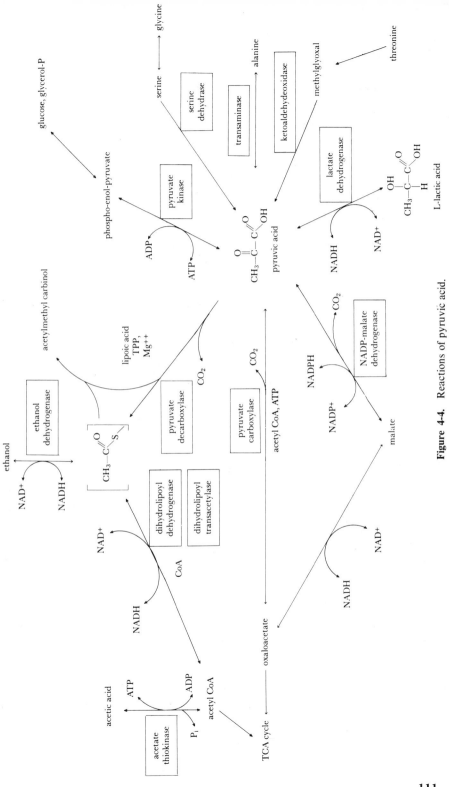

Figure 4-4. Reactions of pyruvic acid.

the 6th and 8th carbons, then accepts the aldehyde to form an acyl thioester. Reaction of the latter with the sulfhydryl group of CoA in a "thioacyl exchange" produces acetyl CoA and the disulfhydryl form of lipoate, which is then dehydrogenated with the concomitant reduction of NAD^+. Probably FAD, which is closely attached to the enzyme, is an intermediary in electron transfer to NAD^+. There is a macromolecular enzyme complex (as isolated from heart and kidney mitochondria) which incorporates at least three distinct functions: decarboxylation of pyruvate, transfer of the resultant acetaldehyde from lipoic acid to coenzyme A, and dehydrogenation of the dihydrolipoic acid. The complex can be dissociated into three components with these distinct properties and reconstituted from them. A phosphorylated form of pyruvate dehydrogenase is less active than a dephosphorylated form. An equilibrium between the two may depend upon the availability of ATP.

Under oxidative conditions the over-all process of conversion of pyruvate to acetyl CoA appears to be irreversible. Under anaerobic conditions it has been shown that the step of acetyl lipoic acid to acetyl CoA is reversible. Moreover, anaerobically there are some other possible products of the enzymatic activity, e.g., acetoin (acetylmethylcarbinol), a condensation product of two molecules of acetaldehyde. This reaction is much more prevalent in microorganisms, however. The formation of ethanol from acetaldehyde by coupling of alcohol dehydrogenase and NADH together with pyruvate decarboxylase is appreciated as "fermentation" mainly in lower forms, but this has been seen to occur also in rat liver slices anaerobically. There are variable trace levels of ethanol in mammalian blood and other tissues. Since some dietary carbohydrate is fermented to ethanol by bacterial flora in mammalian intestine, occurrence of traces of ethanol in peripheral blood could derive from this source.

A deficiency of thiamine (vitamin B_1) is particularly known to be associated with inhibition of utilization of pyruvate. Heart muscle appears to be especially sensitive to this abnormality, since among the clinical manifestations of deficiency of B_1 the heart affliction in beriberi is outstanding. Heart muscle is known to depend on utilization of pyruvate and fatty acids rather than glucose for its energy metabolism. Indeed, at high concentrations of pyruvate (1 to 5 mM) almost the entire respiration of rat heart can be accounted for by complete oxidation of pyruvate. However, at physiological concentrations (.05 to 0.1 mM) pyruvate provides a minor contribution to CO_2 compared with fatty acids.

Brain is another tissue which reacts to an interference with utilization of pyruvate. When the oxidation of pyruvate is reduced to 30 per cent of normal in rat brain by the in vivo administration of the antithiamine analogue, pyrithiamine, "polyneuritic" convulsions take place. Studies of conversion of pyruvate-2-^{14}C to amino acids in calf liver and to CO_2 in the calf in vivo have suggested that in thiamine deficiency a larger proportion of pyruvate was incorporated in the TCA cycle via CO_2 fixation (Fig. 4-4) and less through acetyl CoA. However, similar studies in rats have demonstrated no such effect.

Magnesium is required whenever TPP is, and the production of Mg^{++} deficiency leads to symptoms in animals similar to those of vitamin B_1 deficiency.

There is some indication that the product acetyl CoA inhibits the overall process of pyruvate oxidation. If this is so, it may represent a control mechanism of possible physiological importance. The acetyl CoA formed from pyruvate appears to mix metabolically with that formed (also in mitochondria) from the catabolism of fatty acids. Substrate competition between the two sources of acetyl CoA has been observed both in vivo and in vitro. Primarily the utilization of acetyl CoA formed from pyruvate occurs via the TCA cycle (see below), but it may also be a source of acetylation of various compounds or react with itself to form acetoacetyl CoA ("ketone bodies," as produced in the liver). The CoA derivatives of these 4-carbon acids (acetoacetic and β-hydroxybutyric) may be further reduced and polymerized to long-chain fatty acids.

Free acetate is seldom produced in significant concentration in animal tissues, but there is an enzyme, acetate thiokinase, which can react with ADP, Pi, and acetyl CoA to produce ATP, CoA, and acetate (Fig. 4-4). By this mechanism pyruvate can act as an alternate substrate for glucose in the production of substrate-linked high-energy phosphate. Restoration by pyruvate of the capacity of polymorphonuclear leukocytes for phagocytosis, under circumstances of inhibition of glycolysis and the pentose cycle, has been explained in this way.

In addition to ready utilization by heart muscle of pyruvate by oxidative decarboxylation, cardiac myofibrils, which have been separated from mitochondria, cause (with the aid of Ca^{++} and thiamine) the decarboxylation of pyruvate anaerobically while at the same time contraction of the muscle protein, actin, occurs in the presence of ATP. On this basis, it was suggested that decarboxylation of pyruvate may serve as an energy source in addition to ATP for the contractile mechanism in heart muscle. Under these circumstances the designation could be made of "high-energy carbonate" in analogy to "high-energy phosphate," although the energy transfer may ultimately involve ATP. Such a special function of decarboxylation of pyruvate might help to explain the particular susceptibility of heart muscle to deficiency of the coenzyme thiamine.

In vitro skeletal muscle oxidizes pyruvate readily, but the extent of utilization in vivo is not clearly known.

Kidney in general utilizes lactate, pyruvate, and fatty acids much more extensively for fuel than glucose, and studies with ^{14}C-labeled pyruvate in slices of renal cortex (of sheep, dog, and goat) suggest that a major fraction of carbon dioxide may be contributed by the carboxyl carbon of pyruvate. The medullary part of the kidney, which is in a partially anaerobic environment in vivo and is primarily glycolytic, does not oxidize pyruvate as readily as does the kidney cortex.

Liver also utilizes pyruvate readily, but the percentage of that utilized which is oxidized through decarboxylation is probably much less than in other tissues because of the greater activity of competing reactions such as transamination to alanine and carboxylation to oxaloacetate and malate. Of the two dicarboxylic acids the one which is probably formed more directly and extensively from pyruvate is oxaloacetate through the action of mitochondrial pyruvate carboxylase. This reaction uses acetyl CoA as an obligate cofactor as well as ATP. It is strongly sensitive to the ATP/ADP ratio and can be markedly inhibited by addition of ADP. The reductive carboxylation of pyruvate to malate is carried out by an extramitochondrial enzyme known as "malic enzyme" or $NADP^+$-linked malate dehydrogenase. It may serve to be coupled with other $NADP^+$- and NAD^+-linked enzymes in transhydrogenation mechanisms, e.g., in liver and adipose tissue (see Fig. 3-10).

Another tissue which is known to use pyruvate is skin; according to studies with isolated respiring rat skin, about one-third of the carbon dioxide may be formed from pyruvate (^{14}C-labeled). Another feature of its metabolism in skin is the extensive formation of acetoacetate in addition to acids of the TCA cycle. This raises the interesting possibility that the common occurrence of skin lesions in diabetes could be related to a local production of ketone bodies.

TRICARBOXYLIC ACID CYCLE

The oxidation to carbon dioxide and water of acetyl CoA, derived either from pyruvic acid or from catabolism of fatty acids and some amino acids, involves a closed cycle of reactions known as the tricarboxylic acid (TCA), citric acid, or Krebs cycle (Fig. 4-5). Its existence was first intimated thirty to forty years ago when Thunberg, Knoop, Wieland, Szent-Gyorgi and others observed catalytic effects on the aerobic oxidation of pyruvate and other compounds by the addition of various four-carbon dicarboxylic acids (succinate, fumarate, L-malate, and oxaloacetate). Working largely with pigeon breast muscle, Krebs noted stimulation also by citrate, 2-ketoglutarate, glutamate, and aspartate. Martius and Knoop studied the interconversion of citrate and isocitrate through the supposed intermediate, cis-aconitate, by an enzyme, aconitase, and also noted the formation of 2-ketoglutarate from these tricarboxylic acids. Already known was the fact that 2-ketoglutarate is oxidized to succinate by muscle tissue, so a connection was established between the tricarboxylic acid reactions and those earlier recognized for interconversion of dicarboxylic acids.

Krebs in 1937 provided the link which closed the cycle when he demonstrated that minced pigeon breast muscle can convert oxaloacetic acid (OAA) to citric acid. He suggested that pyruvic acid condensed with OAA simultaneously with decarboxylation of the pyruvic acid. Later it became clear that decarboxylation of pyruvate to form acetyl CoA is a separate reaction, and that acetyl CoA is the substrate which reacts with OAA to form citrate.

Reactions

The enzyme which catalyzes condensation between acetyl CoA and OAA to form citric acid is called citrate synthase (formerly "condensing enzyme"). Its free energy change is strongly in favor of formation of free citrate and CoA. Citryl-CoA exists only as an enzyme-bound intermediate. The reaction is thus a direction control point, but probably not rate-controlling, since OAA does not accumulate in cells, nor does acetyl CoA except under special circumstances in the liver. Long-chain fatty acid esters of CoA are noncompetitive inhibitors of citrate synthase in vitro, but whether this is actually significant under physiological conditions is not known.

Whereas citrate synthase, located almost entirely in the mitochondria, directs the TCA cycle toward utilization and oxidation of acetyl CoA, there is another enzyme, called "citrate cleavage enzyme", or ATP-citrate lyase which is a property of the cytoplasmic portion of the cell and which cleaves citrate to OAA and acetyl CoA. The reaction is driven by the simultaneous conversion of ATP to ADP and Pi. Acetyl CoA does not move freely across the mitochondrial membrane, but citrate does. The coordinated formation of citrate in the mitochondria, migration to the cytoplasm, and cleavage there to acetyl CoA and OAA is thus a means of channeling pyruvate and other precursors of acetyl CoA to the formation of fatty acids, which occurs outside the mitochondria (Fig. 3-10).

The enzyme aconitase, which is found in both mitochondria and cytoplasm, brings about an interconversion between citrate and isocitrate by removal and addition of water. The intermediate, which is presumably cis-aconitic acid, is bound to the enzyme. Aconitase is activated by cysteine and Fe^{++}. The H which is removed from citric acid is always removed from the $-CH_2-$ group which was originally part of the OAA molecule, and not from that group which derived from acetyl CoA. The same stereospecificity is shown by the cleavage enzyme. Thus, although citric acid has a plane of symmetry, it reacts in an asymmetric manner in biological systems. Ogston explained this as being due to an asymmetric 3-point attachment of such "symmetrical" molecules to an enzyme surface. The specificity is indicated for the individual carbons of the intermediates depicted in Figure 4-5.

For the next part of the cycle there are two types of isocitrate

116 PHYSIOLOGICAL CHEMISTRY OF CARBOHYDRATES IN MAMMALS

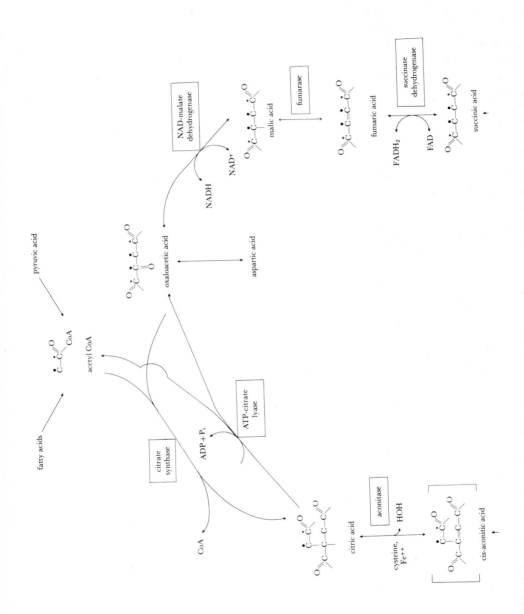

INTRACELLULAR UTILIZATION OF MONOSACCHARIDES 117

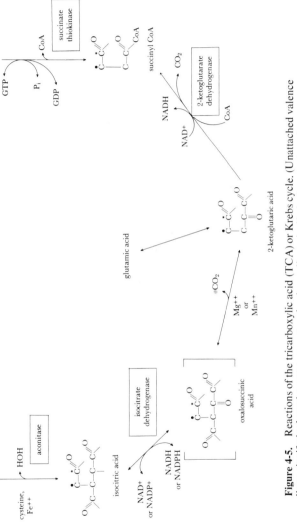

Figure 4-5. Reactions of the tricarboxylic acid (TCA) or Krebs cycle. (Unattached valence lines signify hydroxyl groups; unentered valence lines signify hydrogen atoms.)

dehydrogenase, one utilizing NAD^+ and the other $NADP^+$ as coenzyme. That which uses NAD^+ is found only in mitochondria and is not readily reversible (due to a strong inhibition by NADH). It is more active in skeletal and heart muscle than in kidney, brain, or liver. This enzyme is activated by ADP and requires Mg^{++} or Mn^{++}. The $NADP^+$-specific enzyme is found in mitochondria but also to a considerable extent in the cytoplasm (85 per cent in liver, 45 per cent in brain). The intra- and extramitochondrial enzymes differ from each other in electrophoretic mobility and immunological properties. Mg^{++} and Mn^{++} activate the NADP-type enzymes also.

The end product of isocitrate dehydrogenase (either type) is not oxalosuccinate, which is enzyme-bound, but 2-ketoglutaric acid (2-KGA), since the same enzyme which effects the dehydrogenation also causes decarboxylation of oxalosuccinate. This reaction is analogous to that of $NADP^+$-specific malate dehydrogenase or malic enzyme (Fig. 5-1), which forms CO_2 and pyruvate from malate, or vice versa. Studies of incorporation of $^{14}CO_2$ suggest that reversal of isocitrate dehydrogenase ($NADP^+$ type) may be quite active — even more so than the comparable malate dehydrogenase.

The complete TCA cycle is a property of the mitochondria, where the energy from various oxidations is coupled to synthesis of ATP. However, more recent information indicates that a number of cycle reactions occur also in the soluble portion. In fact, the entire sequence of reactions from acetate and OAA to 2-KGA has been demonstrated to occur in rat liver cytoplasm. Forward (oxidative) operation of $NADP^+$-isocitrate dehydrogenase may provide another source of the reducing coenzyme, NADPH, in the soluble phase of the cell. Studies with tritium-labeled isocitrate (^3H at C-2) have indicated a relatively active transfer of the radionuclide in vivo to fatty acids of the mouse liver and remaining carcass.

This portion of the TCA cycle, i.e., between OAA + acetyl CoA and 2-KGA + CO_2, may also operate in a "backward" (reductive) direction both in mitochondria and in the cytoplasm. This route can be utilized for the new formation both of fatty acids and of glucose (in the liver) from such precursors as glutamic acid (Fig. 4-5).

Oxidative decarboxylation of 2-KGA to form succinyl CoA is an irreversible mitochondrial reaction, which is catalyzed by 2-ketoglutarate dehydrogenase with the aid of TPP, NAD^+, CoA, lipoic acid, and Mg^{++}. The complex reaction is quite analogous to that which oxidizes pyruvate to acetyl CoA. As in the latter reaction there is intermediation of enzyme-bound FAD in the reduction of NAD^+. (In some bacteria the same enzyme can react on both pyruvate and 2-ketoglutarate, but there are separate enzymes in animal tissues.)

The deacylation of succinyl CoA to form free succinate, by succinate thiokinase, is coupled to phosphorylation of the purine nucleotide, guanosine diphosphate (GDP). The resultant GTP can further react with ADP to form ATP. Thereby high-energy phosphate for metabolic work becomes available from one of the Krebs cycle reactions by "substrate-linkage" in a manner similar to that occurring in glycolysis. Also forming succinate from succinyl CoA are two other enzymes, a deacylase and a transferase, which transfers CoA to acetoacetate. The latter enzyme does not occur in liver; consequently a low capacity for oxidation of ketone bodies is one reason for their large release to the circulation under some circumstances.

Free succinate is oxidized to fumarate by a dehydrogenase which contains FAD and nonheme iron. This enzyme is more tightly bound to mitochondria than any of the other Krebs cycle enzymes and is exclusively found in mitochondria. Succinate dehydrogenase does not require a pyridine nucleotide as cofactor; reducing equivalents are transferred directly to the cytochrome system, or possibly to coenzyme Q (a semiquinone). As a result of this by-pass there is a lesser yield of high-energy phosphate per mole of O_2 than for dehydrogenation catalyzed by NAD^+. Though NAD^+ is not a direct component of the succinate dehydrogenase system, nevertheless several studies indicate accumulation of NADH during oxidation of succinate by mitochondria or slices of liver. Also tritium has been shown to be transferred from succinate via pyridine nucleotides to other metabolic products. Circumstances indicate that this is not attributable to the oxidation of malate subsequent to that of succinate. The physiological significance of the energy-requiring reduction of NAD^+ by enzyme-bound $FADH_2$ is not yet understood. Succinate dehydrogenase has been observed to act reversibly as fumarate reductase.

The enzyme fumarase adds water stereospecifically to fumarate to form L-malate. A preponderance of the latter is favored by equilibrium conditions, but the enzymatic action is freely reversible. The apparent affinity of fumarase for fumarate is decreased by ATP, an effect which is reversed by Mg^{++} ion, but whether this represents any physiological control is not known.

The oxidation of L-malate to OAA is catalyzed by malate dehydrogenase, an enzymatic function found about equally in cytoplasm and mitochondria. The enzymes from the two parts of the cell are distinct proteins with different physical properties and composition. The coupled action of these two enzymes may accomplish electron transport between the two cellular areas (Chap. 3). The mitochondrial enzyme is inhibited by high concentration of OAA, citrate, or fumarate; that of the cytoplasm is not. The oxidation of L-malate to OAA is linked to the reduction of NAD^+ for both enzymes, and the reaction is readily reversible. The equilibrium of the reaction is mainly toward reduction of OAA to malate, but normally low cellular concentrations of OAA permit the reaction to proceed in the oxidative direction.

With the regeneration of OAA the cycle can be repeated by the entrance of another molecule of acetyl CoA. As indicated by Figure 4-5, the OAA regenerated by the cycle does not contain the same carbons as that OAA which was the substrate in the previous turn of the cycle. Two of the four carbons of the regenerated OAA are provided by the acetyl CoA joined during the previous turn. Because of this fact, it is possible for the carbons (including labeled carbons) of acetyl CoA to be incorporated into various amino acids, glucose, and other products, which are formed from intermediates of the cycle, but not for net synthesis of such compounds from acetyl CoA to occur by this route, since two other carbons are detached as CO_2 during each turn of the cycle.

Activities and Functions in Different Organs and Species

The TCA cycle is one of the most vital and ubiquitous metabolic systems in virtually all organs of mammalian species. It is situated in mitochondria in juxtaposition to the electron transport system, which accounts for the major quantitative formation of ATP in the cell. From the NADH and $FADH_2$ formed in the Krebs cycle oxidations, a stepwise passage of electrons finally to form water by combination with O_2 is linked to the generation of ATP, thus making some of the energy from these oxidations available to the cell in usuable form. From one mole of glucose completely oxidized to CO_2 and H_2O there is a yield of 30 moles of ATP from the reduced pyridine nucleotides and succinyl CoA formed in the Krebs cycle, 6 moles of ATP derived from NADH formed by GAl-3-P dehydrogenase in glycolysis, and a net of 2 moles of ATP formed from substrates in the glycolytic pathway. Thus each mole of glucose oxidized provides approximately 38 moles of ATP.

Besides serving for general oxidation and energy provision the TCA cycle forms intermediates which are precursors for synthesis of several important cell constituents. From its dicarboxylic acids are formed glucose and other carbohydrates via 3-carbon acids (Chap. 5). Oxaloacetate and 2-ketoglutarate are converted to aspartate and glutamate, respectively. From the latter two amino acids other amino acids are formed; so are parts of purine and pyrimidines. Succinyl CoA is a "building block" for porphyrins. By way of the citrate cleavage enzyme the tricarboxylic acids provide the substance acetyl CoA for synthesis of fatty acids.

The extent of this synthetic activity can be gauged by comparing the proportionate conversion to certain products of the cycle (e.g., CO_2, glucose) of ^{14}C deriving either from the 1-carbon or the 2-carbon of acetate. During metabolism of the labeled acetate if there is "isotope dilution" by influx of unlabeled carbon into the cycle (balanced by an equal outflow in the steady state), then there will be, at isotopic equilibrium, a lesser concentration of ^{14}C in the CO_2 derived from acetate-2-^{14}C than from acetate-1-^{14}C. The magnitude of difference will depend on the relative amounts of labeled acetate and non-labeled, nonacetate carbon passing into, and out of, the cycle. If introduction of non-acetate carbon occurs at half the rate of condensation of acetyl CoA with OAA, then the calculated ratio will be 2.0; if the rates are equal it will be 3.0, and so forth. Such ratios have been measured for several tissues as well as the whole body (see below). Conversely, the amount of ^{14}C from acetate-2-^{14}C in a product such as glucose will be higher than from acetate-1-^{14}C to an extent depending on the relative rates of introduction of labeled acetyl CoA and other non-labeled carbon.

The TCA cycle can in a certain way operate anaerobically by linking some of its oxidative reactions to the reversal of other oxidations in or related to the cycle. Thus animal tissues have been observed to form anaerobically from oxaloacetate such products as citrate, 2-ketoglutarate, succinate, and CO_2, which is dependent on reduction of much of the OAA to malate with NADH formed in other oxidative reactions. No substrate other than OAA can be reduced at adequate speeds to serve this function. Whether this coupling is employed under any physiological circumstances is not known. It could be part of a device for transporting hydrogen from mitochondria to cytoplasm. Reductive amination (with NH_4^+ and NADH) of 2-ketoglutarate to glutamate occurs in some animal tissues (liver, kidney, brain, heart muscle) and may also be linked to the anaerobic utilization of citrate.

Concentrations of intermediates of the TCA cycle have been measured in some tissues—mostly of the rat. They are present at levels of about 10^{-4}M; the following have been found in liver, kidney, and muscle in order of decreasing concentration: fumarate, 2-ketoglutarate, succinate, malate, citrate, and oxaloacetate—the latter considerably lower than the others. According to data on O_2 consumption about 10^{-4} moles of each intermediate is formed per kilogram of rat liver about once every 10 seconds and in kidney every 5 seconds. Therefore the average lifetime of most intermediates in these tissues is only a few seconds.

LIVER. In the liver the enzymes of the TCA cycle are found in high concentration and many of them in both the mitochondria and the cytoplasm. The fairly high respiratory rate of isolated liver further suggests an active forward (oxidative) operation of the cycle. However, liver is an organ in which the TCA cycle (or parts of it) is used for important synthetic functions, among which are gluconeogenesis via dicarboxylic acids, amino acid formation via transamination and reductive amination (glutamate), and fatty acid synthesis via cleavage of citrate. The activity of these synthetic pathways relative to the rate of introduction of acetyl CoA into the cycle is reflected, as discussed earlier, by the ratio of labeling of CO_2 by acetate-1-^{14}C and acetate-2-^{14}C. In rat liver slices this ratio ranges from about 2 to 5, which suggests that on the average TCA cycle reactions are used to a greater extent for synthesis than for oxidation. (The ratio which would be obtained with the liver in situ has not been measured.)

Probably the major synthetic activity occurring via the TCA cycle in liver is gluconeogenesis. If this process becomes excessive, as in severe diabetes or superlactation of dairy cows, the development of ketosis is thereby promoted, because the available OAA is utilized to such an extent for gluconeogenesis that not enough is available for condensation with acetyl CoA. The oxidation of the latter via forward operation of the cycle is hampered. This occurs, moreover, at a time when there is usually an excessive burden of fatty acids delivered to the liver from peripheral tissues. The result is diversion of acetyl CoA to the formation of acetoacetate and beta-hydroxybutyrate ("ketone bodies").

Another, more recently recognized, synthetic activity is the conversion of citrate to OAA and acetyl CoA in the cytoplasm, where the acetyl CoA thus becomes available for fatty acid synthesis and other reactions. This process is markedly decreased in fasting or diabetes. It is much higher in rat liver than in liver of ruminants; in the latter species acetate is supplied to the liver from the rumen, so there is less need for the citrate cleavage reaction.

The fixation of CO_2 with 2-ketoglutarate followed by cleavage of citrate to OAA and acetyl CoA is another pathway of gluconeogenesis, which is increased in diabetes. This "backward" operation of the cycle (in the mitochondria or as a joint function of mitochondrial and cytoplasmic reactions) may account for as much glutamate carbon (labeled) appearing in glucose or fatty acids as does the "forward" oxidative pathway through succinate.

MUSCLE. Enzymes of the TCA cycle were early defined in muscle (of pigeon breast) and are found generally in skeletal and heart muscle. In muscle the TCA cycle is probably used mainly for oxidative purposes — gluconeogenesis through dicarboxylic acids is not operative in this tissue. However, the citrate cleavage enzyme is present and the backward reaction of the cycle may be particularly high in heart muscle, where the activity of NADP-isocitrate dehydrogenase is high and there is prominent fixation of $^{14}CO_2$ into citrate via this cytoplasmic reaction.

BRAIN. Levels of TCA cycle enzymes in brain are similar to those in liver and muscle. Rates of oxidation to $^{14}CO_2$ of variously labeled glucose, added to the circulation through human brain, are consistent with operation of the cycle. However, a special characteristic of the oxidative cycle in brain, as determined by studies with labeled glucose, glutamate, and gamma-aminobutyric acid (GABA), may be the inclusion directly in the cycle of a small "active" pool of glutamate followed by decarboxylation to GABA, deamination of the latter to succinic semialdehyde, and then oxidation to succinate.

After administration in vivo to rats of glucose and of acetate labeled with ^{14}C in different positions, the relative labeling of glutamine, glutamate, and aspartate of brain is different for glucose and for acetate. The differential labeling of these three amino acids has suggested possible heterogeneity of the TCA cycle in brain mitochondria. Other studies, also, of the enzymes for metabolism of these amino acids in relation to GABA and other metabolites have indicated a complex compartmentation of the TCA cycle, corresponding perhaps to different subcellular anatomical structures.

In brain, as in other tissues, there is evidence for a relation between ionic movement (Na^+ and K^+) and operation of the TCA cycle. Both in peripheral (sciatic) nerve and in brain there occurs an active fixation of CO_2 into citrate, which does not occur mainly via malate but (as found lately in other tissue sites) by fixation with 2-ketoglutarate to form oxalosuccinate. As suggested by Waelsch, this carboxylation reaction could control the concentration of citrate, which, by affecting the rate of cleavage to OAA and acetyl CoA, may play a significant role in the formation of acetylcholine. Thus the long-known effect of CO_2 on specialized nervous structures (respiratory center and carotid body) could possibly find a biochemical explanation in this CO_2 fixation reaction.

MAMMARY GLAND. Radioisotopic studies show that in isolated perfused goat udder about 35 per cent of citrate carbon comes from glucose. Much of the remainder probably comes from acetate, propionate, and butyrate, i.e., short-chain fatty acids delivered through the circulation from the rumen. (Metabolism of propionate is linked to the TCA cycle by a mechanism of carboxylation to methylmalonyl CoA, which rearranges to form succinyl CoA). Further such studies with labeled bicarbonate indicate that up to 25 per cent of the carbon in citrate which originates from glucose enters via carbon dioxide. This indicates the large extent to which CO_2 fixation reactions are responsible for formation of TCA cycle intermediates. From 15 to 20 per cent of the CO_2 in citrate is incorporated via carboxylation of 2-KGA. Both for the perfused goat udder and for slices of rat mammary gland the rate of the backward operation of the cycle (2-KGA $\longrightarrow$ ICA $\longrightarrow$ CA $\longrightarrow$ OAA + acetyl CoA) may be up to one-third or one-half that of the forward operation. For the rat this rate of reversal occurs only in the presence of glucose; in its absence only 5 per cent of incorporated glutamate follows the backward route. Possibly glucose metabolism provides the NADPH needed to drive the backward (reductive) reaction. Very likely this route of metabolism of citrate is important as a source of acetyl CoA for the synthesis of milk fatty acids. The effect of glucose to increase the synthetic flow through the TCA cycle is also indicated by an increase in the $^{14}CO_2$ ratio from 1- and 2-labeled acetate (in lactating sheep mammary gland slices).

KIDNEY. Oxygen uptake by isolated glomeruli of the rat is stimulated by all substrates of the Krebs cycle, which indicates operation of the intact cycle in this portion of the kidney. However, activities of certain enzymes (isocitrate dehydrogenase, malate dehydrogenase) are 10 to 20 times higher in the proximal tubule, distal tubule, and collecting duct of human nephron than in the glomerulus. Perfusion of the dog kidney in situ with ^{14}C-labeled succinate followed by analysis of malate and citrate in the urine has further indicated the occurrence of several active TCA cycle enzymes in the tubular cell.

Whereas gluconeogenesis via dicarboxylic acids occurs in the kidney, the extent of this process, or other synthetic activities of the TCA cycle, is less in relation to the oxidative function that it is in the liver, as judged by the $^{14}CO_2$ ratio for rat kidney slices utilizing octanoate-1-^{14}C or octanoate-2-^{14}C.

BONE. Particular attention may be given to the TCA cycle in bone because citrate is present there in appreciable concentration and is believed to be important in the metabolism of calcium. According to the incorporation of ^{14}C from labeled acetate into various intermediates, there is considerable activity of the TCA cycle in epiphysis, spongiosa, and trabecular bone. The decrease of concentration of citrate in the bone of vitamin D–deficient rats is accompanied by a decrease in the amount of acetate-^{14}C appearing in citrate (of isolated bone) but an increase in the amount of ^{14}C in intermediates formed from citrate. Therefore, vitamin D appears not to promote formation of citrate, but possibly to decrease its conversion to other intermediates of the TCA cycle.

OTHER ORGANS AND TISSUES. Oxidation of various labeled substrates indicates an active TCA cycle in adipose tissue. In rat epididymal fat pad there is an active occurrence of citrate cleavage enzyme, but in human peripheral adipose tissue this enzyme appears to be very low or absent. Again this illustrates the hazards of assuming metabolic activities or patterns for one species based on evidence from another.

Many of the Krebs cycle enzymes have been found in thyroid tissue. The activities of several of the enzymes in sheep thyroid were lower per mg. K^+, DNA, or RNA than the levels in sheep liver by a factor of two to five. According to the ratios of yield of $^{14}CO_2$ from different labeled carbon positions in lactate and alanine (C-2 vs. C-3) and acetate and succinate (C-1 vs. C-2), the TCA cycle operates in sheep thyroid tissue (slices and isolated cells) and has a synthetic function, i.e., an output into other metabolic products besides CO_2, which is about half the rate of the oxidative function.

The ciliary body of the eye uses the TCA cycle, as well as a very active aerobic glycolysis, to drive the sodium transport (and formation of aqueous humor) by provision of ATP. About 75 per cent of the cellular ATP is obtained from operation of the cycle and the rest by glycolysis.

Although the TCA cycle does not function as a whole in mature mammalian red cells, certain enzymes of the cycle—fumarase, malate dehydrogenase, aconitase, and isocitrate dehydrogenase—are found in such cells. Their physiological action, if any, in the absence of operation of the cycle is not known.

WHOLE BODY. Comparison of the different rates of appearance of ^{14}C in the expired carbon dioxide of rats, cows, and human subjects after administration of glucose, lactate, or acetate labeled in specific carbon positions gives evidence for the occurrence of the TCA cycle as the common final oxidative pathway for the body as a whole, much as does similar evidence for the most active isolated organs and tissues. The integral synthetic flow through the TCA cycle for the whole body is similar to that in separate organs (liver, kidney, lactating mammary gland, thyroid), as indicated by a $^{14}CO_2$ ratio of about 2.0 on the average for acetate-1-^{14}C vs. acetate-2-^{14}C (mildly diabetic human subjects or lactating cows). Long-fasted rats, however, showed a ratio of about 1.0, which suggested little synthetic activity. In the mildly diabetic patients the $^{14}CO_2$ ratio decreased to 1.0 or less after 3 to 4 hours. This indicates a fairly early return to the TCA cycle of the carbon diverted to "synthetic" pathways. Further such studies may reveal characteristic patterns in different physiological or disease conditions. The advantages of double carbon tracer labeling, e.g., ^{14}C plus ^{13}C, in such situations are obvious.

UTILIZATION OF FRUCTOSE

As a component of the disaccharide sucrose, fructose is a common constituent of the diet of contemporary man in large parts of the world, particularly Western, as already mentioned. Fructose is also present in most fruits and honey, but this is a quantitatively minor source compared with sucrose. Fructose is produced by some mammalian tissues, e.g., the male reproductive organs and the placenta. It is found in high concentration in semen and in fetal blood. This suggests that it has special value for rapidly dividing tissues with high energy and/or nucleic acid requirement, perhaps for the synthesis of pentose.

Fructose is formed by hydrolysis of sucrose at the brush border membrane of the jejunal mucosa. By this derivation or after ingestion as such, fructose is absorbed at moderate and varying rates (Chap. 2). While passing through intestinal mucosal cells some of the fructose is metabolized before entering the portal circulation. The extent of this metabolism is different with different species and depends also upon the load of fructose absorbed. The larger the load, the greater the percentage which traverses the intestinal wall unaltered. In the guinea pig with a load up to 100 mg. about 90 per cent of fructose is converted to glucose and much of the remainder to lactic acid. However, in rats after the same load about half of the fructose is found as such in the portal vein, 25 to 40 per cent is converted to lactic acid, and about 15 per cent to glucose. In the dog about 40 per cent of an oral dose (50 g.) of fructose is absorbed as such, more than half converted to glucose, and very little to lactic acid. With the same dose in man some investigators find that most of the fructose is absorbed as such with some formation of lactic acid. Others have observed that with similar doses up to 70 per cent of the extra hexose appearing in the mesenteric vein is glucose.

As indicated in Figure 4-1, there are alternative ways by which fructose can be converted to glycolytic intermediates and to glucose. In intestinal mucosa of rat, dog, sheep, cow, and horse, there occurs the enzyme fructokinase (ketohexokinase), which transphosphorylates from ATP to form fructose-1-phosphate (Fr-1-P). This enzyme is also found in the liver of various species, and there is inferential evidence for its occurrence in kidney and in PMN leukocytes. The presence of fructokinase in muscle has never been established; yet there is an enzyme in muscle which converts Fr-1-P directly to Fr-1,6-diP. Fructokinase in liver and intestinal mucosa is strongly inhibited by ADP. It requires K^+ and Mg^{++} as well as ATP.

In muscle the enzyme phosphofructokinase may initiate phosphorylation of fructose (at C-1); muscle utilizes fructose at a relatively slow rate, however, in comparison with liver. Brain also uses fructose poorly. For muscle, brain, liver, and adipose tissue there is the possibility of phosphorylation at C-6 by one of the hexokinases, but for all these tissues the affinity of hexokinase for fructose is far less than for glucose. However, at high concentrations of fructose the utilization by hexokinase of adipose tissue proceeds rapidly and can be even higher than the rate for glucose. When fructose is delivered to the circulation via the oral route, the site of major uptake is undoubtedly the liver. The capacity for hepatic utilization when delivered into the human portal vein is at least twice as high for fructose as for glucose. During continuous gastric intubation of as much as 450 g. of fructose daily in diabetic patients there is no significant amount of fructose detectable in the arterial blood. Even when introduced into the peripheral venous circulation in trace or large amount, fructose may be largely metabolized by the liver, since various evidence indicates a conversion of about $\frac{1}{3}$ to circulating glucose in human subjects. Some glucose production may occur in the kidney cortex, which utilizes fructose readily.

Studies with isotopes and selective inhibitors have indicated that Fr-1-P is largely cleaved to dihydroxyacetone phosphate and D-glyceraldehyde (Fig. 4-1). When fructose-1-^{14}C was administered to rats the ^{14}C was distributed to the extent of 54 per cent and 34 per cent respectively, in carbons 1 and 6 of hepatic glycogen-glucose. When fructose-6-^{14}C was used, the $\frac{1}{6}$ ratio was approximately reversed. This finding, made also with isolated liver, suggests that breakdown to trioses, rather than more direct conversion of the intact hexose chain, is the major, perhaps sole, pathway for conversion of fructose to glucose in the liver.

Besides rendering unlikely the alternative pathways for Fr-1-P of direct conversion to Fr-6-P by a "fructomutase" or to Fr-1,6-diP by PFK (Fig. 4-1), such evidence is against the conversion of fructose to glucose via sorbitol, which is a pathway utilizable in some other organs (Chap. 5). Yet there is other evidence that in the liver, fructose, unlike Fr-l-P, may be converted to glycogen by a pathway not involving the major pools of hexose phosphates. The question of alternative routes for metabolism of fructose in the liver appears to be not yet fully resolved. In fetal rat liver, glucose is formed from fructose without splitting of the molecule. The alternative pathway is not known; that which forms sorbitol as an intermediate could not be detected.

The cleavage of Fr-1-P in the liver is carried out by an aldolase, which by present evidence is the same aldolase which degrades Fr-1, 6-diP in the glycolytic pathway. The activities on the two substrates are approximately the same. By contrast, muscle aldolase is 50 times more active on Fr-1,6-diP than on Fr-1-P.*

One of the products of splitting of Fr-1-P, D-glyceraldehyde, could be either phosphorylated to glyceraldehyde-3-P, oxidized to glycerate, or reduced to glycerol. Either of the last two could be subsequently phosphorylated. There are known enzymes in the liver which could introduce D-glyceraldehyde into the glycolytic scheme by any of these three pathways (Figs. 4-1 and 4-2). Since the metabolism of Fr-6-^{14}C results in relatively low labeling of position C-3 or C-4 of hepatic glycogen-glucose or of position C-1 of glyceride-glycerol, formation of the symmetrical compound, glycerol, appears to be minor. Study of glycogen labeling from 4-^{3}H,6-^{14}C-labeled fructose shows that tritium incorporation is inversely proportional to randomization of ^{14}C between C-1 and C-6 of glycosyl units; this indicates phosphorylation of glyceraldehyde rather than oxidation to glycerate.

Both of the triose units formed from Fr-1-P are in one way or another made available for recombination by aldolase to form Fr-1-6-diP and subsequently glucose, which can occur to an appreciable extent both in liver and in intestinal mucosa, as discussed above. Alternatively, the trioses may follow the glycolytic scheme in the other direction to form pyruvate and lactate. It has long been observed that a load of fructose intravenously in man causes a much higher rise in blood pyruvate and lactate than an equivalent load of glucose, which suggests substantial metabolism of fructose by the oxidative route.

Hepatic fructokinase is known not to be depressed in diabetes or stimulated by insulin as is hepatic glucokinase, and fructose loads disappear from the blood as rapidly in diabetic as in non-diabetic patients. On

---

*Muscle and liver aldolase differ in other respects, e.g., response to various adenine nucleotides. Muscle aldolase, like PFK, is inhibited by high concentrations of ATP, so that both enzymes may be involved in any regulatory action of ATP. Hepatic aldolase is inhibited by AMP to a greater extent than by ADP, and not by ATP.

the basis of this and earlier encouraging evidence, fructose has been tried periodically for almost 100 years as a treatment or substitute for other carbohydrate in diabetic patients, particularly those with the more severe form on occasions of ketoacidosis. Fructose has been shown to normalize certain metabolic parameters in some diabetic patients, but its efficacy is diminished in proportion to the severity of the diabetic state and this therapy has not gained widespread use (see Chap. 8).

In severe diabetes the metabolism of fructose is shifted toward formation of glucose and away from conversion to oxidative products. This is consistent with current knowledge of enzymatic changes in diabetes at various points in the glycolytic and gluconeogenic pathways. Even though fructokinase is not affected by diabetes or insulin, this enzyme as well as other pertinent ones, e.g., aldolase and triokinase, is adaptive to fasting and refeeding with fructose in the rat.

There is an uncommon human genetic disorder known as hereditary fructose intolerance (HFI) characterized by high blood fructose curve after oral or intravenous load. The condition is accompanied by hypophosphatemia, chronic defects in kidney tubular reabsorption, signs of liver damage, and signs and symptoms of severe hypoglycemia after ingestion of sucrose or fructose. This disorder is related to a severely depressed Fr-l-P aldolase activity in liver and kidney. The splitting of Fr-1,6-diP is considerably less affected than that of Fr-l-P. The renal accumulation of Fr-l-P appears to underlie the renal tubular dysfunction (acidosis and certain aminoacidurias). The hypoglycemia is due to a virtual cessation of glucose supply to the blood — as indicated by a plateauing of the slope of specific activity of glucose-^{14}C in the blood after injection of fructose. A recent demonstration of inhibition by Fr-l-P of phosphohexose isomerase acting in the direction of formation of Gl-6-P from Fr-6-P provides a possible biochemical explanation for decreased gluconeogenesis and for the typical accumulation in the blood of pyruvate and lactate in patients with HFI. Another possibility is that both glycogenesis and gluconeogenesis are impaired by depletion of both ATP and Pi in the liver cells. The rare condition of "essential fructosuria," with hyperfructosemia as well as fructose in the urine, is due to a defect in fructokinase and is not accompanied by any symptoms.

There is considerable evidence that chronic fructose or sucrose feeding in rats and human subjects predisposes to an elevation of plasma triglycerides (TG) more than does equivalent glucose feeding. Liver TG are also increased by fructose and sucrose, but not glucose, in the rat. Furthermore, there is a steadily increasing adaptation to conversion of more (labeled) fructose and sucrose to liver and serum TG in the rat as feeding continues, whereas glucose feeding causes no adaptation. The mechanism of causation by fructose of increased concentration and presumably synthesis of plasma TG may relate to an excessive amount and rate of formation of α-glycerol phosphate, which could promote esterification of fatty acids. Possibly fatty acids are formed in excess from fructose, also. The fact that lactic acidemia is produced more readily by a load of fructose than one of glucose suggests an exorbitant rate of input into the glycolytic scheme. Hepatic enzymes intermediate in the formation of fatty acids from fructose increase in activity more on fructose feeding than on glucose feeding in the rat.

Another metabolic effect of fructose, recently discovered, is production of a sharp hyperuricemia after either oral or intravenous fructose load in normal children and in patients with HFI (but not in one with essential fructosuria). The magnitude of increase of uric acid is such that it can hardly be accounted for by inhibition of disposal and is more likely due to an increase in formation. Rapid depletion of hepatic ATP and Pi during fructose utilization leads to transient increase in level of AMP, which is then deaminated and catabolized excessively to uric acid. Possibly fructose could promote hyperuricemia through an active formation of NADPH, which is used in the reductive synthesis of purines . Uric acid is an end product of purine metabolism.

UTILIZATION OF GALACTOSE

Galactose is the other hexose, besides glucose, of the disaccharide lactose, which is the main sugar in milk. In addition to being a source of energy fuel—mainly, of course, in infancy and early childhood—galactose is a constituent of various heteroglycans and glycolipids (Chaps. 1, 6 and 7). The unique metabolism of galactose for the formation of these complex structures is discussed in these other chapters. As is evident from consideration of the relevant enzymatic reactions (Fig. 4-1), galactose is not required in the diet, i.e., it can be formed from glucose and in amounts adequate for its special role in glycoproteins and glycolipids.

Galactose is rapidly absorbed from the gastrointestinal tract at a rate comparable to that of glucose (Chap. 2). There is little evidence that it can be metabolized by the intestinal mucosa. Indeed, enzymatic and other assays in general suggest that other organs and tissues, except possibly kidneys, are quite inferior to the liver in regard to the rate of utilization and metabolism of galactose. Galactose is introduced by a series of enzymatic steps into the glycolytic sequence at the point of glucose-l-phosphate (Fig. 4-1). Further metabolism in the liver can provide free glucose (and, of course, fatty acids and other products), so galactose may serve as fuel for the body as a whole via hepatic metabolism if not directly in other tissues.

Metabolism of galactose is initiated by a special kinase, galactokinase, which uses ATP to phosphorylate at position C-1 with formation of galactose-l-phosphate (Gal-l-P). Besides being in liver this enzyme occurs appreciably in intestine, kidney, brain, and erythrocytes. Muscle shows much lower activity than the other tissues. In liver of adult rats the galactokinase activity is about twice as high as in intestine or kidney and three times as high as in brain. Adult male rats show higher hepatic activity than females.

Hepatic galactokinase activity in rats rises sharply after birth, reaches a maximum at 5 days, and declines slowly to adult levels at 36 days. It is several times as high at peak as the stable level in the adult. Hemolysates of newborn human infants also show three times the activity of galactokinase as those of adults. The red cells of newborns also show greater utilization of oxygen after addition of galactose than those of older children. The RBC of newborn infants produce $^{14}CO_2$ from galactose-l-^{14}C three times as rapidly as those of adults, and from glucose-l-^{14}C twice as rapidly. Leukocytes of newborn infants showed an average of 50 per cent greater production of $^{14}CO_2$ from galactose-l-^{14}C than from glucose-l-^{14}C, whereas those of adults were approximately equal for the two hexoses.

The K_m of hepatic galactokinase of the newborn rat is four times greater and the V_{max} (maximal velocity) is six times greater than that of the adult rat. This appears to provide flexibility for handling large amounts of galactose during the nursing period. Inhibition of galactokinase by the product, Gal-l-P, is much more effective in the newborn rat than in the adult — and of a different type. Also, substrate inhibition by galactose occurs at a much lower concentration (near the K_m) in the newborn than in the adult. This has the result of controlling the accumulation of Gal-l-P, which is considered to be at least one of the causes of toxicity in galactose-fed animals and patients with galactosemia. Gal-l-P inhibits several enzymes of carbohydrate metabolism, among which are phosphoglucomutase, UTP;Gl-1-P uridyltransferase, Gl-6-P dehydrogenase, and glucose-6-phosphatase.

Besides the above differences with age in activity and inhibition characteristics, galactokinase of young rats is more adaptive in response to high galactose feeding than that of adult rats. The various differing properties of adult and newborn hepatic galactokinase suggest that there may be two different molecular species of the enzyme.

At 10 minutes after intravenous injection of equivalent amounts (per kg. body weight) of galactose, infants show only about half as much blood concentration of galactose as that of older children and adults. A possible interpretation is that the volume of distribution is roughly twice as high in infants. However, since the increments of blood glucose after galactose injection are two to three times greater at 10 minutes and after in infants than in older children and adults, there may be more rapid and extensive conversion of galactose to glucose in infants. Over-all utilization of galactose in rat liver slices likewise is greater in newborn than in adult animals.

The correlation of changes with age of galactokinase and of general utilization of galactose suggest that initial phosphorylation is the rate-limiting step for galactose metabolism. In hemolysates the rate constant for galactokinase is only one-tenth of that for the next step in utilization of galactose, i.e., the reaction catalyzed by UDP-Gl;Gal-1-P uridyltransferase (Fig. 4-1). In this reaction galactose enters into a complex with the nucleotide uridine diphosphate (UDP). The mechanism involves an exchange of glucose in UDP-Gl for galactose, thus forming UDP-Gal and Gl-l-P. The nucleotide derivative is split between the two phosphates with reattachment of the other hexose-1-P to the UMP moiety in a reaction which is freely reversible. It is this step which is blocked in congenital galactosemia (see below).

There is an alternative reaction to form UDP-Gal which is catalyzed by UTP;Gal-1-P uridyltransferase (also called UDP-Gal pyrophosphorylase). UTP is the reactant instead of UDP-Gl and PPi is the product instead of Gl-l-P. This reaction is also reversible. It has been observed in rat and human liver; in the rat the activity is about $\frac{1}{6}$ that of the UDP-G4;Gal-1-P uridyltransferase. While the reaction using UTP may be a minor pathway normally, it could be important in galactosemic individuals who have lost the capacity for formation of UDP-Gal via the more active uridyltransferase. Both of these enzymes increase markedly with age (there is a sevenfold difference between neonatal and adult rat liver), so it is possible that the improved tolerance of some galactosemic patients is due to the growing activity of this alternate pathway of galactose metabolism.

Galactose is an epimer of glucose because of the difference of configuration at the fourth carbon atom. A conversion of UDP-Gal to UDP-Gl is brought about by the action of an enzyme called UDP-Gal 4-epimerase (Fig. 4-1). This reaction probably involves an oxidation and reduction, since NAD^+ is a coenzyme. Although the latter is rather firmly bound to the enzyme, nevertheless the enzymatic activity is strongly inhibited by NADH or, more accurately, by an increase in the ratio, $NADH/NAD^+$, in the milieu of the enzyme. The inhibition of galactose metabolism by ethanol has been explained on the basis of the known effect of ethanol to increase the $NADH/NAD^+$ ratio. Decreased galactose tolerance during metabolism of ethanol has been demonstrated in rats and human subjects. Agents (e.g., menthol and progesterone) which decrease NADH have a stimulating effect on the oxidation of UDP-Gal-1-^{14}C in rabbit liver homogenates and galactose-1-^{14}C in vivo in prepubertal galactosemic patients.* Perhaps the rate of oxidation of galactose-^{14}C would reflect the redox potential of the liver in other pertinent abnormal situations (see Chap. 8).

---

*No stimulation occurs with normal adult subjects, however. The observation with galactosemics is further curious in that the enzymic lesion is presumably confined to the uridyltransferase step, which precedes the epimerase reaction. By other evidence the latter is not affected in galactosemia.

The 4-epimerase, as isolated from calf liver and adult human hemolysates, shows a requirement for addition of NAD^+. However, hemolysates from newborn infants show a substantial epimerase activity without the addition of NAD^+. Whether this is due to a qualitative difference in the epimerase of infants and adults or to some factor not relating to the enzyme per se is not known. Operating in the other direction 4-epimerase is, of course, responsible for the formation of UDP-Gal from glucose.

Through the joint operation of the successive steps of galactose metabolism (Fig. 4-1) there is a net production of Gl-l-P from Gal-l-P with UDP-Gl acting catalytically. UDP-Gl may be converted directly to Gl-l-P by reversal of UTP;Gl-1-P uridyltransferase or it may be a reactant in formation of glycogen. Thus galactose may be more disposed to formation of glycogen than is glucose.

An interesting possible difference between glucose and galactose in ultimate disposal by main routes of metabolism relates to a difference in anomery, i.e., configuration at the first carbon atom (Chap. 1). The α and β anomers of Gl-6-P are produced in about equal abundance by hexokinase. However, the enzymes which further direct the metabolism of Gl-6-P into the pentose cycle or the E-M-P pathway, i.e., G6PDH and phosphohexose isomerase, respectively, have different anomeric specificities. G6PDH utilizes only the β anomer, while phosphohexose isomerase acts only on the α anomer. The enzyme which finally converts galactose to the "pool" of Gl-6-P, i.e., phosphoglucomutase, is also specific for the α anomer. If the interconversion of α and β forms (by spontaneous or enzyme-catalyzed mutarotation) were not complete, then galactose would follow the glycolytic pathway more extensively than the pentose cycle in comparison with glucose. This has been tested in human RBC by comparing the ratio of conversion of glucose-1-^{14}C to CO_2 and glycolytic products with the ratio obtained with galactose-1-^{14}C as precursor. No difference was found, so a rapid conversion of the α anomer to the β anomer evidently occurs, at least in RBC, after formation of α-Gl-6-P from galactose.

The rare disease congenital galactosemia stimulated much study of galactose metabolism with rewarding outcome for understanding of individual biochemical steps, specific enzyme lesion involved, nature of toxic effects, and management and prognosis of the condition. Loss of the specific enzyme (the uridyltransferase mentioned above) is complete for homozygotes who inherit from both parents the autosomal recessive genes. Heterozygotes, who are carriers of the recessive gene, show, as expected, about one-half of the normal activity of UDP-Gl;Gal-1-P uridyltransferase in their erythrocytes. This group may comprise a reservoir of 1 to 1½ per cent of the population with little if any galactose intolerance and no symptomatology referable to it. Theoretically, if other tissues, particularly liver, show the same ratio of activities of galactokinase and uridyltransferase as do lysed RBC (see above), then the heterozygote would still have ample uridyltransferase activity for handling the amounts of Gal-l-P formed. This would not apply, however, to heterozygotes for the kind of galactosemia due to galactokinase deficiency (see Chap. 8).

True galactosemic patients (homozygotes for uridyl transferase deficiency) when fed milk as infants soon show diarrhea and a failure to thrive, then develop cataracts, jaundice, hepatosplenomegaly, and mental retardation. Galactosuria, aminoaciduria, and proteinuria accompany the galactosemia. Hypoglycemia may also occur after ingestion of milk and be related to ketosis and vomiting. (Hypoglycemia in one galactosemic girl as late as age 16 was found to account for behavior disturbances). Much of the pathology in the liver, spleen, lens, brain, and kidney has been ascribed to the accumulation of Gal-l-P, which has inhibitory effects on carbohydrate metabolism, as mentioned earlier. The administration of food with as low a galactose content as 0.2 per cent can cause a toxic accumulation of Gal-l-P.

A further effect appreciated more recently is the excessive reduction of galactose to galactitol (dulcitol) by aldose reductase with the aid of NADPH. Particularly in lens, brain, and kidney, this could be another cause of the disorders of those sites in galactosemia. In fact, high feeding of galactose (35 per cent in the diet) to rats results in accumulation of galactitol in nearly every tissue of the body.

The defect of congenital galactosemia can be detected by oral galactose tolerance test (with measurement of galactose in blood and urine), measurement of UDP-Gl;Gal-1-P uridyltransferase in red cells, or by measurement of oxidation of Gal-l-^{14}C to $^{14}CO_2$ in hemolysates. Comparison of amounts of ^{14}C in Gal-l-P and UDP-Gal upon incubation of RBC with Gal-l-^{14}C helps to determine levels of the specific enzymes of galactose metabolism in RBC (e.g., in heterozygotes). Curiously, although the defect is clearly expressed in RBC, no early hematologic difficulties are seen. However, Gal-l-P inhibits respiration of erythrocytes in the presence of glucose. Besides being in liver and RBC, uridyltransferase deficiency has been found also in lens, leukocytes, skin, and bone marrow.

Another method of detection or follow-up employs the rate of oxidation of ^{14}C-labeled galactose to $^{14}CO_2$ in vivo. After early childhood most galactosemic patients show only about 1/6 of the rate of normal oxidation of Gal-l-^{14}C. Studies with both Gal-l-^{14}C and Gal-2-^{14}C have suggested that even this residue of galactose utilization is due to a direct oxidative pathway analogous to that for oxidation of glucose in the pentose cycle; such a pathway would by-pass uridyltransferase. A minority of patients show normal or nearly normal rates; nevertheless, they continue to display the typical defect in the red cells.

There are few tests which can measure slight or even moderate decreases in liver function, because this organ has a notable reserve capacity for its various physiological functions. Since there is relatively easy saturation of capacity of the liver for galactose utilization, galactose tolerance has been used as one test of mild liver dysfunction. Galactokinase in the liver has a fairly low K_m (0.1 mM), so if the concentration of galactose in the blood is raised considerably above this value (e.g., to about 2.0 mM or 35 mg./dl.), then the enzyme is virtually saturated, the elimination rate by the liver is essentially constant, and the rate of elimination is a measure of the "Lm," or liver maximum, a term which is comparable to the Tm or tubular maximum as applied to capacity of the kidney for excretion or reabsorption. Oral administration of galactose has been used in the past, but intravenous load has been used more recently in attempts at more quantitative results (Chap. 8). Excretion in the urine as well as elimination from the blood must be evaluated. Because of certain unknown factors and assumptions (e.g., about exclusive hepatic utilization), the test remains somewhat empirical. When galactokinase is saturated, then the rate of subsequent enzymes, e.g., UDP-Gal 4-epimerase, may further determine the rates of total utilization of galactose, including oxidation to CO_2.

UTILIZATION OF MANNOSE

Mannose is an isomer of glucose owing to the difference of configuration at the second carbon atom. It is a constituent of various heteroglycans of mammalian tissues. When ingested as such it forms an insignificant portion of the diet from a nutritional standpoint. Mannose-6-phosphate (Man-6-P) can be formed from more common phosphorylated hexoses in the body (Fig. 4-1), so it is not needed in the diet. Once mannose was considered to be passively transported through the intestinal mucosa. Later studies have shown that at low concentrations mannose is actively transported.

In view of the quantitative insignificance in the diet and the absence of appreciable concentrations of mannose in the blood, there is surprising capacity of several tissues for utilization of mannose—more than for galactose or fructose in certain extrahepatic sites. Mannose is utilized almost—and sometimes equally—as well as glucose in the following organs of the rat: diaphragm, mammary gland, liver, adipose tissue, brain, erythrocytes, and bone marrow. Also bone marrow of the rabbit, leukocytes and kidney of rabbit and dog, ox retinal extract, calf cartilage, and skin from guinea pig and man have all shown similar utilization rates for mannose and glucose.

In both liver and adipose tissue of the rat the metabolism of mannose (^{14}C-labeled) to CO_2, fatty acids and glycogen is about 60 to 90 per cent that of glucose. Metabolism of mannose is depressed in alloxan-diabetic rats to an extent similar to that of glucose, and insulin has a similar effect on stimulating metabolism of both hexoses. Mannose seems to stimulate production of insulin by the β cells of the pancreas in some species (rat, rabbit) but not others (dog). Since transport is not limiting in liver slices, the effect of diabetes and insulin on utilization of mannose seems to occur at the phosphorylation level.

Mannose is phosphorylated by hexokinase in various tissues, including glucokinase (Type IV hexokinase) in liver. In liver the K_m of glucokinase is about twice for mannose as for glucose. A similar ratio holds for hexokinase of adipose tissue. The effective K_m for over-all utilization in adipose tissue, however, is considerably higher than that for hexokinase and three times higher than the respective K_m for glucose. This may be due to the fact that in adipose tissue transport of hexoses rather than phosphorylation is rate-limiting. Another special—and possibly rate-limiting—factor is the dependence of mannose utilization on mannose phosphate isomerase. The K_m of brain hexokinase is only about one-half for mannose as for glucose.

In the lactating rat mammary gland, certain comparisons of the rate of utilization of mannose with that of phosphorylated sugars indicate that mannose is phosphorylated at C-6 rather than at C-1 and that subsequently it is readily isomerized to Fr-6-P, or perhaps to Gl-6-P. There is supposedly a special mannose phosphate isomerase. However, there is presumably also a phosphomannomutase, since activation of mannose to a nucleotide derivative appropriate for introduction of mannose into oligosaccharides involves Man-l-P as a reactant. Man-l, 6-diP is uniquely found in mammalian red cells. GDP-mannose is formed from GTP and Man-l-P by a guanidyltransferase in lactating rat mammary gland and liver. The activity is six times higher in mammary gland than in liver. Also formed (or degraded) by the same enzyme are IDP-Man (IDP = inosine diphosphate) and ADP-Man. Activities in the direction of pyrophosphorylysis for GDP-Man, IDP-Man, and ADP-Man are in the ratio of 1.0:0.8:0.1. The GTP; Man-l-P guanidyltransferase probably has a role in the biosynthesis of GDP-fucose and milk oligosaccharides (Chaps. 1, 5 and 6).

Some data from incorporation of ^{14}C labeled fructose and mannose into glycogen-glucose of PMN leukocytes of rabbit suggest that there may be a link between these two hexoses via the symmetrical compound mannitol, but there is no enzymatic proof for this reaction as yet. A further suggestion has been made that versatility in carbohydrate metabolism by the neutrophil may be useful in protecting the cell from harmful compounds liberated in the course of digestion of bacterial cell wall polysaccharides.

REFERENCES

General

Bell, G. H., Davidson, J. N., and Scarborough, H.: Textbook of Physiology and Biochemistry, 7th ed. E. and S. Livingston, Ltd., 1968.

Fruton, J. S., and Simmonds, S.: General Biochemistry. John Wiley and Sons, Inc., New York, 1958.
Harper, H. A.: Review of Physiological Chemistry, 14th ed. Lange Press, 1973.
Karlson, P.: Introduction to Modern Biochemistry, 2nd ed. Academic Press, New York, 1965.
Landau, B. R., Katz, J., Bartsch, G. E., White, L. W., and Williams, H. R.: Hormonal regulation of glucose metabolism in adipose tissue in vitro, Ann. N.Y. Acad. Sci. *131*:43–58, 1965.
Oser, B. L.: Hawk's Physiological Chemistry, 14th ed. McGraw-Hill, Inc., 1965.
Walker, D. G.: The nature and function of hexokinase in animal tissues. Essays in Biochemistry *2*:33–67, 1966.

For Phosphorylation of Glucose

Ballard, F. J.: Glucose Utilization in Mammalian Liver. Comp. Biochem. and Physiol. *14*:437–443, 1965.
Di Pietro, D. L., Sharma, C., and Weinhouse, S.: Studies on Glucose Phosphorylation in Rat Liver. Biochemistry *1*:455–462, 1962.
Figueroa, E., and Pfeiffer, A.: Incorporation of ^{14}C-glucose 6-phosphate into glycogen and CO_2 by rat liver slices. Nature *204*:576–577, 1964.
Katzen, H., and Schimke, R. T.: Multiple forms of hexokinase in the rat: tissue distribution, age dependency, and properties. Proc. Nat. Acad. Sci. *54*:1218–1225, 1965.
Landau, B. R., and Sims, E. A. H.: On the existence of two separate pools of glucose 6-phosphate in rat diaphragm. J. Biol. Chem. *242*:163–172, 1967.
Lauris, V., and Cahill, G. F., Jr.: Hepatic Glucose Phosphotransferases. Variations Among Species. Diabetes *15*:475–479, 1966.
Neufeld, E. F., and Ginsburg, V.: Carbohydrate metabolism. Ann. Rev. of Biochem. *34*:297–312, 1965.
Nordlie, R. C., and Arion, W. J.: Liver microsomal glucose 6-phosphatase, inorganic pyrophosphatase and pyrophosphate-glucose phosphotransferase. J. Biol. Chem. *240*:2155–2164, 1965.
Sharma, C., Manjeshwar, R., and Weinhouse, S.: Hormonal and dietary regulation of hepatic glucokinase. Adv. in Enzyme Regulation *2*:189–200, 1964.
Smith, E. E., Taylor, P. M., and Whelan, W. J.: Hypothesis of the mode of conversion of glucose into α-glucose-1-phosphate. Nature *213*:733–734, 1967.
Sols, A., Salas, M., and Viñuela, E.: Induced biosynthesis of liver glucokinase. Adv. in Enzyme Regulation *2*:177–188, 1964.

For Glycolysis

Boxer, G. E., and Devlin, T. M.: Pathways of intracellular hydrogen transport. Science *134*:1495–1501, 1961.
Helmreich, E., and Cori, C. F.: Regulation of glycolysis in muscle. Adv. in Enzyme Regulation *3*:91–107, 1965.
Keitt, A. S.: Pyruvate kinase deficiency and related disorders of red cell glycolysis. Am. J. Med. *41*:762–785, 1966.
Lee, J. B., Vance, V. K., and Cahill, G. F., Jr.: Metabolism of C^{14}-labeled substrates by rabbit kidney cortex and medulla. Am. J. Physiol. *203*:27–36, 1962.
Robinson, N., and Phillips, B. M.: Glycolytic enzymes in human brain. Biochem. J. *92*:254–259, 1964.
Shaw, W. N., and Stadie, W. C.: Two identical Embden-Meyerhof enzyme systems in normal rat diaphragms differing in cytological location and response to insulin. J. Biol. Chem. *234*:2491–2496, 1959.
Wu, R., Sessa, G., and Hamerman, D.: Pi transport and glycolysis in leukocytes and platelets. Biochim. Biophys. Acta *93*:614–624, 1964.

For Pentose Cycle

Glock, G. E., and McLean, P.: Levels of enzymes of the direct oxidative pathway of carbohydrate metabolism in mammalian tissues and tumours. Biochem. J. *56*:171–175, 1964.

Horecker, B. L.: Glucose-6-phosphate dehydrogenase, the pentose phosphate cycle and its place in carbohydrate metabolism. Am. J. Clin. Path. 47:271–281, 1967.
Hostetler, K. Y., and Landau, B. R.: Estimation of the pentose cycle contribution to glucose metabolism in tissue in vivo. Biochemistry 6:2961–2964, 1967.
Katz, J., Landau, B. R., and Bartsch, G. E.: The pentose cycle, triose phosphate isomerization and lipogenesis in rat adipose tissue. J. Biol. Chem. 241:727–740, 1966.
Long, W. K., Wilson, S. W., and Frenkel, E. P.: Associations between red cell glucose-6-phosphate dehydrogenase variants and vascular diseases. Am. J. Human Genetics 19:35–53, 1967.
Metzger, R. P., Wilcox, S., and Wick, A. N.: Studies with rat liver glucose dehydrogenase. J. Biol. Chem. 239:1769–1772, 1964.
Moss, G.: The contribution of the hexose monophosphate shunt to cerebral glucose metabolism. Diabetes 13:585–591, 1964.
O'Neill, J. J., Simon, S. H., and Shreeve, W. W.: Alternate glycolytic pathways in brain. A comparison between the action of artificial electron acceptors and electrical stimulation. J. Neurochem. 12:797–802, 1965.
Raggi, F., Hansson, E., Simeson, M. G., Kronfeld, D. S., and Luick, J. R.: Pentose cycle activity in various bovine tissues. Res. Vet. Sci. 2:180–183, 1961.
Stuckey, W. J.: Hemolytic anemia and erythrocyte glucose-6-phosphate dehydrogenase deficiency. Am. J. Med. Sci. 251:104–115, 1966.
Wood, H. G., Katz, J., and Landau, B. R.: Estimation of pathways of carbohydrate metabolism. Biochemische Zeitschrift 338:809–847, 1963.
Wood, H. G., Peeters, G. J., Verbeke, R., Lauryssens, M., and Jacobson, B.: Estimation of the pentose cycle in the perfused cow's udder. Biochem. J. 96:607–615, 1965.

For Utilization of Pyruvate

Benevenga, N. J., Baldwin, R. L., Ronning, M., and Black, A. L.: Pyruvate metabolism in thiamine-deficient calves. J. Nutrition 91:63–68, 1967.
Booij, H. L.: Pyruvate-2-^{14}C metabolism in skin. Canad. J. Biochem. 43:1011–1016, 1965.
Diaz de Arce, H., Crevasse, L., and Shipp, D. L.: Pyruvate decarboxylation and its relationship to contraction of cardiac myofibrils. Am. J. Physiol. 210:1396–1400, 1966.
Evans, J. R., Opie, L. H., and Renold, A. E.: Pyruvate metabolism in the perfused rat heart. Am. J. Physiol. 205:971–976, 1963.
Gans, J. H., Baillie, M. D., and Biggs, D. L.: In vitro metabolism of ^{14}C-labelled pyruvate and propionate by ruminant and dog kidney cortex and medulla. Arch. Biochem. and Biophys. 115:192–196, 1966.
Selvaraj, R. J., and Sbarra, A. J.: Phagocytosis inhibition and reversal. II. Possible role of pyruvate as an alternative source of energy for particle uptake by guinea pig leukocytes. Biochim. Biophys. Acta 127:159–171, 1966.
Williamson, J. R.,: Effects of insulin and starvation on the metabolism of acetate and pyruvate by the perfused rat heart. Biochem. J. 93:97–106, 1964.

For Tricarboxylic Acid Cycle

Abraham, S., Kopelovich, L., and Chaikoff, I. L.: Metabolic characteristics of preparations of isolated sheep thyroid gland cells. III. oxidation of substrates involved in carbohydrate metabolism and the Krebs cycle. Endocrinology 77:863–872, 1965.
D'Adamo, A. F., Jr., and Haft, D. E.: An alternate pathway of α-ketoglutarate catabolism in the isolated, perfused rat liver. J. Biol. Chem. 240:613–617, 1965.
Hardwick, D. C.: The incorporation of carbon dioxide into milk citrate in the isolated perfused goat udder. Biochem. J. 95:233–237, 1965.
Krebs, H. A.: The regulation of the release of ketone bodies by the liver. Adv. in Enzyme Regulation 4:339–354, 1966.
Lamdin, E., Shreeve, W. W., Slavinski, R., and Oji, N.: Biosynthesis of fatty acids in obese mice in vivo. II. Studies with DL-malate-2-^{3}H-3-^{14}C, succinate-2, 3-^{3}H-2, 3-^{14}C, and DL-isocitrate-2-^{3}H-5, 6-^{14}C. Biochemistry 8:3325–3331, 1969.
Lowenstein, J. M.: The tricarboxylic acid cycle. In: Metabolic Pathways, 2nd. ed., pp. 146–270. Academic Press, New York, 1967.
Madsen, J., Abraham, S., and Chaikoff, I. L.: The conversion of glutamate carbon to fatty acid carbon via citrate. J. Biol. Chem. 239:1305–1309, 1964.

Norman, A. W., and De Luca H. F.: Vitamin D and the incorporation of (1-^{14}C) acetate into the organic acids of bone. Biochem. J. 91:124–130, 1964.

Sachs, W.: Cerebral metabolism of doubly labeled glucose in humans in vivo, J. Appl. Physiol. 20:117–130, 1965.

Shreeve, W. W., Hennes, A. R., and Schwartz, R.: Production of $^{14}CO_2$ From 1- and 2-C^{14}-acetate by human subjects in various metabolic states. Metabolism 8:741–756, 1959.

Waelsch, H., Cheng, S.-C., Cote, L. J., and Naruse, H.: CO_2 fixation in the nervous system. Proc. Nat. Acad. Sci. 54:1249–1253, 1965.

Weinman, E. O., Strisower, E. H., and Chaikoff, I. L.: Conversion of fatty acids to carbohydrate: application of isotopes to this problem and role of the Krebs cycle as a Synthetic Pathway. Physiol. Reviews 37:252–272, 1957.

For Utilization of Fructose

Froesch, E. R., Wolf, H. P., Beutsch, H., Prader, A., and Labhart, A.: Hereditary fructose intolerance. An inborn defect of hepatic fructose-1-phosphate splitting aldolase. Am. J. Med. 34:151–167, 1963.

Kaufmann, N. A., Poznanski, R. Blondheim, S. H., and Stein, Y.: Effect of fructose, glucose, sucrose and starch on serum lipids in carbohydrate-induced hypertriglyceridemia and in normal subjects. Israel J. Med. Sci. 2:715–726, 1966.

Landau, B. R., and Merlevede, W.: Initial reactions in the metabolism of D- and L-glyceraldehyde by rat liver. J. Biol. Chem. 238:861–867, 1963.

Moorhouse, J. A., and Kark, R. M.: Fructose and diabetes. Am. J. Med. 23:46–58, 1957.

Nikkila, A. E., and Pelkonen, R.: Enhancement of alimentary hyperglyceridemia by fructose and glycerol in man. Proc. Soc. Exp. Biol. Med. 123:91–94, 1966.

Perheentupa, J., and Raivio, K.: Fructose-induced hyperuricemia. The Lancet ii:528–531, 1967.

Spolter, P. D., Adelman, R. C., DiPietro, D. L., and Weinhouse, S.: Fructose metabolism and kinetic properties of rabbit liver aldolase. Adv. in Enzyme Regulation 3:79–89, 1965.

Zalitis, J., and Oliver, I. T.: Inhibition of glucose phosphate isomerase by metabolic intermediates of fructose. Biochem. J. 102:753–759, 1967.

For Utilization of Galactose

Cuatrecacas, P., and Segal, S.: Mammalian galactokinase. developmental and adaptive characteristics in the rat liver. J. Biol. Chem. 240:2382–2388, 1965.

Donnell, G. N., Ng, W. G., Hodgman, J. E., and Bergren, W. R.: Galactose metabolism in the newborn infant. Pediatrics 39:829–837, 1967.

Kalckar, H. M.: Galactose metabolism and cell 'sociology'. Science 150:305–313, 1965.

Segal, S., Blair, A., and Roth, H.: The metabolism of galactose by patients with congenital galactosemia. Am. J. Med. 38, 62–70, 1965.

Tolstrup, N.: Clinical and biochemical aspects of galactosemia. Scand. J. Clin. and Lab. Invest. 18:Supp. 92, 148–155, 1966.

Tygstrup, N.: Determination of the hepatic elimination capacity (Lm) of galactose by single injection. Scand. J. Clin. and Lab. Invest. 18:Supp. 92, 118–125, 1966.

For Utilization of Mannose

Ball, E. G., and Cooper, O.: Studies on the metabolism of adipose tissue. III. The response to insulin by different types of adipose tissue and in the presence of various metabolites. J. Biol. Chem. 235:584–588, 1960.

Esmann, V., Noble, E. P., and Stjernholm, R. L.: Carbohydrate metabolism in leukocytes. VI. The metabolism of mannose and fructose in polymorphonuclear leukocytes of rabbit. Acta Chemica Scandinavica 19:1672–1676, 1965.

Verachtert, H., Bass, S. T., and Hansen, R. G.: The pyrophosphorylysis of adenosine diphosphate glucose and adenosine diphosphate mannose. Biochim. Biophys. Acta 92:482–488, 1964.

Wood, F. C., Jr., LeBoeuf, B., Renold, A. E., and Cahill, G. F., Jr.: Metabolism of mannose and glucose by adipose tissue and liver slices from normal and alloxan-diabetic rats. J. Biol. Chem. 236:18–21, 1961.

CHAPTER 5

BIOSYNTHESIS OF MONOSACCHARIDES

Biosyntheses of monosaccharides in mammalian tissues serve to (1) assure a continuing supply to the tissues (particularly the central nervous system) of the vital fuel glucose, (2) contribute to storage of hepatic glycogen through gluconeogenesis, (3) maintain the formation of lipids by provision of glycerol phosphate (derived from glucose or from small precursors), (4) help regulate electrolyte excretion in the case of renal gluconeogenesis, (5) help dispose of toxic accumulation of lactic acid, as in severe exercise, (6) provide the pentose component of nucleic acids and nucleotides, (7) provide the components of heterosaccharides, e.g., uronic acids and hexosamines, and (8) supply sugars for extraneous demands, e.g., glucose for the fetus or lactose for the nursing infant. All these needs depend ultimately on gluconeogenesis, i.e., formation of glucose from small precursors, or parts of this multienzymatic process.

GLUCONEOGENESIS

Reactions

The pathway of gluconeogenesis from one of the key precursors, pyruvate, proceeds through reactions which partly reverse the glycolytic scheme, but includes also some reactions which circumvent the energy barriers that would occur by direct and complete reversal of glycolysis. Those reactions which circumvent such barriers are logically points of control for determining the balance between catabolism and anabolism of glucose. Where two different enzymes control, respectively, the two different directions of a reaction, regulatory effects are likely to be exerted and often by factors other than substrate availability.

FORMATION OF PHOSPHOENOLPYRUVATE (PEP). The first set of reactions which circumvents the final reaction in aerobic glycolysis (formation of pyruvate from PEP by pyruvate kinase) to some extent involves the coordinated action of enzymes in different cell compartments and the passage of intermediates between them (Chap. 3). In the rat there is necessarily a movement of pyruvate into the mitochondrion where the enzyme pyruvate carboxylase (P-C) adds carbon dioxide, as bicarbonate, to the methyl carbon of pyruvate to form oxaloacetate (OAA) (Fig. 5-1). In other mammalian species, e.g., guinea pig and human, the enzyme P-C is found both in the mitochondrion and in the cytoplasm. This "beta-carboxylation" is a relatively low-energy process, but does utilize the dephosphorylation of ATP to form ADP. Acetyl CoA is an obligatory, allosteric coenzyme of the reaction. The K_m value for activation is similar to the normal concentration of acetyl CoA in the liver. Acetyl CoA can be replaced by propionyl CoA or, in ruminants, butyryl CoA. There is a high affinity of the enzyme for both pyruvate and bicarbonate. Like some other carboxylations this one requires the vitamin biotin. Indeed, biotin-deficient rats show a ten- to twenty-fold decrease in activity of pyruvate carboxylase, while other gluconeogenic enzymes are not affected. The activity of P-C is rapidly restored following administration of biotin in vivo. There is an absolute requirement of the enzyme for divalent cations, such as Mg^{++}, Mn^{++} or Co^{++}. The reaction proceeds in two steps, as indicated in Figure 5-2.

There are inhibitors of this reaction which may have physiological significance. The immediate product, OAA, is an inhibitor. Methylmalonyl CoA, an intermediate in formation of succinate from propionate (Fig. 5-1), inhibits the exchange of ATP and PP_i. Several nucleotide polyphosphates are inhibitors, with the most efficient being CTP and deoxy CTP. A derivative of the "ketone bodies," acetoacetyl CoA, also inhibits by this mechanism, but the reduced derivative, beta-hydroxy-butyryl CoA, is an activator of P-C. Thus a general change in mitochondrial redox potential toward reduction (increased NADH) may facilitate this gluconeogenic reaction.

Within the mitochondrion OAA is relatively freely reduced to malate in a reaction with NADH which is a reversal of a reaction of the TCA cycle (Chap. 4). Malate then moves out of the mitochondrion to be reconverted to OAA by soluble malate dehydrogenase. The ratio, $NADH/NAD^+$, is high in the mitochondrion but low in the cytoplasm, which favors this direction of reactions. The translocation of malate provides not only the carbon substrate to the cytoplasm but also the reducing equivalents which are needed in a subsequent step in gluconeogenesis (see below). An alternative route to provide extramitochondrial from intramitochondrial OAA (which itself does not translocate readily) may utilize the intermediate formation and translocation of aspartate (Fig. 3-10). When aspartate is the intermediate there can also

BIOSYNTHESIS OF MONOSACCHARIDES

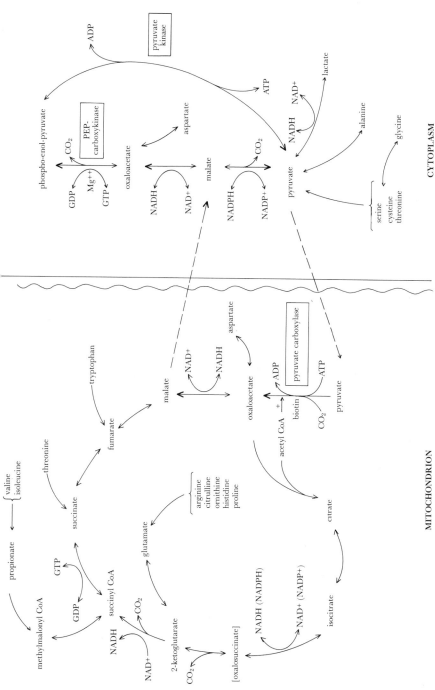

Figure 5-1. Precursors and pathways of formation of phospho-enol-pyruvate (PEP) in first part of gluconeogenesis. Compare Figures 3-10 and 4-5 for other details of reactions, enzymes, and formulae.

1.) $ATP + HCO_3^- + \text{E-biotin} \underset{Mg^{++}}{\overset{\text{acetyl CoA}}{\rightleftarrows}} \text{E-biotin} \sim CO_2 + ADP + P_i$

2.) $\text{E-biotin} \sim CO_2 + \text{pyruvate} \xrightarrow{\text{bound } Mn^{++}} OAA + \text{E-biotin}$

Figure 5-2.

be provision of reducing equivalents during conversion of argininosuccinate into fumarate in the urea cycle (see Physiological Chemistry of Proteins and Nucleic Acids in Mammals).

Oxaloacetate in the cytoplasm (also in mitochondria for some species) is phosphorylated with GTP or ITP as a phosphoryl donor and at the same time decarboxylated to PEP by the enzyme PEP carboxykinase (PEP-CK). The enzyme requires Mg^{++} or Mn^{++} for activity. The catalyzed exchange of the beta-carboxyl of OAA with free CO_2 (the reactive form) is much faster than is the complete reaction. The K_m of the enzyme for both OAA and GTP is far above their normal cellular concentrations, so elevations of either would readily increase the rate. Sulfhydryl groups are significant in its activity, as indicated by a stimulating or stabilizing effect of either cysteine or glutathione in vitro. Tryptophan in vivo has a stimulating effect on the isolated enzymatic activity of PEP-CK, which is paradoxical to its hypoglycemic effect, as discussed later.

The action of the enzyme in mitochondria is markedly enhanced by high ATP/ADP ratio in vitro. A high ATP level may facilitate PEP-CK by maintaining a concentration of GTP; the mononucleotide, GMP, has an inhibitory action on PEP-CK. In various other ways also a high ATP/ADP ratio can promote gluconeogenesis—besides its action as a substrate for P-C, ATP inhibits those enzymes (PFK, citrate synthase, and pyruvate kinase) the activity of which would deflect substrate away from gluconeogenesis. How important this control may be physiologically is not known. However, some evidence suggests that the gluconeogenic potential may in vivo be dependent on the balance of nucleotides in the liver.

As indicated in Figure 5-1, PEP can be converted back to pyruvate by the glycolytic enzyme pyruvate kinase (P-K)—this recycling (sometimes called the pyruvate cycle) would obviously diminish the rate of gluconeogenesis from PEP. Influences which increase the rate of pyruvate kinase (Chap. 4) are therefore quite effective in control of gluconeogenesis.

Mitochondrial OAA and malate equilibrate readily with the symmetrical dicarboxylic acids, fumarate and succinate (Fig. 5-1). In so doing the two halves of the four-carbon di-acid become randomized (see Fig. 4-5, Chap. 4). If pyruvate labeled with ^{14}C in the 2- or 3-carbon position is presented to the tissue (in vitro or in vivo) the randomness of distribution of ^{14}C in formed glucose indicates either the extent of equilibration in the dicarboxylic acid "shuttle" or else passage through the full TCA cycle. Analysis of blood glucose (human and animal) thus indicates that about two-thirds of pyruvate becomes equilibrated with symmetrical di-acids in the course of gluconeogenesis in vivo. Similar studies with the perfused rat liver have shown that about twice as much pyruvate was converted to OAA as to acetyl CoA and that the rate of PEP formation from OAA was three-fold greater than the rate of conversion to citrate. Other such findings are used to gauge the extent of recycling of PEP via P-K.

The beta-carboxylation of 2-ketoglutarate (2-KGA) to form oxalosuccinate, which is a reversal of a "forward" enzymatic reaction of the TCA cycle, is a mitochondrial reaction which is chemically analogous to that of pyruvate carboxylase. The TCA cycle may further proceed "backward" to form OAA and acetyl CoA with the action also of cytoplasmic citrate lyase (Fig. 3-10 and Chap. 4). It is evident from Fig. 5-1 that participation of the TCA cycle in gluconeogenesis permits various amino acids to flow readily into gluconeogenic pathways via intermediates of the TCA cycle. Again, ^{14}C experiments (comparison of incorporation of glutamate-2-^{14}C and glutamate-5-^{14}C into glycogen) have been used to indicate that in the isolated perfused rat liver about half of the glutamate follows the pathway of CO_2 fixation of 2-KGA to citrate, while half takes the oxidative route for 2-KGA through succinate to OAA. In circumstances which predispose to increased gluconeogenesis (e.g., diabetes) the carboxylative route is strongly facilitated.

Two main precursors for gluconeogenesis, lactic acid and alanine, are converted to glucose via pyruvate and subsequent reactions, as already described. The concentration of blood lactic acid for half-maximal rate of gluconeogenesis (in the perfused rat liver) is about 2 mM, which is about twice the normal fasting level of lactate in the blood. Thus an increase above the fasting level of lactate can readily cause an increase in disposal of this common metabolite via gluconeogenesis. Another 3-carbon acid that is gluconeogenic via the TCA cycle is propionic acid, which in liver or kidney is converted to succinate via the branched 4-carbon dicarboxylic acid, methylmalonate. This occurs through beta-carboxylation of propionate (or propionyl CoA), then intramolecular rearrangement to form the symmetrical di-acid, succinate (Fig. 5-1).

The activities of the crucial enzymes P-C and PEP-CK are among the lowest in the gluconeogenic sequence and tend to correlate with the intensity of gluconeogenesis. Various hormonal influences on gluconeogenesis are expressed through these enzymes and are described more fully later.

FORMATION OF TRIOSE PHOSPHATE. All the enzymatic reactions from PEP to triose phosphate (Fig. 5-3) are readily reversible, are the same reactions used in glycolysis, and are likely not rate-limiting for gluconeogenesis under ordinary circumstances. However, in this

144 PHYSIOLOGICAL CHEMISTRY OF CARBOHYDRATES IN MAMMALS

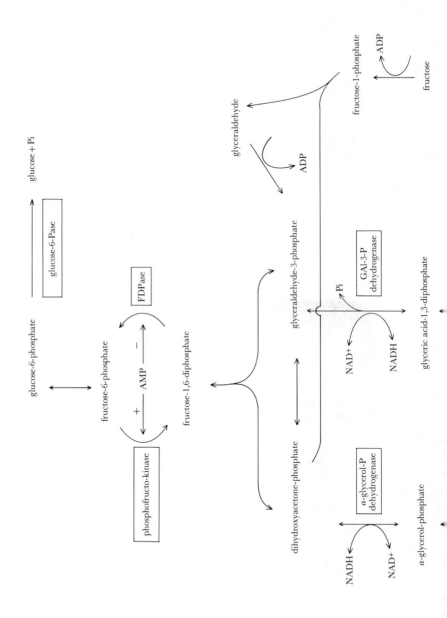

BIOSYNTHESIS OF MONOSACCHARIDES

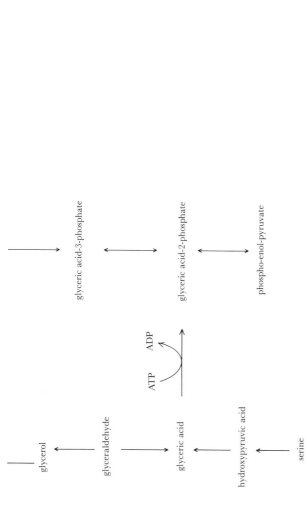

Figure 5-3. Formation of glucose from phospho-enol-pyruvate (PEP) in second part of gluconeogenesis. Compare Figures 4-1, 4-2, and 4-3 for other details of reactions, enzymes, and formulae.

sequence is the reductive step from glyceric acid-3-phosphate to glyceraldehyde-3-phosphate (GA1-3-P). A high $NADH/NAD^+$ ratio would obviously promote the reaction and there are certain indications that the availability of NADH governs the rate of this reaction toward gluconeogenesis. Thus, lactate and malate when added to kidney or liver preparations are converted more rapidly to glucose than are pyruvate or OAA, presumably because oxidation of lactate or malate to pyruvate or OAA, respectively, provides the NADH needed to drive the reductive step toward formation of triose. That such oxidation and reduction reactions are well coupled to each other is indicated by a relatively high transfer efficiency in vivo for tritium which is attached originally to C-2 of lactate or malate and later found in blood glucose (or hepatic fatty acids). Moreover, a condition which increases gluconeogenesis, i.e., administration of glucocorticoids, is accompanied by prompt and sharp increase in transfer of tritium from these substrates to blood glucose and liver glycogen.

Glycerol enters the gluconeogenic scheme at the triose level by the consecutive actions of glycerol kinase (found mainly in the liver) and glycerol phosphate dehydrogenase. Both of these enzymes can be substrate-adapted; the second enzyme in liver is adaptive not only to high glycerol feeding but also to fructose and glucose.

The 3-carbon amino acid, serine, has another route to glucose, in addition to that via alanine, by means of its conversion by transamination (or deamination) to hydroxypyruvate, which is reduced to glycerate and then phosphorylated to GA-2-P (Fig. 5-3).

FORMATION OF HEXOSE PHOSPHATE. The combination of DHA-P and GA1-3-P to form fructose-1,6-diphosphate (FDP) is catalyzed by aldolase in a direct reversal of the glycolytic reaction. There is little energy change, the reaction is not rate-limiting, and no cofactors are involved. There can be an inhibition of aldolase by Fr-1-P when this intermediate accumulates markedly, as in hereditary fructose intolerance, which, together with the inhibition of hexose isomerase, may help explain hypoglycemia due to decreased gluconeogenesis in this condition (see Chaps. 4 and 8).

The next enzyme in the gluconeogenic pathway, fructose-1,6-diphosphatase (FDPase), by-passes the phosphofructokinase (PFK) reaction in glycolysis and forms Fr-6-P and inorganic phosphate (P_i) from FDP (Fig. 5-3). FDPase requires Mg^{++} or Mn^{++} for activity. The nucleotide AMP strongly inhibits FDPase non-competitively (allosterically). Since AMP (or cyclic AMP) activates PFK, the level of this nucleotide affects the direction of substrate flow by two mechanisms which reinforce each other. FDPase can be inhibited down to 20 per cent of the maximal rate by concentrations of its substrate, FDP, greater than 0.1 mM; again, this inhibition is accompanied by an activating effect on PFK. Thus, control mechanisms with these two enzymes operate mutually to switch the metabolic pattern from glycolysis to gluconeogenesis or vice versa.

FDPase exhibits properties of enhanced catalytic activity when it is treated in certain ways to change its normal structure. Combination of some of its sulfhydryl groups with organic compounds (e.g., fluorodinitrobenzene) increases its rate of hydrolysis of FDP, unless the substrate is present during treatment with the sulfhydryl agent. Chelating agents (e.g., EDTA) also activate the enzyme. Further, there are naturally occurring activators of FDPase in both liver and muscle, which are protein in nature and which, in muscle, appear to be identical with PFK. PFK by combination with FDPase could maintain FDPase in a more active state and also protect it from inhibition by AMP. It has been suggested that when glycolysis is active PFK may "dissociate" from FDPase in some way, thereby "shutting down" gluconeogenesis. Pogell and co-workers have speculated on these interrelationships as indicated by Figure 5-4.

The formation of glucose-6-phosphate from Fr-6-P occurs by reversal of another glycolytic reaction, that catalyzed by phosphohexose isomerase. This reaction has appeared usually to be freely reversible and not rate-limiting.

FORMATION OF FREE GLUCOSE. The enzyme glucose-6-phosphatase, which catalyzes the final gluconeogenic step, the hydrolysis of Gl-6-P to glucose and P_i, is associated with lipid components of microsomes. Although a cation requirement is not clearly established, some work suggests that the enzyme contains Zn^{++} or Mn^{++}. The product, glucose, inhibits the hydrolysis of Gl-6-P, as do the organic products of hydrolysis

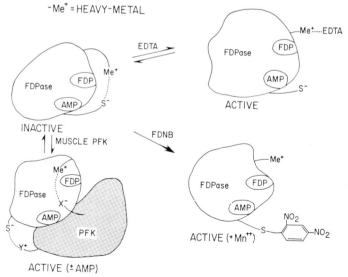

Figure 5-4. Activation of FDPase by various means: FDP = fructose-1,6,diphosphate; AMP = adenosine monophosphate; PFK = phosphofructokinase; Me^{++} = heavy metal; FDNB = fluorodinitrobenzene; EDTA = ethylene diamine tetra acetic acid. (From Pogell et al.: J. Biol. Chem. 243:1356, 1968.)

of other hexose and pentose phosphates which are acted upon by the enzyme. Gl-6-Pase may be inhibited by citrate; pyrophosphate (PP_i) or orthophosphate competitively inhibits the enzyme. Whether these inhibitions have physiological meaning is not known. As described in Chapter 4, the enzyme can act to form Gl-6-P by transfer of a phosphoryl group from PP_i or various nucleoside di- or triphosphates. Carbonyl phosphate is another possible donor. Owing to the lack of specificity of glucose-6-phosphatase the number of possible interactions with substrates and products is very high and the enzyme could be important in the regulation of the concentration of many phosphorylated and free sugars. Like some other gluconeogenic enzymes the activity of this one is readily increased by certain hormonal and nutritional states which favor gluconeogenesis.

Activities and Functions in Different Organs and Species

LIVER. Under ordinary circumstances the liver is the organ which chiefly synthesizes glucose and provides 80 to 90 per cent of the blood glucose in the postabsorptive state. While mainly being responsible for avoidance of hypoglycemia in ordinary, brief periods of fasting, the gluconeogenic activity of the liver can serve to dispose of excessive amounts of lactic acid in the blood and further helps to maintain a balance of amino acid levels in the blood through gluconeogenesis from amino acids released by peripheral tissues. Also hepatic gluconeogenesis from circulating glycerol is part of another metabolic cycle which is linked to the continual turnover of triglycerides by adipose tissue and other peripheral sites.

The main quantitative sources of carbon for hepatic gluconeogenesis are circulating lactate, pyruvate, glycerol, and amino acids. Among amino acids, alanine and glycine are the major precursors. The perfused rat liver also forms glucose rapidly (> 0.5 μMoles/min./wet g.) from serine, proline, fructose, dihydroxyacetone, sorbitol, xylitol, and oxaloacetate. Glutamate, aspartate, and most acids of the TCA cycle react slowly, probably owing to permeability barriers. Circulating glutamine and asparagine form glucose faster than the corresponding acids. Rates of gluconeogenesis from common precursors by the isolated perfused rat liver are shown in Table 5–1. The rates are remarkably additive when all substrates are combined, which suggests that all are major contributors physiologically. Increase of the plasma concentration of any of these constituents can cause increases in glucose production, since the physiological values shown in Table 5–1 are much below saturating levels. In a number of physiological and pathological conditions (see below) such increases no doubt help account for increased gluconeogenesis.

Various studies indicate that liver can form glucose more rapidly

Table 5–1. *Gluconeogenesis from Substrates Perfused at Physiological Concentrations Through Isolated Rat Liver**

SUBSTRATE	CONCENTRATION (mM)	RATE (μmoles/100 g. body wt./hr.)
Lactate	1.0	51
Pyruvate	0.1	12
Mixed amino acids	4.1	96
Glycerol	0.3	18
		177
Combination of all above substrates	as above	177

*From Park et al., Proc. Sixth Cong. Internat. Diab. Fed., 354, 1967.

from those substrates—glycerol, dihydroxyacetone, fructose—which enter the gluconeogenic scheme at or beyond the triose level. The rate-limiting step prior to triose formation may occur at different points—pyruvate translocation into mitochondria, formation of PEP, reduction of phosphoglyceraldehyde to triose—and the point of limitation may shift, depending on the metabolic circumstances of substrate type or level, hormonal influences, species differences, and other factors.

A particular difference among species is the intracellular location of hepatic PEP-CK. In the mouse and hamster this enzyme is present only in the soluble portion of the cell; in the rat 80 to 85 per cent is found there. In the guinea pig and other species half is in the mitochondrion and the remainder either in nuclei or in cytoplasm. Ruminants have PEP-CK in both mitochondria and cytoplasm. Conceivably mitochondrial PEP-CK has the advantage of ready availability of its cofactor, GTP, from substrate phosphorylation during oxidation of 2-ketoglutarate and succinate (Fig. 5-1).

KIDNEY. The kidney cortex of many species has a substantial capacity for gluconeogenesis, as shown by rates of at least 0.2 to 0.5 mMoles/hr./dry g., from several substrates, e.g., lactate, pyruvate, fumarate, glutamate, proline, fructose, dihydroxyacetone, and D-glyceraldehyde (at supraphysiological concentrations). Various polyols (e.g., sorbitol, xylitol) may be converted to glucose by kidney. Smaller animals show generally higher rates of renal gluconeogenesis than large ones. Lactate gives high rates in mouse and rat and rather low rates in guinea pig, hamster, and rabbit. As expected, propionate is a very good precursor for renal gluconeogenesis in sheep but oddly not in cattle.

Maximal rates of gluconeogenesis from glycerol and pyruvate are similar in perfused rat liver and kidney (per g. dry wt. of organ); the rate from lactate is about five times higher in liver, while the dicarboxylic acids (fumarate, malate, succinate) are 10 to 20 times more rapidly converted in kidney and the amino acids glutamate and aspartate about three times more rapidly. The higher utilization of these substrates by perfused kidney reflects mainly the limited permeability of liver, since enzymes for their metabolism are not significantly lower in liver than in kidney, except for PEP-CK, which is two and one-half times more active in kidney. Renal gluconeogenesis from pyruvate appears to occur via the dicarboxylic acid shuttle, but there is much less interaction with the TCA cycle in the kidney than in the liver, as indicated by studies of ^{14}C distribution in glucose and glutamate.

Control by hormones (see below) seems to be exerted on renal gluconeogenesis in much the same way as on glucose production by the liver. A singular difference in kidney, however, is its response to acidosis (metabolic or respiratory) by an increase in gluconeogenesis. This change, which is inversely related to the pH in kidney, is not exhibited by liver. The mechanism of this gluconeogenic increase is not presently clear, but its physiological consequences are so convenient as to appear quite teleological. With the increase in gluconeogenesis there is a cellular depletion of 2-KGA and glutamate. The latter amino acid is an inhibitor of glutamate transaminase and of glutaminase I, which forms NH_3 and glutamate from glutamine. Increased formation of NH_4^+ (derived from NH_3) minimizes loss of electrolytes with organic acids in the urine during just those conditions (e.g., diabetes and fasting) in which organic acids are produced in superabundance. The increased renal gluconeogenesis not only facilitates ammoniagenesis but can absorb some of the glucogenic organic acids, e.g., lactate.

The formation of glucose and ammonia production by rat renal cortex slices, as well as a fall in concentration of glutamate, is promoted by cyclic AMP (cAMP). Glucose production from glutamine, glutamate, 2-KGA, fumarate, malate, and OAA is increased but not from glycerol or fructose, which tends to localize the effect at formation of triose phosphate from OAA and probably at the PEP-CK reaction. The similarity of the effect of cAMP to that of acidosis has led to the suggestion that there is possibly a common mechanism.

MUSCLE. Some kinds of skeletal muscle can synthesize hexose phosphate and glycogen from small molecular precursors, so in that sense are capable of gluconeogenesis but not of formation of free glucose, because of the absence of glucose-6-phosphatase. Therefore, muscle cannot contribute to circulating glucose. However, glycogenesis in situ from 3-carbon precursors might be significant under some circumstances and could utilize some circulating lactate, glycerol, and so forth. The enzyme PEP-CK is not present in skeletal muscle, and in the conversion of ^{14}C-labeled lactate or pyruvate to glycogen of muscle there is no randomization of the outer carbons as there is in liver. The action of pyruvate kinase in the direction of phosphorylation (and against a seeming energy barrier) must occur in this tissue to account for glycogenesis from certain small-molecular compounds. ATP must be available to drive this reaction. In skeletal muscle pyruvate kinase is far more active than in liver and in kidney, and under optimal conditions may be three times as active as FDPase. The latter enzyme is not present in heart muscle, which therefore cannot make its glycogen from metabolic intermediates and must utilize circulating glucose for the accumulation of cardiac glycogen.

OTHER ORGANS AND TISSUES. During lactation in mammals there is an acute need for formation of the constituent monosaccharides of milk lactose. The evidence indicates that the glucose moiety comes from circulating glucose rather than endogenous synthesis in the mammary gland. However, PEP-CK and FDPase have been found in bovine lactating mammary gland; FDPase is present also in the gland of the rabbit and shows weak activity in mammary gland of the guinea pig and rat. There is some evidence that these enzymes may be involved in the synthesis of galactose at the mammary gland site.

Gluconeogenesis does not occur in adipose tissue because of the absence of FDPase and Gl-6-Pase. However, glyceride-glycerol is readily formed from pyruvate in adipose tissue, and isotope studies of randomization of label during transit from pyruvate to glycerol indicate that, as in liver and kidney, the dicarboxylic acid pathway is utilized. Both P-C and PEP-CK are found in adipose tissue but the latter enzyme so far only in rats. In the absence of glucose this pathway is active (in rats) and utilizes 40 per cent of an amount of pyruvate presented to the tissue in vitro. In the presence of glucose and insulin, when glycerol phosphate becomes available from glucose, only 3 per cent of pyruvate is channeled to glyceride-glycerol. In contrast to the response of P-C and PEP-CK in liver or kidney, these enzymes in adipose tissue are increased by adrenalectomy and glucocorticoids suppress activity. However, as in liver or kidney, the adipose tissue responds to fasting by an increase in activity of PEP-CK and in glycerogenesis. Both adrenalectomy and fasting decrease the supply of glucose to the adipose tissue, so the increase of PEP-CK in both cases helps to supply α-glycerol phosphate, which maintains some degree of re-esterification of fatty acids and may act as a brake on net lipolysis. Increases in PEP-CK by fasting and adrenalectomy, which are additive, appear to be due to changes in enzyme synthesis and degradation, respectively. Dietary and hormonal fluctuations of PEP-CK occur in the cytoplasmic rather than the mitochondrial form.

In the intestinal mucosa the gluconeogenic enzyme, glucose-6-phosphatase, is present in considerable amount. Possible roles for this enzyme in absorptive processes have been suggested but not proved (Chaps. 2 and 3). A varying amount of fructose is transformed to glucose during absorption, which would utilize this enzyme in a gluconeogenic capacity.

GLUCONEOGENESIS IN RUMINANTS. The conversion of hexose to

acetic, propionic, and butyric acids by bacteria in the rumen of ruminants accentuates the importance of gluconeogenesis in these species. During feeding some 20 to 50 per cent of hepatic gluconeogenesis derives from propionate via reactions shown in Figure 5-1. Microbial protein provides another large fraction of newly formed glucose. Under fasting circumstances the turnover of plasma glycerol from peripheral fat stores increases markedly and then glycerol largely takes the place of propionate as a gluconeogenic precursor.

More than other mammals the ruminants show a distribution of the two hepatic enzymes, P-C and PEP-CK, into mitochondrial and extramitochondrial compartments. The advantages of such a dual pathway for PEP synthesis are not clear, but acetyl CoA is readily available in the cytoplasm from acetic acid, as are propionyl CoA and butyryl CoA, which are also activators of P-C in ruminants. Butyryl CoA activation shows cooperativity, wherein the binding of one molecule by the enzyme facilitates the binding of a second; this tends to amplify small changes of butyryl CoA concentration into large changes in P-C activity. Probably the gluconeogenic effect of butyrate in cows is to be explained by an activation of P-C.

In the fed state ruminants are actively utilizing the OAA-PEP pathway for gluconeogenesis from propionate. Accordingly, PEP-CK does not increase with fasting as it does in rats, although the levels of some other gluconeogenic enzymes, P-C, FDPase, Gl-6-Pase, do increase. Likewise, glucocorticoids produce no increase in hepatic PEP-CK in sheep.

In lactating cows there may occur bovine ketosis, which has been explained as being due to the large drain on available hepatic OAA to form glucose in large amounts for lactose production. A supposed depletion of OAA reduces TCA cycle activity and acetyl CoA is converted to ketone bodies instead of being oxidized. However, the balance of changes in P-C and PEP-CK, as indicated above, does not suggest a depletion of OAA with increased gluconeogenesis. Another theory holds that a large increase in hepatic uptake and oxidation of fatty acids with an excess of acetyl CoA causes increases in both gluconeogenesis (activation of P-C) and ketogenesis. For reasons still obscure the most profound ketosis may occur in cows with hypoglycemia. Unlike the fetal rat, the liver of the fetal cow or sheep has the capacity for gluconeogenesis; this makes the fetus less dependent on maternal glucose, for which there is relatively little direct supply from the diet in ruminants.

Physiological and Pathological Changes

HORMONAL EFFECTS. Several hormones which dispose to hyperglycemia are known to increase gluconeogenesis. These include epinephrine from the adrenal medulla, steroids (cortisone, hydrocortisone) from the adrenal cortex, glucagon from the pancreatic islets, and thyroxine from the thyroid. All have a common tendency to increase the flow of free fatty acids (FFA) to the liver by a lipolytic action at peripheral sites, e.g., adipose tissue. Moreover, these hormones could increase hepatic fatty acids by a similar lipolytic increase (or decrease in fatty acid esterification) in the liver. A persuasive argument, based on considerable evidence, can be made that these hormones promote gluconeogenesis by the following course of events: (1) increased fatty acid oxidation in the liver increases the concentration of acetyl CoA, which activates pyruvate carboxylase and inhibits pyruvate dehydrogenase, (2) increased oxidation of acetyl CoA through the TCA cycle increases ATP and citrate, which have multiple enzymic influences that increase gluconeogenesis, as already described, (3) increased NADH due to fatty acid oxidation promotes formation of mitochondrial malate, its transfer to the cytoplasm and provision of reducing equivalents for the reductive step to triose phosphate. Furthermore, increased NADPH would shift 2-KGA toward formation of citrate and OAA via increased carboxylation and reduction to isocitrate.

In accordance with these postulates, the effect of adding oleate to the perfused rat liver is an increase in concentration of acetyl CoA with such effects on pyruvate metabolism that the ratio of carboxylation to decarboxylation may increase as much as tenfold. Reduced pyridine nucleotides are concurrently increased in rats, though not in guinea pigs. The rates of formation of glucose from alanine, lactate, or pyruvate are doubled. "Cross-over" studies (comparison of cellular concentrations of intermediates with or without oleate) suggest points of acceleration at P-C, FDPase, and (with alanine) GAl-3-P dehydrogenase. Oleate does not, however, increase ATP or citrate.

Although increased concentrations of plasma FFA could have the above effects in vivo, opposing this is a possible decrease in gluconeogenesis via stimulation by FFA of secretion of insulin from the pancreas. Insulin reduces the flow of FFA from peripheral fat stores and has other actions to decrease gluconeogenesis. In the intact dog this latter effect prevails when plasma FFA levels are raised acutely; hepatic glucose output decreases, as indicated by direct arteriovenous measurements across the splanchnic bed.

There are indications that the gluconeogenic actions of various hormones are exerted in additional, and perhaps more important, ways than via the "FFA effect." The pronounced effect of the adrenal cortical steroids seems to occur primarily by an increase in transaminases in peripheral tissues, liver, and kidney. This initiates a flow of amino acids from the tissues through the circulation into the liver or kidney and there into the TCA cycle. There are also increases in activities of rate-limiting gluconeogenic enzymes (PEP-CK, FDPase, Gl-6-Pase) and of some of the enzymes with higher activity. To what extent the gluconeogenic effect of glucocorticoids operates by direct action on enzyme or RNA synthesis is not yet settled. The effect of hydrocortisone on hepatic purine and RNA synthesis can be reproduced by injected amino acids or casein feeding. Some of the gluconeogenic (particularly early) effect of glucocorticoids may be expressed via changes in substrate availability, e.g., hepatic cellular uptake of amino acids and intracellular translocation of amino acids or metabolic intermediates. Glucocorticoids increase incorporation into glucose of radionuclides from several substrates prior to PEP, except for succinate. Incorporation of isotopic label from glycerol, which need not pass through PEP for gluconeogenesis, is not stimulated by adrenal cortical steroids.

Glucocorticoids, and also glucagon and epinephrine, may promote formation of PEP by increasing the effective cellular concentration of cyclic AMP, which is such a fundamental action of these and other hormones in various tissues. The effect of glucocorticoids on hepatic cAMP seems to be indirect, i.e., they have a "permissive" role which allows expression of the glucagon stimulation of gluconeogenesis via cAMP. An increase of cAMP by epinephrine, on the other hand, does not require the presence of glucocorticoids. The liver responds promptly to both glucagon and epinephrine with increased gluconeogenesis; the response to glucocorticoids, in vivo or in vitro, is somewhat less prompt (30 to 60 minute delay). Epinephrine increases acetyl CoA; glucagon and glucocorticoids may or may not do so. By increasing cAMP the hormones epinephrine and glucagon presumably promote hepatic lipolysis, thereby increasing acetyl CoA (increased P-C activity) and long-chain acyl CoA, which inhibits citrate synthase, thus directing OAA toward gluconeogenesis. Besides, there may be a more direct effect of cAMP to stimulate PEP-CK — this is not clearly established.

In any case, glucagon seems to have an effect on gluconeogenesis which is not attributable to increased lipolysis. Oleate added to perfused rat liver increases ketogenesis manyfold but only moderately increases gluconeogenesis; glucagon increases gluconeogenesis more prominently than indicated by its ketogenic effect. Moreover, the gluconeogenic effects of fatty acids and glucagon are additive under certain conditions. Possibly glucagon acts mainly to increase conversion of liver protein to glucose rather than to stimulate gluconeogenesis via lipolysis. Indeed, increase of gluconeogenesis by glucagon is accompanied by increased urea production.

Insulin can have both direct and indirect actions to counteract the gluconeogenic hormones and to decrease gluconeogenesis. At peripheral sites the effect of insulin on net lipolysis is generally an inhibitory one, so that plasma FFA fall. Hepatic and renal clearance of FFA are reduced with metabolic results in the liver and kidney as suggested above. More directly in the liver insulin may have a braking effect on protein breakdown or transamination of amino acids, since urea as well as glucose production is decreased. Insulin counteracts the increased gluconeogenesis caused by glucagon or cAMP. Whether this action of insulin is exerted essentially via reduction of cAMP is not yet known. However, large amounts of glucagon or cAMP can overcome the effect of insulin. The direct, hepatic effect of insulin on gluconeogenesis can be evident, either in vivo or in vitro, within a short period. Within 10 minutes of intravenous administration of insulin to human subjects the conversion of ^{14}C-labeled pyruvate to blood glucose (and probably liver glycogen) is reduced to about half of the usual rate. This rapid action indicates some effect on substrates or cofactors rather than enzyme amounts. On the other hand, over longer periods insulin seems to repress synthesis of gluconeogenic enzymes, thus opposing the later, sustained effect of glucocorticoids. The effect of absence of insulin is seen in the two- to threefold increase in gluconeogenesis repeatedly observed by various techniques in the diabetic animal or human (see Chap. 8). Whereas this degree of abnormality is noted with most substrates and in toto, the conversion of glutamate via the carboxylation route (reverse TCA cycle) appears to be particularly accelerated in diabetes—up to eightfold.

FEEDING AND FASTING. The effects of dietary changes on gluconeogenesis are probably expressed to a large extent through their effect on insulin production. Not only glucose from the diet but fatty acids and certain amino acids stimulate the production and release of pancreatic insulin with the effects noted above. Fed rats have hepatic gluconeogenic rates only one-half to one-third those of fasted rats. Although increase of insulin is a prime regulator of gluconeogenesis after feeding, studies (again with the isolated perfused rat liver—surely winner of the Purple Heart in gluconeogenesis research) indicate that in the absence of hormonal regulation an elevation of glucose above 150 mg./100 ml. reduces gluconeogenesis from alanine-^{14}C by about one-third while net glucose release is reduced to about half. Urea formation and amino acid uptake by the liver are reduced, so the effect may be located at the transaminase stage or on the translocation of amino acids into the liver.

The increase of gluconeogenesis with fasting may be principally a result of the increased mobilization of peripheral lipids with increased hepatic or renal uptake of FFA, or it may be a more complex phenomenon not yet fully appreciated. Studies of isotope distribution in glutamate and glucose suggest that the "classical" shift of hepatic pyruvate metabolism from oxidation to gluconeogenesis rapidly occurs upon fasting. Hepatic acetyl CoA increases with the expected gluconeogenic consequences. Citrate decreases during starvation, so citrate cannot serve to regulate by decreasing PFK. However, PFK may become more sensitive to inhibition by ATP during starvation by virtue of a decrease in the substrate, Fr-6-P. This substrate has the property of reducing ATP inhibition and could be a factor in control of the PFK-FDPase cycle.

Increase in gluconeogenesis from glycerol occurs quite early in fasting and is probably not dependent on the "FFA effect."

Hepatic gluconeogenesis, after the phase of early fasting and initial "swell," gradually diminishes on more prolonged fasting, but renal gluconeogenesis continues unabated. In obese patients after starvation for 5 to 6 weeks, the hepatic glucose output has lessened to the point that the liver is producing only about one-half of the blood glucose and the kidneys the remainder. Approximately all the lactate, pyruvate, glycerol and amino acid carbon taken up by these tissues is then being converted to glucose. Such metabolic effects of starvation can become medically significant in various situations. Anorexia due to advanced cancer, irradiation, and other conditions may cause a virtual starvation with marginally adequate gluconeogenesis. In kwashiorkor, due to protein starvation, low blood glucose levels (20 to 40 mg./100 ml.) are commonly observed but are not usually the major clinical problem. Occasionally a profound hypoglycemia in kwashiorkor is associated with coma, hypothermia, and severe bacterial or parasitic infection.

EFFECTS OF AMINO ACIDS. As already mentioned, the administration to rats in vivo of L-tryptophan causes an apparent increase in the activity of hepatic PEP-CK, when the latter is measured in the isolated tissue in vitro. However, the effect of this amino acid in the intact animal is a paradoxical accumulation of all substrates prior to PEP in the gluconeogenic scheme, which furthermore is accompanied by hypoglycemia in fasted animals. There is some kind of blocking in vivo of the catalytic function of the enzyme. Other investigation has shown that administration of a tryptophan metabolite, indole-3-acetic acid or the analogues, indole-3-propionic acid or indole-3-butyric acid, can cause profound hypoglycemia in both normal and alloxan-diabetic mice. Another tryptophan metabolite, quinolinic acid, may be inhibitory for gluconeogenesis. In rats the amounts of tryptophan required to cause maximal accumulation of gluconeogenic precursors are well within the limits of the usual daily intake of tryptophan, which suggests that the latter could play a significant role in the postprandial state. The effect of tryptophan has a very interesting and possibly etiologic relationship to a curious medical syndrome recognized within the past quarter century—that of intermittent but occasionally profound hypoglycemia associated with large neoplasms, usually retroperitoneal and sarcomatous in type. Patients with such neoplasms have been found to have increased amounts of tryptophan and its metabolites in the blood and urine, and the increase correlates with the periods of hypoglycemia.

Leucine and arginine in certain sensitive human individuals, particularly infants and children, causes hypoglycemia when administered intravenously or even by mouth. Mainly this seems to be due to a stimulation of insulin production or release by the pancreas. However, studies with isolated guinea pig liver have also suggested that leucine, and its 2-

keto analogue, can reduce gluconeogenesis by some intrinsic hepatic action. Other observations indicate that the 2-keto acid products of deamination of certain amino acids, e.g., phenylalanine, tyrosine, and methionine, inhibit gluconeogenesis from a variety of precursors in rat kidney cortex slices.

In spite of these indications of particular inhibitory influences on gluconeogenesis of certain amino acids and their products, the general effect of an abundance of amino acids more often may be to support gluconeogenesis, not only as substrate but as precursors for synthesis of gluconeogenic enzymes when called upon, for example, by glucocorticoids.

EFFECTS OF COLD EXPOSURE. The rate of hepatic gluconeogenesis, as indicated by various enzyme activities and by conversion of ^{14}C-labeled substrates to glucose, is increased in cold-exposed rats. During early cold stress, when the animal is shivering, the rate from dicarboxylic acids may be increased approximately fivefold according to the ^{14}C results. Muscle activity during shivering, or for that matter, general exercise, probably causes such an increase in blood lactate concentration from glycolysis that the gluconeogenic organs take up much larger than normal amounts of lactate and reconvert it to glucose (the Cori cycle). Increases in actual enzyme amount are at the most 175 per cent (for PEP-CK), so the five-fold increase in substrate conversion suggests some activation of pre-existing enzyme or increased availability of substrate or both. Probably there is also mobilization of fat, which stimulates via acetyl CoA, and so forth, and mobilization of protein. The capacity for renal gluconeogenesis from some substrates was increased by about 50 per cent by exposing rats or hamsters to a cold environment for 2 weeks.

EFFECT OF ETHANOL. Ethanol is well known to cause hypoglycemia under appropriate conditions (mainly fasting or semistarvation) and part of this effect is due to a decrease in gluconeogenesis. Because the metabolism of ethanol in the liver involves consecutive dehydrogenation of ethanol and acetaldehyde by ethanol dehydrogenase with reduction of NAD^+ to NADH in each case, there is an abundant formation and often accumulation of NADH during ethanol metabolism. This tends to reduce dihydroxyacetone phosphate to α-glycerol phosphate, OAA to malate, and pyruvate to lactate, thus removing or diverting key substrates from the gluconeogenic pathway. Glutamate dehydrogenase is blocked by NADH, which further inhibits conversion of this and other amino acids to glucose. Nevertheless, in perfused rat liver the increased $NADH/NAD^+$ ratio caused by ethanol infusion appears to be maximally developed at ethanol concentrations below those which are required to show an inhibition of gluconeogenesis. This and other evidence indicates that at higher concentrations of ethanol there is an interference with hepatic uptake of other substrates (glycerol, lactate), which largely accounts for the decreased gluconeogenesis and hypoglycemic effect.

The nutritional and hormonal state of the animal or man is highly significant in determining the effect of ethanol on hepatic gluconeogenesis. When substrate is abundant, as in feeding, when alcohol concentrations are moderate, and, particularly if the gluconeogenic substrate does not itself readily yield reducing equivalents (e.g., pyruvate or alanine), then the gluconeogenic rate may actually be increased by ethanol, which provides reducing equivalents needed to form triose phosphate from GA-3-P. Stimulation by ethanol of release of a gluconeogenic hormone, e.g., epinephrine, may also override an inhibition of gluconeogenesis, with resultant hyperglycemia following ethanol. In any case, even with reduction of gluconeogenesis hypoglycemia does not supervene until hepatic glycogen is depleted. Kidney cortex has a low activity of ethanol dehydrogenase, and renal gluconeogenesis is not inhibited by ethanol.

BIOSYNTHESIS OF PENTOSES

The major pentoses of mammals are D-ribose and D-deoxyribose. Ribose is an integral component of the nucleotides, vital functions for which include electron transport, high-energy phosphate transfer, and substrate activation and transfer in polymerizing reactions. Also ribose phosphate is, of course, the repeating unit which forms the skeletal "backbone" of ribonucleic acids (RNA). Correspondingly, deoxyribose phosphate, lacking an oxygen function at C-2 of the carbon chain, is contained in DNA. D-xylose is another pentose contained in some heteroglycans.

Ribose

Reactions

There are two possible major routes for biosynthesis of ribose-5-phosphate (Rib-5-P): (1) from Gl-6-P via the oxidative reactions of the pentose cycle (PC) pathway, or (2) by reversal of the non-oxidative reactions of this cycle, i.e., by formation of Rib-5-P from fructose-6-phosphate.

The two enzymatic reactions which oxidize Gl-6-P successively to 6-phosphogluconate and 3-keto, 6-phosphogluconate, followed by spontaneous decarboxylation of the latter to ribulose-5-P (Rbu-5-P), have already been described in Chapter 4. Ribulose-5-P is readily isomerized by a specific enzyme, Rib-5-P ketoisomerase, to ribose-5-P. The oxidative as well as the non-oxidative enzymes of the pentose cycle are found not only in the cytoplasm but also in the mitochondria of rat liver, though to a lesser extent than in the cytoplasm. This potentiality for pentose synthesis may be significant for the autonomy of mitochondria.

BIOSYNTHESIS OF MONOSACCHARIDES

The non-oxidative reactions of the pentose pathway have also been described in Chapter 4. To summarize, there are four enzymes—transketolase (TK), transaldolase (TA), Rib-5-P ketoisomerase, and Rbu-5-P epimerase (inversion at C-3)—which are involved in carrying out the separate reactions (a-e) of Figure 5-5 (see also Fig. 4-3). These are all reversible and, if operating to the right in Figure 5-5, pentose is synthesized from hexose and triose as the sum of these reactions, which are all at the same oxidative level and involve no electron transfer. In a possible variant of these reactions FDP may form sedoheptulose-1,7-diphosphate, which can be dephosphorylated by a phosphatase to form sedoheptulose-7-P. The latter reaction makes the over-all pathway irreversible.

Other unusual biosyntheses of ribose-5-P have been described. For instance, enzymes of rat liver can condense formaldehyde with D-glycerotetrulose-1-P to form pentose which seems to be mainly ribose-5-P. FDP and dihydroxymaleic acid (tautomer of 3-hydroxy-OAA) will react in minced rabbit muscle to form ribose-5-P. Presumably beta decarboxylation of the dicarboxylic acid yields 3-hydroxypyruvate, from which an "active glycolaldehyde" unit is condensed with glyceraldehyde-3-P (formed from FDP) simultaneously with a second decarboxylation. Hydroxypyruvate may also be derived from serine (Fig. 5-3). There is no clear evidence for the natural occurrence of dihydroxymaleate in mammals.

The uronic acid pathway (Fig. 5-8) leads to the formation of free pentoses, one of which, D-xylulose, can be phosphorylated to xylulose-5-

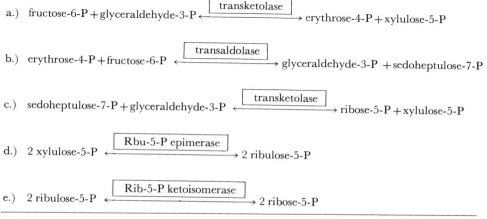

Figure 5-5. Non-oxidative reactions of the PC pathway for biosynthesis of ribose-5-P. Compare Figure 4-3 for formulae.

P, thus interlocking with the PC pathway. By this route ribose-5-P would be derived from the first 5 carbons of glucuronic acid—and glucose. Studies with ^{14}C-labeled glucose and glucuronolactone have been done to evaluate this pathway (see below).

Although DNA serves as a template for RNA, there is no known reversibility of any of the reactions which form deoxyribose from ribose derivatives (see below), so ribose-containing nucleotides or RNA do not arise by direct conversion of deoxyribose counterparts.

Activities and Functions in Various Organs and Species

As discussed in Chapter 4, both of the oxidative enzymes for pentose synthesis utilize the coenzyme NADP to transfer electrons from Gl-6-P and form 2 moles of NADPH for each one of pentose. It is probably of fundamental biological significance that two components for growth, i.e., reducing equivalents and pentoses, are together provided by the pentose pathway. As mentioned before, it seems that provision of NADPH is more critical than that of pentose in terms of the over-all cellular need. When glucose-^{14}C is presented in vitro to heart muscle, diaphragm, or abdominal muscle (of the mouse) the percentage found in ribose derivatives does not exceed 0.5 per cent. Even though the contribution of the pentose pathway to total glucose catabolism in muscle is only about 2 to 5 per cent, the quantity of ribose formed by the oxidative route should be well in excess of the amount required—unless the turnover of some ribonucleotides is much higher than supposed. Nevertheless, in growing tissues the pentose pathway is more active. For instance, conversion of glucose-^{14}C to ribose, largely by the oxidative route, in abdominal muscle of baby mice is about 10 times higher than that of adult mice.

The alternative, non-oxidative route of ribose synthesis could be used under anaerobic conditions—it may therefore be a more primitive metabolic pathway. The extent to which this pathway is ordinarily used in the direction of pentose synthesis is still unsettled. If the activity of the oxidative reactions is well in excess of the pentose requirements, the net flow of the non-oxidative reactions should be from pentose phosphate to fructose phosphate. Since the non-oxidative reactions are all reversible, there can be exchange of carbon from hexose to pentose without net synthesis of pentose. Some studies with ^{14}C in vivo indicate little exchange between ribose phosphate and the hexose-phosphate pool, while others indicate that the molecular distribution of ^{14}C in glucose newly formed from three-carbon precursors is affected to some extent by exchange.

The distribution of ^{14}C in the ribose molecule following formation from a position-labeled precursor glucose, e.g., glucose-2-^{14}C, has been analyzed in order to help determine the relative activities of, and contributions by, the oxidative vs. the non-oxidative routes. By the oxidative route glucose-2-^{14}C becomes ribose-1-^{14}C. Further transformation of this ribose through the full pentose cycle and repetitive cycling will distribute a minor amount of ^{14}C into C-2 of ribose (Fig. 4-3). However, by reversal of the non-oxidative reactions another amount of glucose-2-^{14}C will be converted to ribose-2-^{14}C.

In heart and abdominal muscle tissues more glucose-2-^{14}C appears in C-1 than in C-2 of ribose, indicating a preponderance of the oxidative route over the non-oxidative reactions. In diaphragm muscle, on the other hand, there is more activity in C-2 of ribose, which suggests greater activity of synthesis (or exchange) via TK and TA. Predominant activity of the non-oxidative route is also indicated for synthesis of ribose in both acid-soluble nucleotides and RNA of liver, spleen, kidney, and intestine. The difference between the pattern in other tissues and that in skeletal muscle (except for diaphragm), indicates an in situ origin of ribose in muscle rather than transport from other tissues.

A technique used to good advantage by Hiatt has been the administration in vivo to animals or humans of imidazoleacetic acid, which is conjugated to a riboside form in the liver and excreted as such in high yield in the urine. Ribose so excreted shows the same labeling pattern as the visceral nucleic acids. Analysis of the ribose from the riboside in human urine suggests about equal contributions of the oxidative and nonoxidative pathways. This is a greater relative proportion for the oxidative route than is found in the rat. Similar differences were found between human and rat erythrocytes. Possibly this technique could be further exploited to investigate the possible effects of various pathological states, and of certain drugs on the formation of pentoses, provided the conjugation reaction does not itself disturb the balance and activities of pathways.

Studies with the imidazoleacetic acid–ribose conjugation after administration of glucose-2-^{14}C indicate that the uronic acid pathway is not—at least in the liver—a significant source of ribose in normal man or animals.

Physiological and Pathological Changes

As mentioned in Chapter 4, thiamine is a coenzyme for transketolase. In thiamine-deficient rats the normal 3:1 preponderance of the non-oxidative route for hepatic ribose synthesis is replaced by a 2:1 predominance of the oxidative route, which suggests that the decrease of TK activity in such rats has a marked effect on the synthetic pathways.

Starvation or feeding of glucose has had no effects on the ratio of labeling via the two pathways, nor have hormonal influences such as alloxan diabetes, hypophysectomy, or treatment with ACTH, glucagon, or thyroxin in rats. However, severe diabetes has resulted in quite low total labeling from glucose-^{14}C. According to measurement of enzymatic activities, a 48-hour fast in rats causes lowering of the total activities of all the hepatic enzymes of the pentose cycle, both oxidative and non-oxidative. Alloxan diabetes also has caused a decrease to 60 or 70 per cent of normal values for all of the involved enzymes except G6PDH and transaldolase. Treatment with insulin restored the depressed enzymes. In adrenalectomized rats the activities of TK and Rbu-5-P epimerase fall and are restored by cortisone. In thyroidectomized rats Rib-5-P isomerase and TK activities are decreased, while transaldolase does not change significantly. Hypophysectomy causes a 50 per cent fall in TK, which is partly reversed by thyroxine and almost fully by treatment with growth hormone. Thus, of the non-oxidative enzymes transketolase appears to be the most consistently responsive to hormonal changes. Among the various ribogenic enzymes only G6PDH has shown the "overshoot" effect of refeeding a high carbohydrate diet after fasting.

The widespread occurrence of genetic G6PDH deficiency in several organs of human individuals from various races has been discussed (Chap. 4). The pathologic consequences of this deficiency (mainly susceptibility to hemolytic anemia) are probably due to a paucity of NADPH rather than ribose phosphate. However, G6PDH deficiency appears to be accompanied by other general abnormalities in glucose metabolism, so further investigation in this condition of ribose synthesis by either the oxidative or non-oxidative reactions seems warranted.

Deoxyribose

The synthesis of deoxyribose is less well understood than that of ribose. A number of studies indicate that reduction of a ribose derivative provides deoxyribose congeners. However, other studies have suggested that another possible route may be a reversal of a catabolic reaction catalyzed by deoxyribose-5-P aldolase, which splits deoxyribose into a triose phosphate and acetaldehyde.

Most evidence supports the concept of reduction of a ribose derivative. Ribose-containing purine and pyrimidine nucleosides, which are labeled in the sugar and the aglycone moieties, are converted to deoxyribonucleosides with nearly the same ratio of activity in the two moieties, which suggests direct reduction, although it does not preclude other types of synthesis. In most cases there is little dilution of label in the sugar moiety during incorporation of labeled uridine and cytidine into DNA; this indicates that direct reduction of ribose is the major, perhaps the only, route of formation of deoxyribose.

To achieve this direct reduction some animal tissues are known to possess a reductase which converts cytidine diphosphate (CDP) to deoxycytidine diphosphate (dCDP) with NADPH, Mg^{++} and Fe^{++} (or

BIOSYNTHESIS OF MONOSACCHARIDES

CO^{++}) as cofactors. ATP is thought to be an allosteric activator and not involved in the reaction per se. Via a specific enzyme, thioreductase, electrons are transferred from NADPH to a more direct reducing agent, thioredoxin, a cystine-containing polypeptide (Fig. 5-6). Thioredoxin contains FAD; in this it resembles glutathione reductase and dihydrolipoamide dehydrogenase. Reduced lipoic acid in some cases can substitute for thioredoxin, but not as effectively. Presumably after phosphorylation to diphosphonucleotides CMP, AMP, GMP and UMP can also be reduced by the catalytic system for CDP or by corresponding reductases. The reduction of UDP is strongly inhibited by dUDP, which may help account for the absence of dUMP in DNA. However, there are similar feedback inhibitions by products among the other nucleotides, as well as some cross-over feedback inhibitions (e.g., inhibition of reduction of GMP by dAMP). Control mechanisms for deoxyribonucleotide synthesis thus seem to be complex and are as yet poorly understood. Reduction of ribonucleotides appears to be irreversible.

The reversible reaction catalyzed by the enzyme deoxyribose-5-P aldolase (Fig. 5-7) is widely distributed in animal tissue and microorganisms. If this reaction were to operate synthetically in animal tissues, acetaldehyde would need to be provided by some possible precursor like pyruvate (Chap. 4) or threonine (via threonine aldolase). Even if so, the enzyme has a poor affinity for acetaldehyde. The evidence in favor of such a synthesis includes observations that, when position-^{14}C-labeled

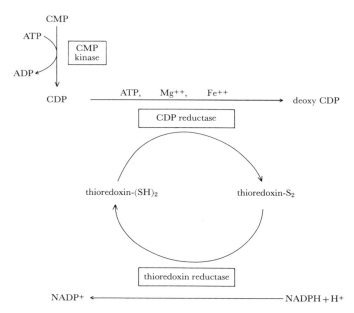

Figure 5-6. (From Sable, H. Z., Adv. in Enzymology, 28:391–460, 1966.)

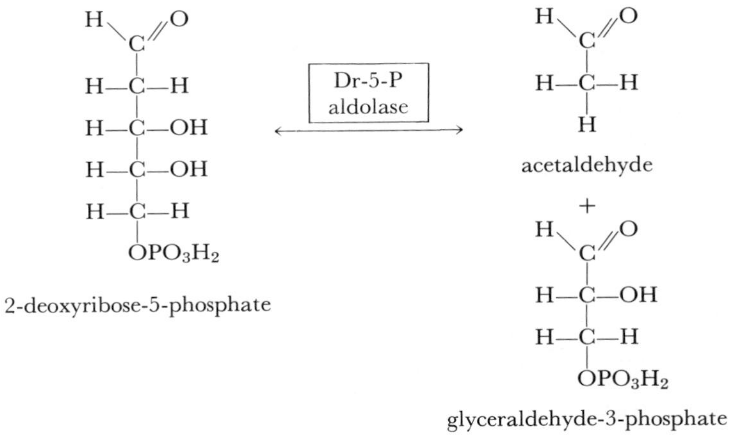

164 PHYSIOLOGICAL CHEMISTRY OF CARBOHYDRATES IN MAMMALS

$$
\begin{array}{c}
\text{H}-\text{C}(=\text{O})-\text{H} \\
\text{H}-\text{C}-\text{H} \\
\text{H}-\text{C}-\text{OH} \\
\text{H}-\text{C}-\text{OH} \\
\text{H}-\text{C}-\text{H} \\
\text{OPO}_3\text{H}_2
\end{array}
\quad \xrightleftharpoons{\text{Dr-5-P aldolase}} \quad
\begin{array}{c}
\text{H}-\text{C}(=\text{O})-\text{H} \\
\text{H}-\text{C}-\text{H} \\
\text{H}
\end{array}
$$

2-deoxyribose-5-phosphate ⇌ acetaldehyde

$$
+ \quad
\begin{array}{c}
\text{H}-\text{C}(=\text{O})-\text{H} \\
\text{H}-\text{C}-\text{OH} \\
\text{H}-\text{C}-\text{H} \\
\text{OPO}_3\text{H}_2
\end{array}
$$

glyceraldehyde-3-phosphate

Figure 5-7.

precursors such as glucose, glycine or acetate or $NaH^{14}CO_3$ are administered in vivo, the molecular labeling patterns of ribose and deoxyribose (in liver RNA and DNA) are often different. In some cases the different patterns in deoxyribose could be explained by the Dr-5-P aldolase reversal and in other cases by an exchange of labeled triose phosphate with the 3, 4 and 5 carbons of bound deoxyribose. However, the differences could also be due to compartmentation of the precursor pools or to asynchrony of synthesis of RNA and DNA. In either case, the relative contributions of the oxidative and the non-oxidative portions of the pentose cycle to the synthesis of deoxyribose might differ from those to ribose synthesis.

THE URONIC ACID PATHWAY

The uronic acid pathway provides UDP-D-glucuronic acid (UDP-GA) and UDP-L-iduronic acid, important precursors for the synthesis of common heteropolysaccharides in connective tissue and skin. UDP-GA is further used to combine with a variety of natural and foreign compounds, including noxious agents, to form glucuronides, which are then excreted in the bile or urine. In addition, this pathway to some extent provides ascorbic acid for some mammalian species. Further, there is the provision of UDP-xylose for heteroglycans by one route of metabolism of UDP-glucuronic acid.

Reactions

The conversion of glucose-6-phosphate to UDP-GA proceeds by initial conversion of Gl-6-P to Gl-l-P by the enzyme phosphoglucomutase,

(Fig. 5-8). This reaction commits glucose-6-P in an anabolic direction—to glycogen as well as to heteropolysaccharides. The equilibrium of the reaction in vitro lies mostly (95 per cent) toward Gl-6-P. An intermediate in the reaction is α-Gl-1,6-diP. The diphosphate acts like a coenzyme. Although phosphoglucomutase can isomerize other sugar phosphates the activity is less than 1 per cent of that on glucose phosphate. Magnesium ion is required by phosphoglucomutase; imidazole or similar compounds (a naturally active one has not been identified) act in conjunction with Mg^{++}. There is an alternate route for synthesis of Gl-1-P from glucose, as described in Chapter 4, which utilizes a kinase and a dismutase and also has α-Gl-1,6-diP as an intermediate.

Glucose-1-P reacts with the nucleotide UTP under the influence of the enzyme UTP;Gl-1-P uridyltransferase (also called UDP-glucose pyrophosphorylase) to form UDP-glucose (UDP-Gl) and pyrophosphate. The concentration of this uridyltransferase is high—0.3 per cent of the extractable protein in liver. Mg^{++} is also required by this enzyme. There is a high specificity for the reactants—substitution of other nucleotides or hexoses reduces activity to 4 per cent or less. Both phosphate and UDP inhibit the uridyltransferase; thus, when the cell's store of high energy phosphorylation is low, synthesis of UDP-Gl and subsequent polymers is reduced. A factor which tends to drive the reaction toward synthesis of UDP-Gl (and other nucleotide sugars) is the widespread occurrence in cells of pyrophosphatases, which could continually remove the product, PP_i. Another reaction which could utilize PP_i is the reversal of glucose-6-phosphatase to form Gl-6-P (Fig. 4-1). There is a suggestion that the phosphatase and uridyltransferase reactions could be "coupled."

The first unique reaction in the uronic acid pathway is the conversion of UDP-glucose to UDP-glucuronic acid (Fig. 5-8). This involves a double oxidation at the C-6 position of glucose in which two equivalents of NAD^+ are reduced. The same enzyme (UDP-Gl dehydrogenase) seems to catalyze both oxidations without evident existence of an intermediate aldehyde. UDP-glucose dehydrogenase can be inhibited in vitro by UDP-galactose and UDP-xylose, and the synthesis of glucuronides in liver slices is likewise inhibited by UDP-galactose, D-galactose, D-galactosamine and UDP-xylose. The mechanism may be various, however—either inhibition of the UDP-GA-forming enzymes or depletion of UTP by the free sugars—and there is no knowledge of any physiological significance. UDP-glucuronic acid can be degraded by animal enzymes to glucuronic acid-l-P and then to glucuronic acid. Alternatively a glucuronyl transferase may form free glucuronic acid. UDP-L-iduronic acid may also be formed from UDP-GA; this reaction is catalyzed by an epimerase changing the configuration at C-5. Another product of UDP-GA by decarboxylation is UDP-D-xylose, which is another precursor for heteroglycans.

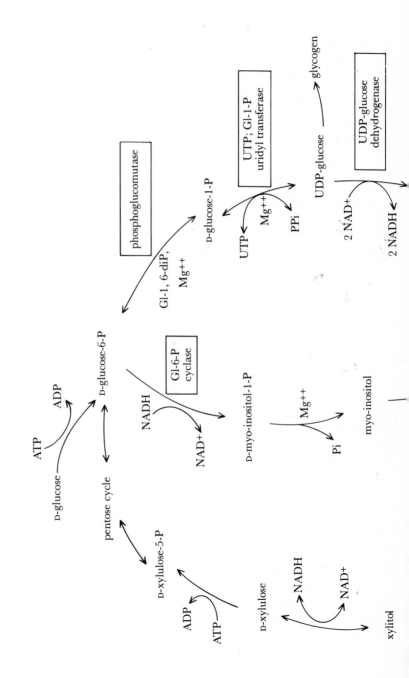

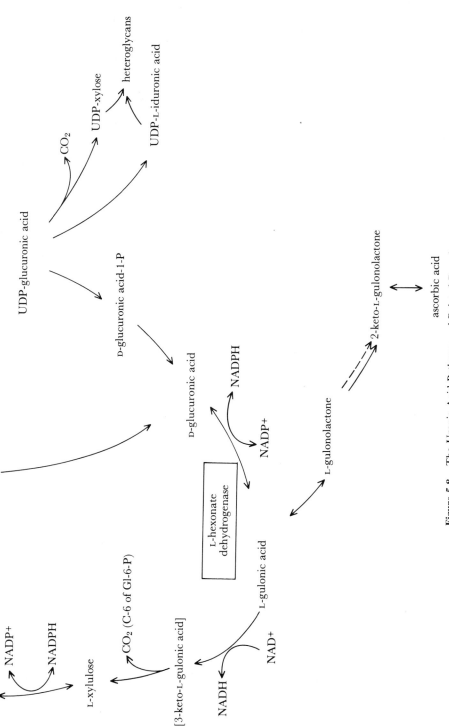

Figure 5-8. The Uronic Acid Pathway and Related Reactions.

In further reactions of this pathway the aldehyde at C-1 of glucuronic acid is reduced to a primary alcohol with the aid of the coenzyme NADPH. This compound, when inverted, has the structure of L-gulonic acid (Fig. 5-9). The enzyme for this reaction is known as L-hexonate dehydrogenase (as well as gulonate;NADP oxidoreductase) since it is rather non-specific; e.g., it can reduce L-iduronic acid to D-idonic acid.

After lactone formation by a lactonase found in liver of several mammalian species, L-gulonic acid can be oxidized to 2-keto-L-gulonolactone by most mammals; enolization of the latter compound in the liver forms reduced ascorbic acid or vitamin C (Fig. 1-5). This is readily oxidized to yield carbonyl groups on both C-2 and C-3 and thereby may function in biological electron transfer in ways not yet elucidated. In man, other primates, and the guinea pig, L-gulonolactone cannot be oxidized to the 2-keto derivative and for these species ascorbic acid is an essential vitamin. In the rat biosynthesis of ascorbic acid is greater in males than in females and found to be dependent on androgenic activity.

Another carbohydrate which is related tangentially to the uronic acid pathway is the cyclic hexahydric alcohol, myo-inositol. This compound, a component of some phospholipids and possibly functional as a "lipoptropic agent," is known to derive from glucose-6-P via D-myo-inositol-1-P (Fig. 5-8). The reductive formation of D-myo-inositol-1-P by Gl-6-P cyclase requires NADH, and the specific phosphatase forming free myo-inositol requires Mg^{++}. Both enzymes are widely distributed among various organs (brain, liver, kidney, lung, spleen, heart, and testis) of the rat. Activity of the cyclase is considerably higher in the testis (seminiferous tubules) than in other organs. Other species (pig, hamster, guinea pig) evidently cannot synthesize inositol, at least in adequate amounts, because in such species it is an essential dietary constituent. The catabolism of myo-inositol involves conversion to D-glucuronic acid in an oxidative cleavage of the cyclic alcohol, such that C-2 of myo-inositol (Fig. 1-5) becomes C-6 of glucuronic acid.

The further metabolism of L-gulonate occurs by oxidation of C-3 with an NAD-linked enzyme to form 3-keto-L-gulonic acid, which decarboxylates to L-xylulose (Fig. 5-10). This enzyme catalyzes such reactions with other sugars; e.g., L-idonic acid is converted to L-xylulose and D-gluconic acid to D-ribulose.

L-xylulose is transformed to D-xylulose via the polyalcohol, xylitol, as an intermediate. One dehydrogenase catalyzes the reduction of L-xylulose to xylitol with NADPH as coenzyme; the other utilizes NAD^+ in the oxidation of xylitol to D-xylulose. Since NADPH and NAD^+ are more predominant in tissues relative to $NADP^+$ and NADH, the reactions tend to flow from L-xylulose to D-xylulose. In this series of reactions the carbons one to five of D-xylulose can be seen to originate from carbons one to five of D-glucuronic acid and therefore of glucose (Figs. 5-9 and 5-10).

BIOSYNTHESIS OF MONOSACCHARIDES

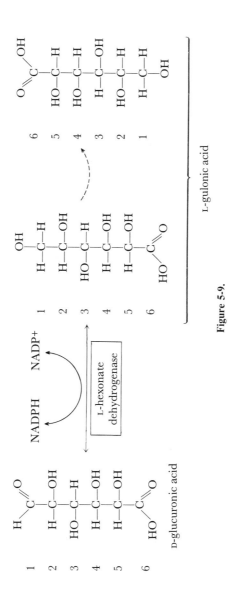

Figure 5-9.

170 PHYSIOLOGICAL CHEMISTRY OF CARBOHYDRATES IN MAMMALS

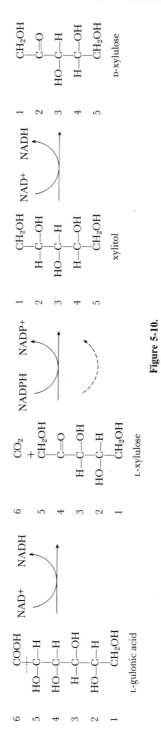

Figure 5-10.

In the liver there is a kinase which phosphorylates D-xylulose to D-xylulose-5-P, which thereby becomes a link between the uronic acid pathway and the pentose cycle pathway (Figs. 5-8 and 4-3).

Activities and Functions in Different Organs and Species

The occurrence of the full uronic acid pathway with all its enzymatic reactions seems to be best established for liver and kidney. In the liver, and possibly the kidney and intestine, a prime function of the pathway is the production of UDP-glucuronic acid, which is a substrate for the action of glucuronyl transferases in detoxification of various compounds with elimination of the glucuronides in urine or bile. A wide assortment of drugs, including carcinogens, antipyretics, hypnotics, and antimalarials, are known to stimulate generally the capacity of various enzymes, e.g., UDP-Gl dehydrogenase and UDP-glucuronyltransferase, which lead to formation of glucuronic acid and its conjugation with these and other drugs. For each of the inducing drugs there tends to be an organ specificity of induction. Free glucuronic acid is not itself a physiological precursor of glucuronides. A detoxifying effect of administered glucuronic acid or glucuronolactone is explained as being due to a conversion in vivo of glucuronolactone to glucaro-1,4-lactone, which is a powerful inhibitor of β-glucuronidase. Hydrolysis of glucuronic acid conjugates by β-glucuronidase would be inhibited.

It is probable that the uronic acid pathway operates according to Figure 5-8 in the connective tissue cell. UDP-G1, its dehydrogenase, and UDP-Gl pyrophosphorylase have been found in cartilage and skin (of rabbits) and other connective tissue (of rats). The C-5 epimerase forming UDP-L-iduronic acid from UDP-GA has also been noted in extracts of rabbit skin. When glucose labeled with ^{14}C has been presented to isolated cartilage (guinea pig) or skin (rat) the uronic acid moieties of hyaluronic acid, chondroitin-4-sulfate, or dermatan sulfate have become labeled.

The extent of occurrence of the uronic acid pathway in adipose tissue is still unsettled. Phosphoglucomutase, UDP-Gl pyrophosphorylase and, more significantly, L-hexonate dehydrogenase have been demonstrated to exist. However, the procedure of analyzing molecular distribution of ^{14}C in products of metabolism of position-labeled glucose indicates that no more than a few per cent, if any, of the metabolized glucose follows the uronic acid pathway in adipose tissue.

Physiological or Pathological Changes

In view of the association of hormonal changes with overt connective tissue diseases and with less well-defined disturbances in perivascular connective tissues, for example, questions naturally arise regarding hormonal effects on the uronic acid pathway and particularly the synthesis of UDP-glucuronic acid. However, information is quite fragmentary and inconclusive.

The adrenal cortical hormones (glucocorticoids) no doubt play a role in controlling the level of enzymes involved in the synthesis and metabolism of glucuronic acid, but the nature of this effect is quite uncertain. Some glucocorticoids, when given to "sham-operated" rats, cause increases in the enzymes phosphoglucomutase, UDP-Gl pyrophosphorylase and UDP-Gl dehydrogenase, in the connective tissue, while results with adrenalectomized rats are variable. Isotope studies are also inconsistent in suggesting increases or decreases in conversion of substrates to heteropolysaccharides as effects of glucocorticoids. Whether or not growth hormone, as studied with liver or adipose tissue, has any specific effect on the uronic acid pathway is also controversial.

A carbohydrate load stimulates glucuronide formation in animals and man. This could be due to a non-specific increase in the disposal of glucose through the uronic acid pathway, among others, but some evidence suggests that insulin, released in response to glucose, particularly increases the activity of UDP-glucose dehydrogenase. In diabetic rats UDP-glucose accumulates in the liver and the concentrations of UDP-GA and UDP-G1 dehydrogenase are reduced. Activity of UDP-GA glucuronyltransferase is not significantly changed. Other evidence indicates that in alloxan-diabetic or starved rats the synthesis of ascorbic acid is impaired, whereas the formation of xylulose from L-gulonate is enhanced. The enzymatic activity of L-hexonate dehydrogenase is moderately increased. In diabetic humans the serum level of L-xylulose is increased to three-fold the normal value and the concentration of glucuronic acid may also be increased. On the basis of such findings Winegrad and others have postulated hyperactivity of the uronic acid pathway in diabetes mellitus, presumably a "passive" result of diversion of glucose from main pathways which are decelerated in the disease. However, as indicated above, not all evidence can be easily reconciled with this hypothesis. The effects of insulin, glucocorticoids, and other hormones on the uronic acid pathway deserve further study.

A special genetic (probably autosomal recessive) and rare condition, occurring almost exclusively in Semitic races, is known as "essential pentosuria." The pentose is L-xylulose, of which relatively large amounts (about 5 g./day) may be excreted. There appears to be a defect in the NADPH-linked L-xylulose reductase; a marked increase in the K_m of the enzyme for NADPH has been demonstrated. No pathological effects are known to accompany this enzymatic defect.

BIOSYNTHESIS OF HEXOSAMINES AND SIALIC ACIDS

The C-2 amine derivatives of the hexoses, glucose and galactose, with an acetyl group almost always attached to the nitrogen of the amine group, are widely contained as components of the major heteropolysaccharides and the oligosaccharides of many glycoproteins of diverse functional nature. Widely distributed also among the glycoproteins (but not usually found in heteropolysaccharides) are the acidic nonosamines known as sialic acids or neuraminic acids. These amino sugars are found commonly at all evolutionary stages of development and their presence must be considered fundamental in life processes. Their value in conferring structural and functional specificities has been discussed (Chap. 1).

Reactions

The amino sugar initially formed is D-glucosamine-6-P (NGl-6-P) by the enzyme glutamine; Fr-6-P amidotransferase, which occurs widely in animal tissues. The transfer of the amide group from L-glutamine to Fr-6-P (some consider the direct reactant to be Gl-6-P) by this enzyme is irreversible and without the need of any cofactors. Another possible synthesis of glucosamine-6-P is by reversal of a deaminase which hydrolyzes NGl-6-P to Fr-6-P and ammonia (Fig. 5-11). This enzyme is found in kidney and brain. An equilibrium strongly toward deamination may be overcome by acetylation of the amine, since the acetylation is irreversible. Probably the reaction involving L-glutamine is the primary biosynthetic route. Another reaction leading to NGl-6-P is the phosphorylation of free glucosamine, which is possibly catalyzed by one of the same hexokinases that phosphorylate glucose with ATP. This may be useful in reutilization of glucosamine. In the liver the presence of glucose inhibits this phosphorylation, so acetylation is a more favored and possibly prior reaction (Fig. 5-11). In the brain phosphorylation is the more avid reaction. Acetyl CoA is a reactant in the formation of NAcGl-6-P by an enzyme found in liver, kidney or muscle. The same enzyme can acetylate D-galactosamine-6-P.

The C-6 phosphates of amino sugars must be transformed to the C-1 phosphates in their further course toward incorporation into glycosides. The same phosphoglucomutase which interconverts Gl-6-P and Gl-l-P can also form NGl-l-P reversibly from NGl-6-P. Another mutase works on NAcGl-6-P to form NAcGl-1-P. A 1,6-diphosphate of N-acetylglucosamine has been detected as a probable intermediate in the latter reaction, which therefore appears quite analogous to that of phosphoglucomutase. Both mutases use Mg^{++} as a cofactor. N-acetylgalactosamine is phosphorylated by a kinase (in liver or kidney) directly to the C-1 phosphate derivative.

The next type of reaction is the familiar prelude to polymerization for sugars, i.e., activation by formation of a nucleotide. In typical fashion the reactant nucleotide is UTP, which together with N-acetylglucosamine-1-P forms UDP-NAcGl and PP_i.* Possibly NAcGl-1-P will react

---

*Any of these nucleotide-sugar complexes, considered as a nucleoside-P-P-glycose, has the same general structure as the coenzymes NAD^+ and FAD. In an analogous reaction NAD will react with PP_i to form ATP and nicotinamide mononucleotide. The structural resemblances could be significant for competitive interactions among the coenzymes and the nucleotide sugars.

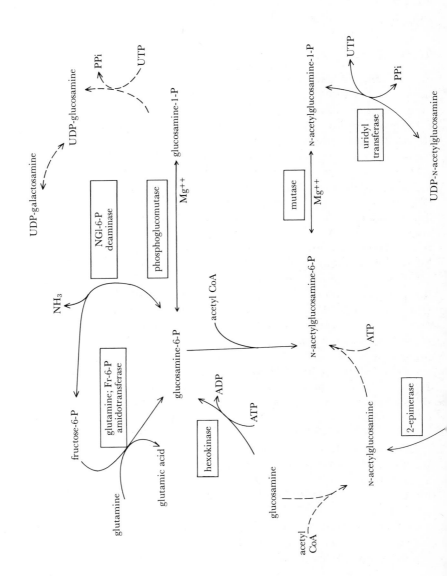

BIOSYNTHESIS OF MONOSACCHARIDES

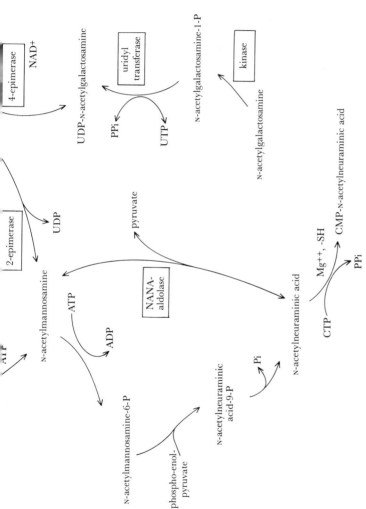

Figure 5-11. Biosynthesis of free and reactive forms of hexosamines and neuraminic acids.

similarly. The non-acetylated NGl-l-P can also be so transposed by a uridyl transferase to UDP-NGl, but the latter compound has never been found to occur naturally as does the acetylated analogue. There is a similar formation of UDP-NGal.

Epimerization of UDP-NAcGl or UDP-NGl to the corresponding UDP-N-acetylgalactosamine or UDP-galactosamine occurs enzymatically in a manner quite analogous to the classical interconversion of UDP-Gl and UDP-Gal. However, the enzymes for epimerization of the amines appear to be distinct from those for UDP-Gl, even though NAD^+ is likewise required. Whether UDP-hexosamines, without acetyl groups, are natural intermediates has still not been decided. So far there is no evidence for acetylation in mammalian tissues of hexosamines in linkage with nucleotides.

The biosynthesis of hexosamines is linked to that of the sialic acids by an epimerization occurring at C-2 of the hexosamine carbon chain. UDP-N-acetylglucosamine is converted to UDP and N-acetylmannosamine (simultaneous and irreversible cleavage and epimerization) by an epimerase which, unlike the 4-epimerases, does not require NAD^+ (Fig. 5-11). Free N-acetylglucosamine can also be converted reversibly to N-acetylmannosamine by an epimerase from hog kidney; corresponding N-glycolylhexosamines are also interconverted. ATP is required by this enzyme. The formation of glycolyl groups in glycolylneuraminic acids may depend upon an ascorbate or NADPH-dependent oxidoreductase which hydroxylates acetyl groups attached to neuraminic acids.

N-acetylmannosamine is phosphorylated to NAcMan-6-P by a specific kinase using ATP. NAcMan-6-P then reacts irreversibly with phosphoenolpyruvate to form N-acetylneuraminic acid-9-P, which is dephosphorylated (by an enzyme the specificity of which is not known) to NANA- (Figs. 5-11 and 1-5). N-glycolylmannosamine-6-P can also react with PEP to form NGNA-9-P. The cyclization of NANA-9-P is probably non-enzymatic. A reversible reaction for either the breakdown or formation of NANA from pyruvate and N-acetylmannosamine is catalyzed by an enzyme known as NANA-aldolase. The enzyme also acts on NGNA, for which it is about two-thirds as active (in splitting) as it is for NANA. The equilibrium is about 1:20 in favor of the degradative products, and the biosynthetic role of this enzyme is still somewhat in doubt.

In order to be introduced into heterosaccharides NANA, like other monosaccharides, must first be complexed with a nucleotide. NANA (not the usual phosphorylated derivative) reacts with CTP to form CMP-NANA and PP_i under the influence of an enzyme which requires Mg^{++} and glutathione (or other SH compound). The phosphoryl of the mononucleotide is linked to the hydroxyl at C-2 of the cyclized NANA. The enzyme can attach the nucleotide to glycolylneuraminic acid as well as to NANA.

BIOSYNTHESIS OF MONOSACCHARIDES

Activities and Functions in Various Organs and Species

LIVER. The liver has a much greater capacity to synthesize amino sugars from glucose than have other organs. In this function it serves as a primary factory for the many amino-sugar-containing compounds of the blood and perhaps other tissues. There is a rapid interchange of glucosamine between liver and serum which is considerably more rapid than de novo synthesis. Bound glucosamine is thus recovered for reutilization. There is, in fact, little breakdown of glucosamine to hexose and ammonia in liver, whereas other tissues (brain, kidney, testis, white cells) readily convert glucosamine to non-amine sugars. Possibly the inhibition by glucose of phosphorylation of glucosamine (and its resultant diversion to acetylation) plays a role in protecting the amino sugar from breakdown.

There is an important inhibition of the glutamine; Fr-6-P amidotransferase by UDP-NAcGl, which is an example of end-product inhibition of the first step in an enzymatic pathway. In liver this shows kinetics of a competitive type, i.e., UDP-NAcGl increases the K_m for Fr-6-P without changing the maximal velocity. The K_i of inhibition is definitely within the range of physiological concentration of UDP-NAcGl. There is a similar inhibition of UDP-NAcGl-2-epimerase by CMP-NANA, but higher than over-all cellular concentrations of the latter are required. These feedback inhibitions are not restricted to the liver.

In liver there is an alternative to other mechanisms of formation of UDP-NAcGal, which is a reaction of UDP-Gl with galactosamine-1-P involving a transfer of the uridyl unit to the amino sugar in substitution for glucose.

NANA is rapidly synthesized in rat liver from glucosamine. The specific activity of NANA synthesized from glucosamine-^{14}C is virtually the same as that of protein-bound hexosamine, indicating a high specificity of origin. The amount of NANA formed from the labeled glucosamine is about one-third to one-half of the bound hexosamine itself, so there is also quite an extensive conversion.

OTHER TISSUES. In many other sites, e.g., kidney, brain, intestine, lung, testis, blood cells, skin, mast cells, placenta, the enzymes are found which substantiate the occurrence of hexosamine synthesis much the same as in liver and as summarized in Figure 5-11. Furthermore, various isotope studies support the general scheme. Thus after intraperitoneal injection of glucose-^{14}C into rats, analysis of rat skin at various times indicates a progression from hexose phosphates to acetylhexosamines to nucleotidated amino sugars. An interesting study of the maturation series of red cells (in rabbits) from bone marrow erythroid cells through reticulocytes to mature red cells indicates that certain enzymatic functions are progressively decreased and then lost—in correspondence with a diminishing need for membrane synthesis. In retina the initial amidotransferase, like that of liver, can be inhibited as much as 90 per cent by UDP-NAcGl. However, the mechanism is different from that with the liver enzyme in that the retinal enzyme shows non-competitive inhibition with respect to Fr-6-P.

Physiological or Other Changes

Virtually nothing is known yet of the effect of various influences—neural, hormonal, dietary, infectious or other environmental—which might cause changes in synthesis of amino sugars. Investigators have commented that the feedback control of amidotransferase by UDP-NAcGl would serve to protect this synthetic pathway from the vagaries of change in concentration of hexose phosphate with feeding, fluctuating or abnormal glycolysis. On the other hand, the inhibition of glucosamine phosphorylation by high concentrations of free glucose in the liver, e.g., in diabetes, could mean a possible alteration of metabolism of glucosamine toward acetylation and away from hexose formation. The extent of formation of hexosamines from glucose and glutamine seems to be greater under anaerobic conditions for isolated specimens of lung, skin, liver, and kidney, which provides a hint that hexosamine synthesis is not uninfluenced by changes in cellular concentration of important substrates.

BIOSYNTHESIS OF FUCOSE

The unique hexose L-fucose (6-deoxy-L-galactose) is found almost as commonly in mammalian heteroglycans as the sialic acids, yet its synthesis, which involves some unusual transformations for mammalian carbohydrate metabolism, has been little investigated in mammals, although much more so in bacteria. Present evidence indicates that the primary pathway, like that in bacteria, proceeds via D-mannose-6-P, which is itself formed either from free mannose or from fructose-6-P (Fig. 4-1). A mutase presumably forms D-mannose-1-P which is activated to GDP-α-D-mannose. GDP-D-mannose then undergoes conversion to GDP-L-fucose as shown, in the steric form for clarity, in Figure 5-12.

Another source of GDP-L-fucose occurs throughout the phosphorylation of free L-fucose by a liver enzyme with ATP to form L-fucose-1-P. Thus, fucose in the polysaccharides and glycoproteins in the diet after hydrolysis by fucosidase could be utilized for incorporation into similar compounds endogenous to the mammalian tissue. Likewise fucose from breakdown of endogenous heterosaccharides could be reutilized. L-fucose-1-P subsequently reacts with GTP, as catalyzed by a uridyl transferase, to form GDP-L-fucose and pyrophosphate.

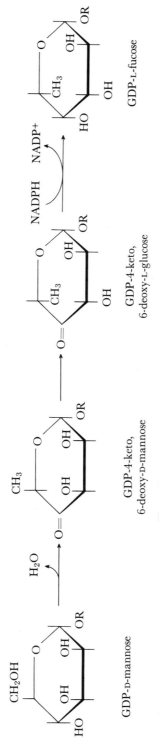

Figure 5-12. Biosynthesis of GDP-α-L-fucose, where R = GDP.

BIOSYNTHESIS OF GALACTOSE

Galactose for synthesis of glycoprotein is derived through the intermediate formation of UDP-galactose. The latter originates by a reversal of that pathway which utilizes galactose by converting it through UDP-Gal, UDP-Gl and Gl-1-P to the mainstream of glycolysis (Fig. 4-1). The activity of UDP-Gal-4-epimerase in the liver of the newborn rat is considerably greater than in the adult organ. Although this is probably utilized mainly for galactose metabolism as needed by the suckling infant, such an increase would permit greater synthesis of UDP-Gal from glucose if that were needed for glycoprotein synthesis. As noted earlier, UDP-Gal epimerase is activated by NAD^+. It is inhibited by various uridine nucleotides, including UDP-Gl, UDP-Man and UMP and also by TDP-Gl, but whether this has any physiological meaning for regulation of synthesis is not known.

Galactose as formed in the mammary gland and converted to lactose is believed to originate from glucose via the same reactions as already mentioned. Most of the galactose formed by the mammary tissue during lactation derives from the blood glucose, according to study in vivo in the cow.

BIOSYNTHESIS OF FRUCTOSE

The occurrence of free fructose in seminal fluid and in fetal blood suggests that metabolism of fructose has some advantage for germinating and embryonic tissues. The advantages of rapidly available energy via the "short-cut glycolysis" used for fructose metabolism (Chap. 4) or of pentose synthesis from fructose, both of which processes can occur anaerobically, are two possibilities. Free fructose derives from glucose via the polyalcoholic compound sorbitol. It is the suggestion of Horecker that fructose is one of the products of primitive enzymes arising in primordial life forms to utilize the polyols that would prevail in the highly reducing atmosphere of the earth when life first developed. Another suggestion for the role of the sorbitol pathway, wherever it may be found, has been that it serves as a transhydrogenation mechanism from NADPH to NADH; this idea is just as speculative as the others mentioned.

Reactions

Since there is no evidence for a phosphatase which can dephosphorylate Fr-6-P, free fructose appears to be formed only by the polyol pathway occurring in certain tissues. By this pathway glucose is reduced to sorbitol by aldose reductase with NADPH as the electron-donating coenzyme (Fig. 5-13). This enzyme, as found in seminal vesicle, pla-

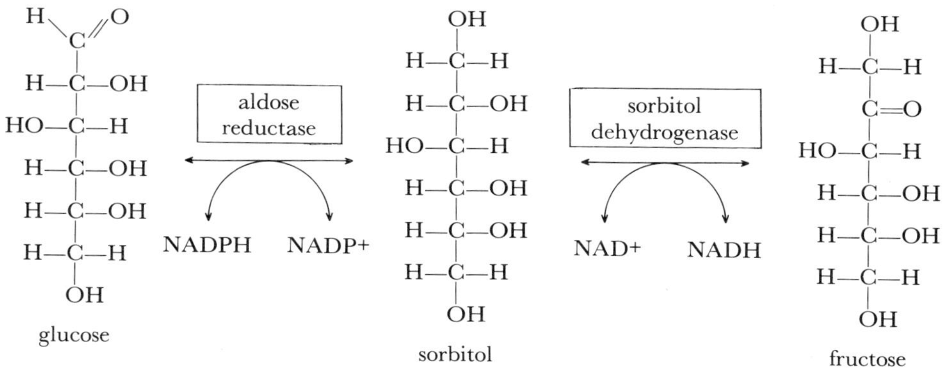

BIOSYNTHESIS OF MONOSACCHARIDES 181

```
  H   O                          OH                                OH
   \ //                          |                                 |
    C                          H-C-H                             H-C-H
    |                            |                                 |
  H-C-OH      ┌──────────┐     H-C-OH      ┌──────────────┐       C=O
    |         │  aldose  │       |         │   sorbitol    │       |
 HO-C-H       │ reductase│    HO-C-H       │ dehydrogenase │    HO-C-H
    |         └──────────┘       |         └──────────────┘       |
  H-C-OH         ⇌             H-C-OH           ⇌               H-C-OH
    |                            |                                 |
  H-C-OH     NADPH   NADP+     H-C-OH       NAD+    NADH         H-C-OH
    |                            |                                 |
  H-C-H                        H-C-H                             H-C-H
    |                            |                                 |
    OH                           OH                                OH
  glucose                      sorbitol                          fructose
```

Figure 5-13. The sorbitol pathway for fructose formation.

centa, lens of the eye and nerve tissue, is non-specific and can reduce several aldoses as well as aliphatic and cyclic aldehydes. Liver also contains a similar enzyme, which, however, is inactive toward glucose. In the same tissues, including liver, there is another and more specific enzyme, sorbitol dehydrogenase, which oxidizes sorbitol to fructose via the coenzyme NAD^+. This enzyme in liver catalyzes the oxidation of galactitol at only 3 per cent the rate of sorbitol. Presumably the enzyme in brain or nerve is at least as specific, since no tagatose, the theoretical product of action by the enzyme on galactitol, is detected in the brain or nerve of galactosemic animals. However, there may also be less specific polyol dehydrogenases which can oxidize sorbitol to fructose. The K_m constants of both reductases and dehydrogenases of this type for their substrates are relatively high.

Activities and Functions in Various Organs and Species

The occurrence of free fructose in seminal fluid, fetal blood (particularly of ungulates), cerebrospinal fluid, peripheral nerves and lower levels of CNS corresponds with the finding of the enzymes of the sorbitol pathway in these tissues or those (seminal vesicles, prostate gland, placenta) contributing to the fluids mentioned. The lens of the eye also has the requisite enzymes.

The calculated rates of conversion of glucose to sorbitol in brain and nerve are about equal. Fructose accumulates much more readily in peripheral nerve than in brain or spinal cord because a mechanism for its removal, phosphorylation by hexokinase followed by glycolysis, is about 65 times faster in brain than in nerve. This is due to differences in both V_{max} and K_m for fructose of nerve hexokinase vs. brain hexokinase. Peripheral nerves and spinal cord of normal rats contain sorbitol and fructose in concentrations only slightly less than that of glucose. In brain L-hexonate dehydrogenase acts like aldose reductase in reducing glucose to sorbitol with NADPH.

The liver is unique in being penetrable to sorbitol from the circulation, and the sorbitol present in the diet in certain fruits (cherries, plums, pears, and apples) is metabolized by the liver. Sorbitol, like fructose, has been used to provide to diabetic patients a carbohydrate which can be readily oxidized in the liver without the insulin-dependent enzyme hexokinase; sorbitol has, indeed, appeared to decrease ketosis under some circumstances. However, sorbitol, like fructose, is largely converted to glucose in diabetic liver and excessive amounts may be undesirable.

There was an earlier hypothesis that the placenta of ungulates (e.g., sheep, horse) forms sorbitol from glucose followed by transport of sorbitol to the fetal liver, where fructose is formed from sorbitol. More likely the reaction sequence takes place entirely within the placenta, and the carbohydrate transported to fetal blood to reach the observed levels of 100 to 150 mg./dl. is fructose. The rate of delivery of fructose from placenta to fetus shows product inhibition by fructose returning to the placenta via the umbilical artery.

Physiological and Other Changes

An increase in fructose in the accessory sex organs (seminal vesicle, prostate gland, and so forth) is known as a chemical indicator of activity of the male sex hormone testosterone. Estrogens counteract this action, at least for the prostate gland. Other hormones of the pituitary, e.g., ACTH and prolactin, may work synergistically with testosterone to increase fructose formation. Fructose in semen is produced mainly, perhaps entirely, by the seminal vesicles, and the prostate produces most of the citric acid in semen. The fructose concentration of human semen shows an inverse correlation with the quality of semen (sperm count, and other factors), whereas citric acid concentration shows a direct correlation with the quality. The concentration ratio for citric acid/fructose is generally below 1.0 for specimens from subfertile and sterile males. Blood glucose serves as the precursor of seminal fructose. In diabetic animals there are found increased amounts of fructose in semen with a decrease after insulin administration. However, in some way the male sex hormone is necessary before hyperglycemia, as in diabetes, can lead to increased synthesis of fructose.

The occurrence of the polyol pathway in nervous tissue and brain is now invoked to explain certain signs and symptoms of neurological disorders in some diseases of carbohydrate metabolism which can lead to excessive accumulation of products of the pathway. Thus, in diabetes the high glucose levels in the blood and tissues (including nerve tissue and aorta) evidently lead to shunting of glucose to the polyol pathway with formation of sorbitol and fructose. The high K_m levels for glucose of the enzymes of the pathway favor this effect. Levels of glucose, fructose, and sorbitol in sciatic nerve of diabetic animals have been seen to be 6, 12, and 23 times the normal values. A harmful excessive osmotic effect might well be exerted by such an accumulation of monosaccharides, which could play a role in diabetic neuropathy or angiopathy. Although the pathway is reversible, the equilibrium is so strongly toward fructose synthesis that high levels of the products are readily reached and maintained despite fluctuations of glucose concentration. Cataract formation in the eye lens in diabetes may also involve activity of the polyol pathway. Similarly galactosemia produced experimentally is accompanied by high levels of galactitol in nerves (seven times the glucose concentration) and in the lens of the eye. This is also discussed in Chapters 4 and 8.

REFERENCES

For Gluconeogenesis

Ballard, F. J., Hanson, R. M., and Kronfeld, D. S.: Gluconeogenesis and lipogenesis in tissue from ruminant and nonruminant animals. Federation Proceedings 28:218-231, 1969.

Eisenstein, A. B.: Current concepts of gluconeogenesis. Am. J. Clin. Nutr. 20:282-289, 1967.

Goodman, A. D., Fuisz, R. E., and Cahill, G. F., Jr.: Renal gluconeogenesis in acidosis, alkalosis, and potassium deficiency: its possible role in regulation of ammonia production. J. Clin. Invest. 45:612-619, 1966.

Heath, D. F., and Threlfall, C. J.: The interaction of glycolysis, gluconeogenesis, and the tricarboxylic acid cycle in rat liver in vivo. Biochemical J. 110:337-362, 1968.

Newsholme, E. A., and Gevers, W.: Control of glycolysis and gluconeogenesis in liver and kidney cortex. Vitamins and Hormones 25:1-87, 1967.

Nishiitsutsuji-Uwo, J. M., Ross, B. D., and Krebs, H. A.: Metabolic activities of the isolated perfused rat kidney. Biochem. J. 103:852-862, 1967.

Oji, N., and Shreeve, W. W.: Acute effects of hydrocortisone on gluconeogenesis from ^{14}C- and ^{3}H-labeled substrates in intact rats. Endocrinology 80:1062-1068, 1967.

Park, C. R., Mallette, L. E., Friedmann, N., and Exton, J. H.: Control of hepatic gluconeogenesis by the supply of substrates and the role of adrenal glucocorticoids. Proc. Sixth Cong. International Diab. Fed., pp. 354-359, 1967.

Pogell, B. M., Tanaka, A., and Siddons, R. D.: Natural activators for liver fructose 1,6-diphosphatase and the reversal of adenosine 5-monophosphate inhibition by muscle phosphofructokinase. J. Biol. Chem., 243:1356-1367, 1968.

Pontremoli, S., and Grazi, E.: Gluconeogenesis. In: Carbohydrate Metabolism and Its Disorders, Vol. I, pp. 259-295, ed. by F. Dickens, P. J. Randle, and W. J. Whelan, Academic Press, New York, 1968.

Rinard, G. A., Okuno, G., and Haynes, R. C., Jr.: Stimulation of gluconeogenesis in rat liver slices by epinephrine and glucocorticoids. Endocrinology 84:622-631, 1969.

Williamson, J. R., Browning, E. T., and Scholz, R.: Control mechanisms of gluconeogenesis and ketogenesis. I. Effects of oleate on gluconeogenesis in perfused liver. J. Biol. Chem. 244:4607-4616, 1969.

For Pentoses

Green, M. R., and Landau, B. R.: Contribution of the pentose cycle to glucose metabolism in muscle. Arch. Biochem. and Biophys. 111:569-575, 1965.

Hiatt, H. H.: Studies of ribose metabolism. V. Factors influencing in vivo ribose synthesis in the rat. J. Clin. Invest. 37:1453-1460, 1958.

Novello, F., Gumaa, J. A., and MacLean, P.: The pentose phosphate pathway of glucose metabolism. Hormonal and dietary control of the oxidative and non-oxidative reactions of the cycle in liver. Biochem. J. *111*:713-725, 1969.

Sable, H. Z.: Biosynthesis of Ribose and Deoxyribose. Adv. in Enzymology: *28*:391-460. Ed. by F. F. Nord, Interscience Publishers, New York, 1966.

For Uronic Acid Pathway

Horecker, B. L.: Pentose phosphate pathway, uronic acid pathway, and interconversion of sugars. *In*: Carbohydrate Metabolism and Its Disorders, Vol. I, pp. 139-167, ed. by F. Dickens, P. J. Randle, and W. J. Whelan. Academic Press, New York, 1968.

Müller-Verlinghausen, B., Hasselblatt, A., and Jahns, R.: Impaired hepatic synthesis of glucuronic acid conjugates in diabetic rats. Life Sciences *6*:1529-1533, 1967.

Winegrad, A. I., and Burden, C. L.: Hyperactivity of the glucuronic acid pathway in diabetes mellitus. Trans. Assn. Am. Phys. *78*:158-173, 1965.

For Inositol

Eisenberg, F., Jr.: D-myo-inositol-l-phosphate as product of cyclization of glucose-6-phosphate and substrate for a specific phosphatase in rat testis. J. Biol. Chem. *242*:1375, 1967.

For Hexosamines and Sialic Acids

Horecker, B. L.: Pentose phosphate pathway, uronic acid pathway, and interconversion of sugars. *In* Carbohydrate Metabolism and Its Disorders, Vol. I, pp. 139-167, ed. by F. Dickens, P. J. Randle, and W. J. Whelan, Academic Press, New York, 1968.

Warren, L.: The biosynthesis and metabolism of amino sugars and amino sugar-containing heterosaccharides. *In*: Glycoproteins: Their Composition, Structure, and Function, ed. by Gottschalk, Elsevier Publishing Co., New York, 1972, pp. 1097-1126.

For Fucose

Ginsburg, V.: Sugar nucleotides and the synthesis of carbohydrates. Advances in Enzymology. *26*:35-79, 1964.

Ishihara, H., Massaro, D. J., and Heath, E. C.: The metabolism of L-fucose III. The enzymatic synthesis of β-L-fucose-1-phosphate. J. Biol. Chem. *243*:1102-1109, 1968.

For Fructose

Gabbay, K. H., Merola, L. O., and Field, R. A.: Sorbitol pathway: presence in nerve and cord with substrate accumulation in diabetes. Science *151*:209-210, 1966.

Stewart, M. A., Sherman, W. R., Kurien, M. M., Moonsommy, G. I., and Wisgerhof, M.: Polyol accumulations in nervous tissue of rats with experimental diabetes and galactosaemia. J. Neurochemistry *14*:1057-1066, 1967.

CHAPTER

6

METABOLISM OF POLYSACCHARIDES

The chemistry of the homosaccharide glycogen and the heterosaccharides, i.e., the glycosoaminoglycans or proteoglycans ("mucopolysaccharides") and glycoproteins, was presented in Chapter 1. This chapter describes the metabolism of these polymeric and macromolecular carbohydrates together with further discussion of their physiological roles.

GLYCOGEN

Chemical Reactions

Glycogen is synthesized and degraded in mammals by means of multiple enzyme systems which are located in various organs and organelles. In this area of metabolism relating to emergency provision of ready fuel, the controls and "back-up" systems are elaborately developed. Many of the enzymes exist in different molecular forms, the interconversion of which allows for rapid increase of enzymatic activity without synthesis of new enzyme. As if further to assure quick responses the substrate glycogen is more or less continually bound to enzymes concerned with its synthesis and degradation. The need for rapid glycogen breakdown or formation is generally acute, intermittent, and fluctuating in response to environmental changes. Controls by both metabolites and hormones are finely exerted. There is an interlocking or cascading sequence of enzyme activations for the primary enzymes affecting glycogen directly. Larner has pointed out an analogy to the cascading series of reactions of blood coagulation. The complexity of control allows a responsiveness to multiple signals from within and outside the cell which assures an appropriate and adequate response of glycogen metabolism at critical times. A further guarantee of rapid and complete "turning-on" of glycogen breakdown, or conversely an abrupt and decisive switch to glycogenesis, is a form of "dual control" of synthetic and degradative reactions. Glycogen metabolism is an outstanding biochemical example of this general biological principle first formulated by Sherrington for muscle physiology, after he observed that the contraction of one set of muscles for a given movement is accompanied by relaxation of the opposing muscles. The Teppermans have recently noted how commonly this principle applies to biochemical phenomena.

SYNTHESIS OF GLYCOGEN. The primary intracellular monosaccharide precursor for glycogen synthesis is glucose-1-phosphate. Formation of Gl-1-P by phosphoglucomutase from Gl-6-P has been described in Chapter 4, as has the alternative combination of two other enzymes, Gl-1-P kinase and Gl-1-P dismutase, to produce Gl-1-P from glucose and ATP without the intermediate formation of Gl-6-P. The further formation of UDP-glucose (UDPG) from UTP and glucose-1-phosphate by the specific uridyl transferase enzyme is also described in Chapter 4.

An enzyme known as glycogen synthetase adds glucosyl units from UDPG to preformed glycogen. The main enzyme of this system (and often itself identified as synthetase) is UDPG:glycogen α-4-glycosyltransferase, which forms α-1 $\rightarrow$ 4 bonds. The reaction is irreversible (Fig. 6-1). The mammalian enzyme can use ADP-glucose half as well as UDPG (starch synthetase in plants prefers ADP-glucose) and it will also catalyze the incorporation of glucosamine from UDP-glucosamine. This enzyme exists in two forms: "I," which is "independent" of Gl-6-P and "D," which is "dependent" on Gl-6-P as a cofactor for significant activity. While Gl-6-P increases manyfold the V_{max} of transferase D, there is also an effect of Gl-6-P on transferase I to lower its K_m for UDPG. Cellular accumulation of Gl-6-P thus imposes a "positive feed-forward" control on the glycogenic system to provide for storage of carbohydrate when it is in excess of cellular needs.

Transferase I, or "synthetase a,"* the active form of the enzyme, and transferase D, or "synthetase b," the inactive, are interconvertible. Synthetase a is phosphorylated with Mg^{++}-ATP to the b form by another enzyme, synthetase a kinase. Conversely the b form is dephosphorylated to synthetase a by a specific phosphatase. Activation of this phosphatase is inhibited by glycogen itself, an example of "negative feedback" control. The deactivating enzyme, synthetase a kinase, also exists in two forms. One of these is dependent on cyclic-3',5'-adenosine-

---

*The terms a and b, in analogy to active and inactive phosphorylase a and b, may be preferable to I and D. For one reason, the inactive transferase D in liver is completely inhibited by physiological concentrations of P_i, no matter what the concentration of Gl-6-P.

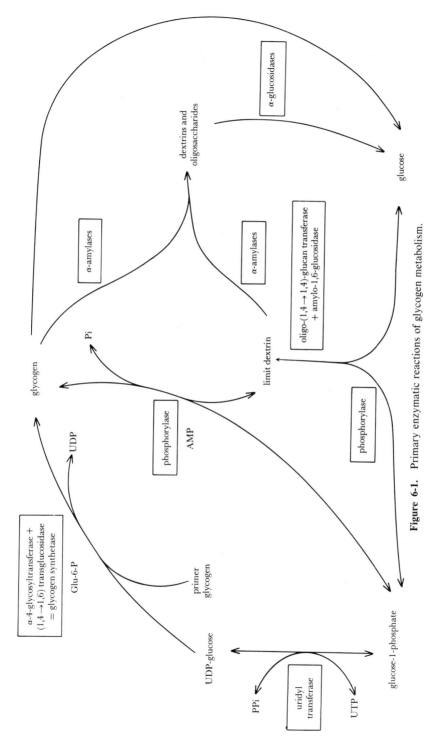

Figure 6-1. Primary enzymatic reactions of glycogen metabolism.

monophosphate (cAMP) as a cofactor and the other is not. At physiological concentrations an increase of cAMP stimulates the inactive synthetase a kinase by markedly decreasing the dissociation constants for the complex between the enzyme and its cofactor, Mg^{++}, as well as for the substrate, ATP. Thus the activity of synthetase a kinase is correlated positively, and that of glycogen synthetase negatively, with the level of the "second messenger hormone," cyclic AMP.

Since epinephrine promotes formation of cAMP (by adenyl cyclase), this adrenal hormone inhibits glycogen formation. Insulin, conversely, can lower cAMP by increasing the activity of the enzyme phosphodiesterase, which degrades cAMP. Although this may be one mechanism by which insulin exerts a notably rapid, direct, and extensive effect to increase glycogen synthesis, some studies suggest that insulin promotes a conversion of active synthetase a kinase (cAMP-independent) to the inactive form. Synthetase a kinase is accelerated not only by cAMP but also by the cyclic-3',5'-monophosphates of guanosine, uridine, and cytidine. Whether these have any physiological role is not yet known, but cyclic-3',5'-GMP has been found in diaphragm and brain of rats.

There are alternative mechanisms of conversion of synthetase a to synthetase b. A certain protein factor plus Ca^{++} will effect an irreversible conversion; so also will trypsin. The physiological significance of these alternate deactivations is in question.

When the α-1,4 chain length in the outer glycogen tier reaches 7 or more (up to 15 or 20) units, the brancher enzyme amylo (1,4 → 1,6) transglucosidase can transfer an outer segment of the 1,4 chain to another point in the same or an adjacent chain to form a branch via an α-1,6 linkage (see Fig. 1-7, Chap. 1). The optimal size of transferred outer segments is 6 or 7 glycosyl residues. This reaction is not known to be reversible, but the possibility of repositioning of a chain at another C-6 hydroxyl is not excluded. Thus UDPG; glycogen transferase and transglucosidase together form the synthetase system (Fig. 6-1). Most likely the brancher enzyme is in excess and the structure of internal glycogen represents the maximal branching limit. Glycogen from several sources has a rather narrow range of chain length for most internal chains.

DEGRADATION OF GLYCOGEN. The main enzyme for breaking down glycogen is phosphorylase, which specifically cleaves α-1,4-linked glucosyl residues with inorganic phosphate to form glucose-1-phosphate (Fig. 6-1). The reaction is reversible and, in fact, the thermodynamic equilibrium favors the synthetic direction. However, the high physiological ratio of P_i to Gl-1-P causes the reaction to be mainly degradative in vivo. At the equilibrium point phosphorylase can extensively rearrange the glucose units in the outer branches and under special circumstances it may act in the synthetic direction. Ordinarily the α-1,4-linked units are sequentially phosphorylyzed until the chain "stub" reaches a limit of 2 or 3 residues before the 1,6-linkage. Glycogen thus degraded to the limit of the outer tier of branches is called "limit dextrin." Since the outer chains constitute over half of the total glucose residues, a relatively large proportion of glycogen is convertible into Gl-1-P by phosphorylase alone. For reasons unknown the enzyme acts faster on outer tiers of residues than on the internal chains. One possible limiting factor is the occurrence in glycogen of glucosamine incorporated by the synthetase system. Phosphorylase cannot act on glucosamine-1-P in the synthetic direction, so presumably cannot phosphorylyze it from glycogen. Phosphorylase contains pyridoxal phosphate as a bound prosthetic group, which does not appear to participate in the enzymatic function, as with other enzymes.

Like glycogen synthetase, phosphorylase exists in active and inactive forms, termed respectively phosphorylases a and b, which are interconvertible. In muscle, phosphorylase b is dependent on AMP while the more active form is not, although phosphorylase a can be further stimulated by the mononucleotide. Liver phosphorylase b is not stimulated by AMP, but the active form is. Inosine-monophosphate (IMP) also activates muscle phosphorylase b but kinetically in a different way than does AMP. IMP affects the V_{max} of the enzyme but not the affinity for Gl-1-P; AMP does both. Since the mononucleotides accumulate when the phosphate energy system is depleted, activation by AMP (or IMP) is a way by which the cell can "call for" more carbohydrate as an energy source. Appropriately, ATP or Gl-6-P competitively inhibits the AMP activation of phosphorylase. Another control (in liver) is an inhibition by UDPG, which is competitive with P_i.

Phosphorylase b → a conversion, like that of synthetase a → b, occurs through phosphorylation of serine residues in the enzyme. This is catalyzed by a "second-stage" enzyme, phosphorylase b kinase, using Mg^{++} and ATP. Thus, the phosphorylating enzymes for synthetase a and phosphorylase b reinforce each other physiologically, because both conduce to a decrease of glycogen and increase of available monosaccharide. There is a curious inference that synthetase (transferase) and phosphorylase, although not identical, may have evolved together, in view of the similar sequence of hexapeptide around the phosphorylated site. Phosphorylase b kinase is inhibited by excess ATP and stimulated by glycogen, which lowers the K_m for phosphorylase b. This is another instance of positive feed-forward control.

In muscle the process of activation of phosphorylase b involves conversion of a dimer form to a tetramer; in liver phosphorylase b does not undergo this change during phosphorylative activation. The muscle enzyme appears to exist in multiple conformational forms, depending on successive interactions with allosteric activators (glucose as well as AMP), and the most active form is another dimer which is tightly bound to glycogen.

Phosphorylase b kinase itself exists in active and inactive forms with activation occurring by a "third-stage" enzyme, phosphorylase b kinase-kinase, which also phosphorylates its substrate with ATP and Mg^{++}. This reaction is inhibited by excess ATP and stimulated by glycogen and by cyclic-3′,5′-AMP. Through the latter cofactor epinephrine and glucagon exert their stimulation of glycogenolysis, since these hormones facilitate higher levels of cAMP by stimulation of adenyl cyclase, the fourth-stage enzyme which forms cAMP from ATP. Again like synthetase a kinase, the activity of phosphorylase b kinase can be stimulated by Ca^{++} plus a protein or by trypsin. These activations are considered secondary or perhaps not even physiological. The interlocking of glycogen synthesis and breakdown is dramatized by the recent establishment of identity of synthetase a kinase and phosphorylase b kinase-kinase.

Phosphorylase a is converted to the b form by a specific phosphatase which removes all four phosphates. This may be a stepwise process with partially phosphorylated forms that are AMP-dependent under some circumstances. Phosphorylase a phosphatase is activated by Gl-6-P and inhibited by ATP. There is a phosphatase also for phosphorylase b kinase.

Acting on limit dextrin is the debrancher enzyme, amylo-1,6-glucosidase (Fig. 6-1). However, closely associated with this enzyme and necessary as a prelude to it is the action of another enzyme, oligo-(1,4→1,4)-glucantransferase, which transfers maltotriosyl (or maltosyl) units from a shortened outer chain to any non-reducing end glucose unit, thereby exposing a single 1,6-linked residue for cleavage by the debrancher enzyme. The acceptor molecule for the glucantransferase may be branched oligosaccharide, glucose, or maltose as well as limit dextrin or glycogen. Branched acceptors are preferred by the enzyme. The glucantransferase reaction is reversible, as is potentially the action of amylo-1,6-glucosidase. The glucantransferase can transfer oligosaccharide segments from outer chains longer than 4 residues. Thus theoretically it, together with the 1,6-glucosidase, could partially degrade glycogen in the absence of phosphorylase. Amylo-1,6-glucosidase seems to consist of two enzymic species, one with an acid and one with a neutral pH optimum. These two species may be part of a multienzyme complex containing also the oligo-1,4 → 1,4-glucantransferase (Fig. 6-1).

A group of hydrolytic enzymes, α-amylases, attack the α-1,4-glycosidic linkages in the internal structure of glycogen (or starch) to form maltose, maltotriose, and dextrins (α-limit dextrins). Some of these amylases occur in the digestive secretions (Chap. 2). However, there are also some poorly characterized amylases in liver and muscle with low activity, which may nevertheless be physiologically functional. Oligosaccharides of glucose are known to occur in muscle, liver, and brown fat; how these are related to glycogen metabolism is not clearly resolved (see below). The oligosaccharides may be metabolized to glucose by α-glucosidases (γ-amylases), which occur mainly in the digestive tract but also in other

METABOLISM OF POLYSACCHARIDES

Table 6-1. *Known Actions of Hormones on Enzymes of Glycogen Metabolism*

ORGAN	ENZYME	Insulin	Epinephrine	Glucagon	Corticoids
Liver	glycogen synthetase	+	(−)	−	+
	phosphorylase	−	(+)	+	−
	α-1,4-glucosidases		+		
Skeletal muscle	glycogen synthetase	+	−	0	
	phosphorylase	−	+	0	0
	amylo-1,6-glycosidase		+		
	α-amylase		+		
	α-1,4-glucosidases		+		

tissues in small amounts. These enzymes can also attack glycogen directly. Depending on pH optima there are in liver and muscle neutral α-glucosidases and acid-α-glucosidases (acid maltases). The latter occur in lysosomes along with other hydrolytic enzymes. Some glucosidases are directed toward glycogen and others toward oligosaccharides (Fig. 6-1). Some may be specific for α-1,4-linkages and others not. Only the lysosomal acid-α-glucosidase has a significant capacity for degradation of glycogen. However, acid glucosidase in lysosomes of liver and other tissues hydrolyzes maltose about 6 times faster than glycogen; neutral glucosidases act on maltose about 25 times faster than on glycogen. The neutral-type enzyme may also catalyze transglycosylations; however, its more important role seems to be to break down malto-oligosaccharides formed by α-amylase in the liver. As with phosphorylase these hydrolytic enzymes are also susceptible to some hormonal effects (see below and Table 6-1).

Activities and Functions in Various Organs and Species

LIVER. Glycogen in the liver serves a special function not shared by other tissues (except for kidney) because the occurrence of glucose-6-phosphatase allows the breakdown of hepatic glycogen to free glucose, which can be utilized to maintain the blood sugar level in periods of brief fasting. The service does not extend to long fasting because the amount of liver glycogen (3 to 5 per cent of wet liver weight in the normal, well-fed animal or human) and its rate of breakdown are such that the store of liver glycogen is essentially depleted in 6 to 12 hours. After that, and even before that for smooth homeostatic control, the body maintains blood sugar by an increase of gluconeogenesis from protein as well as by the mobilization and utilization of fat stores. If there is an impairment in hepatic glycogen breakdown (to be discussed later) the periods of transition from the fed to the fasting metabolic state even in common daily routines may be rough and stressful.

Under normal average circumstances the liver synthesizes and degrades glycogen at a rate of about 0.5 mg./hr./g. of liver weight. After fasting and refeeding of carbohydrate, however, the rate of synthesis may be 5 to 10 mg./hr./g. The known capacity of hepatic glucokinase seems too low to account for this rapidity of glycogen synthesis; possibly the phosphotransferase activity of glucose-6-phosphatase or else the kinase-dismutase route for synthesis of glucose-1-phosphate (Chap. 4) are auxiliary mechanisms. About 15 minutes after an oral glucose load, or within a few minutes after intravenous administration, the level of liver glycogen rises sharply and extensively. The rate of glycogenesis may increase eight- to ten-fold in response to a doubling of the blood glucose concentration.

The important effect of the glucose is not that of "mass action"; glucose specifically increases activity of glycogen synthetase, evidently by promotion of synthetase b → a conversion. Although elevation of blood glucose also stimulates insulin release and insulin itself activates glycogen synthetase (see above), the specific effect of glucose is not mediated via insulin. Nor is it dependent upon an increase of Gl-6-P, since hepatic concentration of the latter as well as of UDPG is low during the period of rapid glycogen synthesis after a glucose load. With insulin the glycogenic effect of a glucose load is obtained at a lower level of blood sugar. However, the glycogenic effect of insulin does not depend upon glucose penetration into the liver cell, i.e., insulin promotes glycogenesis from newly formed or endogenous hexose not in equilibrium with glucose entering the cell.

The glucocorticoids (cortisone, hydrocortisone) are powerful stimulators of hepatic glycogen synthesis also through increased activity of glycogen synthetase (plus enhanced gluconeogenesis), although the effect of glucocorticoids is more delayed than that of glucose or insulin. Some effect of glucocorticoids may appear in one-half to 1 hour, but maximal effect occurs some hours later. A mechanism (perhaps protein synthesis) for an increase of glycogen synthetase by glucocorticoids that is different from that by glucose or insulin is suggested by the difference in time of activation. Furthermore, the effects of glucocorticoids and of glucose (or of insulin) are observed to be additive.

Glucagon can almost completely inactivate glycogen synthetase in the liver, presumably via its effect on cAMP. Whether glycogen in liver limits its own accumulation by inhibition of synthetase b phosphatase, as in muscle, is not known. In the liver phosphorylase a is an inhibitor of synthetase b phosphatase, which, of course, reinforces the effect of phosphorylase to cause glycogenolysis.

Liver phosphorylase generally decreases slightly after a glucose load, although, as noted in the dog, insulin and glucose together are followed by marked decline of hepatic phosphorylase. Conversely, glucagon — which rises in the blood during periods of fasting, or of hypoglycemia—stimulates phosphorylase about twofold. In the liver at physiological concentrations glucagon is much, at least 10 times, more glycogenolytic than epinephrine. (There is some effect of epinephrine to increase the activities of neutral and acid-α-glucosidases.) Both glucose and glucocorticoids stimulate hepatic phosphorylase a phosphatase, thus inhibiting glycogen breakdown. By acting as a cofactor for phosphorylase phosphatase to convert phosphorylase a → b, glucose relieves the inhibition of synthetase b phosphatase, which may be the mechanism by which it promotes synthetase a and glycogenesis. The K_m of phosphorylase phosphatase for glucose is about 130 mg./dl., which is, of course, in the physiological range. The degree to which controls by metabolites and hormones on the dual activities of glycogen synthesis and breakdown are interlocked is indeed striking for liver (Table 6–1).

Liver glycogen concentration manifests a circadian rhythm, as seen in mice, which persists even in starvation. Glycogen concentration is highest in the early morning hours in these rodents. Most likely this phenomenon relates to the important control of liver glycogen metabolism by the adrenal glucocorticoids, which are well known to vary in plasma concentration with circadian periodicity.

SKELETAL MUSCLE. The need for rapid energy provision for the function of skeletal muscle is obvious. Even during light muscular exercise in fasting subjects the muscle's own glycogen store, rather than blood glucose, appears to provide the main carbohydrate substrate for energy. Under these circumstances about 50 per cent of the energy supply for muscle is derived from fatty acid oxidation. With more intense exercise under anaerobic conditions glycogen utilization comes increasingly into play. The capacity for prolonged work or heavy exercise is directly correlated with the prior glycogen content of the exercised muscles. Resting glycogen content of muscles of exercise-trained animals and humans is in general higher than normal. After exercise there is an increase in glycogen synthetase of the exercised (but not the unexercised) muscles, with repletion of glycogen stores to higher levels than those found in the unexercised muscles. Possibly this phenomenon accounts partly for the increased glucose utilization which occurs in normal as well as diabetic individuals after exercise, when storage may be shifted significantly from fat to glycogen.

High-carbohydrate feeding (not necessarily sucrose) enhances the general level of muscle glycogen and the extent of repletion after exercise. In human subjects a high-carbohydrate diet results in accumulation of about 3 g. glycogen/100 g. of muscle (quadriceps femoris), compared with about 0.6 g./100 g. muscle on a fat plus protein diet. During starvation the muscle glycogen also decreases to levels like those found on a low-carbohydrate diet, which are generally not less than half of the content on an average mixed diet. Thus muscle glycogen is considerably better preserved than is liver glycogen during long fasting.

The effect of glycogen concentration in muscle to permit heavy work appears to be more than just the availability of an energy source, since with low muscle glycogen work is impaired even in the presence of plentiful blood glucose and fatty acids. Villar-Pilasi and Larner have suggested a relationship between the cellular retention of K^+ and the steps involved in glycogenesis. The conversion of UDP-Glucose to UDP generates H^+ which may be exchanged for K^+. Conversely, with glycogenolysis and the utilization of P_i to form Gl-1-P there are losses of K^+ or other cation from the cell. These changes could have ramifying effects on various cell functions.

In general, glycogen metabolism in muscle is less susceptible to various hormones (except epinephrine) than is liver glycogen (Table 6–1). Nevertheless, within minutes insulin can increase muscle glycogen by the aforementioned mechanism, i.e., an increase in percentage of synthetase a without change in total synthetase activity. Epinephrine opposes an increase in synthetase, presumably via its effect on cAMP. In muscle, synthetase b is less inhibited by P_i than it is in liver, nor is synthetase b phosphatase inhibited by phosphorylase as in liver.

Epinephrine, rather than glucagon as in liver, can increase the activity not only of phosphorylase a but of debrancher enzyme, α-amylase, and various glucosidases. This truly emergency hormone thus calls for full mobilization to provide the muscles with energy for "fight or flight." Electrical stimulation of muscle, i.e., muscular contraction, leads to a conversion of phosphorylase b to a by stimulation of phosphorylase b kinase. An involvement of the Ca^{++}-activation mechanism seems attractive to explain this effect of muscular contraction, but this theory is presently beset by problems, such as much different K_m values of Ca^{++} for phosphorylase b kinase activation and muscular contraction. The latter does not activate via an increase of cAMP and otherwise also differs from epinephrine in its mechanism. There is a much shorter half-time of activation than with epinephrine. Furthermore, the two effects on glycogenolysis, muscular contraction and epinephrine, can be additive. Epinephrine seems to prepare the muscle for the glycogenolytic action of muscular contraction, since with epinephrine there is a decrease in the lag period as well as an increase in final steady-state levels of phosphorylase a attained with muscular contraction.

Cold exposure decreases muscle (as well as liver) glycogen in rats. Perhaps this occurs via epinephrine or by glucose depletion. In any case, exercise plus cold exposure decreases muscle glycogen in ways which are not only additive but synergistic. The presence of glucose, Gl-6-P, or Gl-1-P decreases the level of phosphorylase a in isolated rat diaphragm, thereby helping to maintain better glycogen levels when these substrates are present.

Another hormonal influence is exerted by testosterone to increase glycogen synthetase, hexokinase, and glycogen content of the sexually specialized muscle, the levator ani. This occurs after a lag time of several hours and is inhibited by actinomycin D and puromycin, which indicates an effect on protein synthesis. The effect is additive to that of insulin. Generalized skeletal muscle is less sensitive to testosterone, so the relevance of this phenomenon to common muscular activity is questionable.

CARDIAC MUSCLE. Glycogen metabolism in heart muscle differs in several respects from that of skeletal muscle. Aerobic metabolism is higher in heart muscle, with ready utilization of pyruvate and lactate as well as fatty acids and ketone bodies. Therefore, heart muscle is less dependent on glycogen for energy. The glycogen is relatively well conserved under different metabolic and work conditions. Indeed, during fasting or in diabetes the cardiac glycogen increases in contrast to the opposite for skeletal muscle. This seems to depend on an increase in circulating fatty acids and ketones, which not only provide substitute fuel but curtail glycolysis, particularly the activity of phosphofructokinase. The general effect of fasting depends on the presence of circulating growth hormone, presumably in order to help initiate the sequence by mobilization of peripheral fat.

Insulin promotes the uptake of glucose and glycogenesis by heart muscle. Activation of synthetase is not opposed by epinephrine as it is in skeletal muscle; in fact, synthetase a may be increased. However, epinephrine activates phosphorylase b kinase and hence glycogenolysis, though more slowly than for skeletal muscle. Glucagon also enhances cardiac glycogenolysis, even though it has little such effect on skeletal muscle. Thyroid hormone potentiates the action of epinephrine on phosphorylase a activity. A mechanism recently suggested is an increased formation of adenyl cyclase under the influence of thyroid hormone additive to the effect of epinephrine to stimulate adenyl cyclase. Epinephrine in cardiac muscle, unlike skeletal muscle, has no effect on degradative enzymes other than phosphorylase.

Cardiac glycogen is rapidly responsive to changes in glucocorticoids, which promote glycogen deposition probably by some effect on the synthesis of glycogen synthetase or conversion of synthetase b to a. There is a diurnal variation in cardiac glycogen of the rat, which probably reflects a similar variation in the circulating level of glucocorticoids. Exercise-conditioned rats show an increase in cardiac glycogen stores, a response resembling that of skeletal muscle. The utilization of cardiac glycogen is increased as mean ventricular pressure develops, which depends on some conversion of phosphorylase b to a. Hypoxia may also trigger a conversion of cardiac phosphorylase b to a; this is in part but not altogether mediated by adrenergic stimulation.

SMOOTH MUSCLE. Little is known about glycogen metabolism in smooth muscle other than uterus. Generally smooth muscle contains glycogen, and the occurrence of glycogen synthetase, phosphorylase a and b, and phosphorylase b kinase has been identified. Whereas hyperglycemia causes increases of uterine glycogen, the glycogen content of this organ is remarkably protected against losses in the face of hypoglycemia produced by fasting or insulin. This trait depends upon estrogen, which is the "insulin of the uterus." Estrogen is able to induce increases in both glucose and glycogen of the uterus even under conditions of low circulating glucose. The female steroid hormone increases glucose transport into the uterine muscle cells (Chap. 3), increases the synthetase system and also the concentration of phosphorylase a. An increased turnover of glycogen is implied, but with a net increase of glycogen content which may be two- or three-fold higher after estrogen administration in various species. In some species this is associated with an increase in amount of intracellular glucose and in others not, so the question of primacy of effect (if there is any) on glucose transport or intracellular enzyme is difficult to answer.

In the uterus estrogens increase cAMP via stimulation of adenyl cyclase. No such effect, or glycogenolysis, is caused by estrogen in skeletal muscle, indicating the receptor specificity of the target organ. The glycogenic effect requires a lag time of a few hours and is inhibited by inhibitors of protein synthesis. This may be related to the effect on cAMP, which itself may be involved with protein synthesis. According to study with the hamster, progesterone can potentiate estrogen in causing an increase of glycogen in the longitudinal layers of uterine muscle; however, other studies with the rat suggest that progesterone as well as testosterone and hydrocortisone partially inhibit the estrogen stimulation of glycogen synthesis in the uterus.

BRAIN. Brain contains about one-tenth the glycogen concentration of muscle and far less than that of liver. However, the turnover of brain glycogen is several times faster than that of liver. Glycogen synthetase in the cerebral cortex of rats is present in both a and b forms, as are the enzymes catalyzing a and b interconversion, i.e., synthetase a kinase and synthetase b phosphatase. Yet glycogen deposition induced by elevation of blood glucose is not associated with increase in synthetase a, as in liver or muscle. Furthermore, hypoglycemia and anoxia, which diminish glycogen levels, are paradoxically accompanied by severalfold increases of synthetase a, though also by considerable increase in its K_m for UDPG. The latter is considered to represent a protective mechanism for brain whereby glycogenesis is minimized during conditions when energy is low and the available glucose must be used maximally for glycolysis. With the presence of large amounts of synthetase a form, though inoperable at low concentrations of glucose, Gl-6-P and UDPG, the system is poised for rapid replenishment of glycogen by mass action when ambient glucose again becomes plentiful.

Insulin increases the synthesis and accumulation of glycogen in the cerebral cortex of normal or diabetic rats, which tends along with other evidence (Chaps. 3 and 4) to refute the former belief that insulin does not directly affect glucose metabolism in brain. The insulin effect can be observed under conditions (e.g., intracisternal administration) wherein the blood glucose levels are not disturbed. Furthermore, the effect of insulin on glycogenesis in cerebral cortical slices of rats seems more direct than can be ascribed to increased transport (Chap. 3), since the conversion of glucose-^{14}C to $^{14}CO_2$ is not increased at a time when glycogenesis from the labeled glucose is promoted.

The normal rate of glycogenesis of cerebral cortex is not diminished in diabetic rats, and the level of brain glycogen in such rats may indeed be higher than normal. There is glucose-6-phosphatase activity in the brain of normal rats (one-fifth as active as in liver), but in diabetic rats the Gl-6-Pase is very low, which is opposite to the usual diabetic change of this enzyme in liver. Possibly the enzymatic defect results in diversion of Gl-6-P to formation of glycogen in the diabetic brain. Anesthetic and tranquillizing agents typically increase levels of glycogen in brain, accompanied by elevations of intracellular glucose and Gl-6-P, though not of UDPG.

BLOOD CELLS. Among blood cells polymorphonuclear (PMN) leukocytes and platelets contain much more glycogen than erythrocytes. Neutrophilic granulocytes account for almost all the glycogen in PMN leukocytes; eosinophils and basophils, besides lymphocytes and monocytes, have much less. The importance of glycogen in leukocytes relates primarily to provision of energy for phagocytosis, when glycogen in the white cells is rapidly consumed. Fasting, on the other hand, does not lower the glycogen level. The glycogen in neutrophilic granulocytes is in a condition of constant turnover and the level is dependent on continual provision of glucose. In vitro studies suggest that neutrophils in 100 ml. (1 dl.) of blood may utilize about 2 or 3 mg. of glucose per hour for adequate renewal of glycogen stores. The total glycogen in formed blood cells of normal humans (leukocytes plus erythrocytes) is also 2 to 3 mg./dl., so the turnover time appears to be about one hour. The content of glycogen in blood cells of diabetics is almost twice as high and with insulin-treated diabetics it is still higher. Neutrophilic glycogen is also increased in infectious leukocytosis, polycythemia vera, sometimes with malignant neoplasm, and in glycogen storage diseases (see below).

In PMN leukocytes of normal humans, glycogen synthetase exists exclusively in the b or D form (dependent on Gl-6-P for activity). Though not detectable in normal human leukocytes, the enzyme synthetase b phosphatase is found in the white cells of diabetics, which may help explain the elevated glycogen levels. In the leukocytes of normal as well as alloxan-diabetic rats the synthetase b → a transformation is found to occur and the K_a of activation of synthetase by Gl-6-P is much less than for human PMN leukocytes. Thus, the latter are more dependent on high intracellular levels of Gl-6-P. For both human and rat leukocytes Mg^{++} decreases both the K_m for UDPG and the K_a for Gl-6-P.

Even though very low in glycogen content the red cells contain all the enzymes required for the synthesis and degradation of glycogen, which is actively turning over in such cells.

Platelets contain about one-fiftieth as much glycogen per cell as neutrophils but in total amount per volume of blood the two types of blood cells have about equal amounts of glycogen. The larger, heavier, and younger fraction of the blood platelet population contains more glycogen than the smaller, lighter, and older platelets. During coagulation glycogen is depleted from the platelets. This has been attributed to the high energy need for the ATPase-like action of the platelet contractile protein, thrombosthenin, during clot retraction. In platelets glycogenolysis contributes about 50 per cent of the carbohydrate flux through the glycolytic pathway in the presence of physiological concentrations of glucose. Whereas phosphorylase of leukocytes is similar to that of liver in certain characteristics (e.g., response to AMP), the phosphorylase of platelets is more like that of muscle, which is in keeping with the existence in both types of cells of a contractile protein dependent upon glycogenolysis and glycolysis for energy.

KIDNEY. The proximal tubular cells of the kidney near the glomerulus show little glycogen deposition compared with the more distal portion of the proximal tubule, the loop of Henle, the distal convoluted tubule, and, particularly, the cortical collecting ducts. However, enzymes for glycogen metabolism in kidney are generally more active in the proximal convoluted tubule, which suggests that glycogen accumulation in the renal epithelial cells is not proportional to the activity of glycogen metabolism.

Glycogen levels in the kidney are directly correlated with the degree of diuresis imposed upon the kidney. Since antidiuretic hormone may increase cyclic AMP, Darnton has suggested that cAMP-induced glycogenolysis provides in distal tubules an energy source for the transport of ions, which is increased during antidiuresis. She further suggests that a thickening of the mucopolysaccharide (MPS) structure of the papilla in antidiuresis could be related to increased availability of MPS precursor through glycogenolysis. In diabetic, hyperglycemic animals glycogen deposition in the kidney papillae is exaggerated and the glycogen is characteristically contained in lysosomes, which suggests inadequacy of disposal (perhaps of an overload) by acid α-glucosidase.

REPRODUCTIVE ORGANS. The glycogen content of the vaginal epithelium is increased by either estrogen or progesterone and most effectively by both together. Estrogen elicits a marked increase when instilled locally. Increases of uterine and cervical glycogen are more characteristic of systemically administered estrogens. The increase in vaginal glycogen appears to precede a period of increased cell division, which could be stimulated by energy provision from the glycogen. The glycogen content of the human placenta increases until about the twentieth week, after which it remains constant. Utilization of this glycogen may be significant for energy supply, particularly when fetal malnutrition occurs.

Glycogen probably is important in the maturation of the germinal cells of the seminiferous tubule. Stages of spermatogenesis which are involved in RNA and DNA synthesis and cell division are characterized by rapid decrease in glycogen, which is probably needed to supply energy and substrate. Some enzymes of glycogen metabolism have been found in seminiferous tubules. The apparent absence of glycogen synthetase and the demonstration of phosphorylase activity in the synthetic direction suggest that glycogen synthesis may be differently catalyzed in the seminiferous tubules than in liver, muscle, and so forth.

PITUITARY AND ADRENAL. The posterior pituitary contains more glycogen than the anterior portion. The content in both neuro- and adenohypophysis is responsive with increase or decrease of glycogen as blood glucose concentrations are varied. Possibly the utilization of glycogen by the anterior pituitary during hypoglycemia plays some role in the increased secretion of growth hormone consequent to hypoglycemia.

The adrenal medulla has a greater concentration of glycogen than the adrenal cortex. In the adrenal cortex glycogen is lowered by both hyper- and hypoglycemia. The effect of ACTH on the adrenal cortex includes a reduction in glycogen; this has been attributed to an increase in cAMP, thereby activating phosphorylase. The glycogen content of adrenal medulla is less susceptible than that of the cortex to changes in the blood glucose level, although there is some decrease during hypoglycemia, which may be related to secretion of catecholamines. For reasons yet unknown, alloxan diabetes in the rat is associated with marked elevations of medullary glycogen and an unresponsiveness of cortical glycogen to ACTH.

ADIPOSE TISSUE. Glycogen in adipose tissue resembles muscle glycogen in its molecular weight and monoparticulate characteristics. Glycogen has a short half-life of only 2 to 4 hours in adipose tissue, and the inner as well as outer tiers show an extensive turnover. Thus, although the concentration of adipose tissue glycogen is low, its high metabolic activity indicates importance to cellular function. Levels are generally higher in meal-fed than in nibbling animals and there are rapid and extensive increases in glycogen synthesis upon refeeding of fasted animals.

OTHER TISSUES. Glycogen is found in several other tissues such as thyroid; beta-cells of the pancreas; skin; intestinal mucosa; and cartilage. Little is presently known of the metabolic characteristics or function of glycogen at these sites. Glucose-6-phosphatase occurs in the pancreatic β-cells and the intestinal mucosa, as in liver and kidney.

Some Pathological Changes

The abnormalities of glycogen metabolism which may accompany hormonal disorders have been discussed in the preceding sections. More specifically there are genetic defects or losses of particular enzymes of glycogen metabolism which are recognized clinically as various glycogenoses or "glycogen storage diseases."

Glycogenoses

Although a properly operating glycogen metabolism cushions against metabolic and nutritional vicissitudes, it is not essential to life, so genetic deficiencies or even losses of glycogen-metabolizing enzymes can be tolerated and are expressed as glycogen storage disease (GSD). In these conditions, which are evident early in life, glycogen may markedly accumulate when the defect is in glycogen breakdown (more common) or be unduly sparse, with loss of an enzyme for synthesis (less common). Most of the glycogenoses are autosomal recessive traits, which are rarely expressed homozygously but for which there are much larger heterozygous populations. These can now often be recognized by laboratory tests among near relatives. Since individuals with GSD show a certain panoply of disturbances in carbohydrate and fat metabolism, the heterozygotes may contribute more commonly than earlier supposed to milder but similar disturbances in the general population. The glycogenoses are frequently designated by type (I, II, and so forth) in general chronological order of description (Table 6–2).

Type I Glycogenosis

Deficiency of glucose-6-phosphatase, which catalyzes the final step in the conversion of glycogen to glucose in liver or kidney is designated

Table 6–2. *Classification of Glycogen Storage Diseases*

TYPE	ENZYME DEFECT	GLYCOGEN STRUCTURE	MAIN ORGAN INVOLVEMENT
I	glucose-6-phosphatase	normal	liver, kidney
II	acid α-1,4-glucosidase	normal	generalized (or skeletal muscle)
III	amylo-1,6-glucosidase	limit dextrin	liver, kidney, cardiac, and skeletal muscle
IV	amylo-(1,4→1.6)-trans-glucosidase	amylopectin-like	generalized
V	muscle phosphorylase	normal	skeletal muscle
VI	liver phosphorylase	normal	liver
VII	glycogen synthetase	normal	liver

Type I glycogenosis (clinically called Von Gierke's disease). Because gluconeogenesis also proceeds through the intermediate, Gl-6-P, the body as a whole is in double jeopardy for available blood glucose without hepatic glucose-6-phosphatase. Clinical symptoms of hypoglycemia, ketonemia, and acidosis upon fasting are pronounced, particularly in the childhood years. The ketosis is largely due to intemperate peripheral fat mobilization evoked by hypoglycemia. Hyperglyceridemia, hypercholesterolemia and elevation of circulating FFA are characteristic. Elevation of serum lactic acid (hyperlacticemia) and uric acid (hyperuricemia) is also typical. Lactic acid may rise because of excessive glycolytic rates in the liver, because another major pathway for Gl-6-P (i.e., dephosphorylation) is blocked. Decreased renal urate excretion may be secondary to hyperlacticemia; however, overproduction of uric acid in the liver associated with abnormal carbohydrate metabolism (e.g., excessive formation of phosphoribosylpyrophosphate, a precursor of uric acid via nucleotides) is another possible cause for hyperuricemia. In conjunction with the latter, gout and renal failure are notably distressing problems of adults with Type I glycogenosis. From an early age there are hepatorenomegaly, retarded growth and often eruptive xanthomata (consequent to hyperlipemia). Whether hyperlipemia is just a phenomenon of increased peripheral mobilization or is due also to increased synthesis in the liver—perhaps by increased formation of substrate or cofactors via glycolysis or the pentose cycle pathway—is not known.

Galactose, fructose, and glycerol are converted poorly to blood glucose but rapidly to lactic acid via glycolysis with sharp rises in serum levels of lactate. Glucagon and epinephrine often fail to produce their usual hyperglycemic response. These tests with carbohydrate loads or hormones are used diagnostically and to distinguish from other types of glycogenoses. Another method for diagnosis of Type I is peroral biopsy of intestinal mucosa, which will also show the enzyme deficiency.

In Type I there are normal or only slightly reduced levels of other enzymes concerned with glycogen metabolism. The high glycogen concentration may nevertheless, by the mechanism earlier discussed, inhibit normal rates of glycogen synthesis and obviate to some extent another pathway for metabolism of Gl-6-P. Since the action of debrancher enzyme, perhaps with the aid of glucantransferase or γ-amylase plus α-1,4-glucosidase, can break down some glycogen to free glucose, there are some avenues for turnover of glycogen not involving glucose-6-phosphatase. Enzymes of hepatic gluconeogenesis, other than glucose-6-phosphatase, seem to be increased in Type I glycogenosis, perhaps in some way to compensate for the disorder of glucose provision to the blood by liver. In the muscles, where Gl-6-Pase normally is absent, there are moderately increased amounts of glycogen. Increased glycogenesis in muscles from pyruvate has been demonstrated, which may represent homeostatic adaptation to raised serum levels of pyruvate and lactate. A peculiar finding in some patients with Type I glycogenosis (and possibly others) is an increased tolerance for alcohol with increased blood clearance, no accompanying impairment of galactose tolerance (Chap. 8), and fall (rather than rise) in serum lactate. The indication of changed ethanol metabolism remains unexplained. Glucose-6-phosphatase is also reduced in blood platelets in Type I glycogenosis, which may be evidence of defective platelet function.

Type II Glycogenosis

In this genetic condition (Pompe's disease) there is absence of the lysosomal acid-α-1,4-glucosidase (acid maltase), which hydrolyzes maltose or the external branches of glycogen to free glucose. Heart, liver, kidney, skeletal and smooth muscle, blood cells, and nerve tissue are all deficient in the enzyme. This type of generalized glycogenesis is clinically devastating, with muscular weakness, cardiac enlargement and failure, and mental retardation. Two general subtypes emphasize cardiac and neuromuscular involvement, respectively. In either case, death usually occurs early in life. Milder cases do occur, however, and there are descriptions of patients with involvement only of skeletal muscle. Besides glycogen, a mucopolysaccharide may also accumulate in the skeletal muscle, which is a clue to other biochemical defects in this condition besides deficiency of lysosomal acid maltase. In the generalized form Type II can be detected by the absence of α-glucosidase in leukocytes or in fibroblasts (from skin biopsy). With this recessive trait parents of affected individuals have only about half of the normal activity of α-glucosidase in these cells. Type II can also be diagnosed in utero by demonstrating deficient α-glucosidase activity or abnormal lysosomes in amniotic fluid cells.

Type III Glycogenosis

Absence of the debrancher enzyme amylo-1,6-glucosidase, usually from several tissues, produces Type III glycogenosis (Forbe's disease). The signs and symptoms (hepatomegaly, retarded growth, "doll's face," fasting hypoglycemia, moderate acidosis, hyperuricemia, hyperlipemia) are much like those of Type I, but are usually noted to be milder. However, affection of cardiac and skeletal muscle, unlike Type I, gives rise to myopathies of varying degree and nature. A diabetic type of glucose intolerance may occur, as indeed may be found in any of the hepatic glycogenoses. Various subtypes of Type III have been described on the basis of predilection for occurrence in liver or muscle or both. The glycogen accumulating in this condition resembles limit dextrin with short outer chains; hence, another name for Type III is limit dextrinosis.

Leukocytes, erythrocytes and fibroblasts are all affected by the enzyme deficiency in its more generalized form. The amount of excess glycogen in these cells varies, but gross accumulations in the red cells are particularly indicative of Type III glycogenosis. An enzyme deficiency in the leukocytes or erythrocytes of heterozygotes (parents and some siblings of homozygotes) can possibly be detected by functional tests such as incorporation of glucose-^{14}C into glycogen in vitro or liberation of glucose from limit dextrin. Fibroblasts in amniotic fluid could presumably be tested for prenatal detection of debrancher enzyme deficiency.

Glycogen-rich red cells from patients with Type III glycogenosis show a more rapid utilization of glucose than normal cells, but the glucose is not incorporated into glycogen or into lactate more rapidly. The fate of the excess glucose and the reason for the increased utilization are not understood.

Type IV Glycogenosis

This very rare condition (Anderson's disease) of deficiency of the brancher enzyme, amylo-(1,4→1,6)-transglucosidase, causes hepatosplenomegaly and severe cirrhosis with progressive hepatic failure and early death. Called also amylopectinosis, this type of glycogenosis shows in several organs accumulation of glycogen which is more like starch in having very long chains without branching. Leukocytes share the deficiency in the brancher enzyme and half-normal amounts are found in the leukocytes of heterozygotes for this recessive trait.

Type V Glycogenosis

A complete lack of phosphorylase in skeletal muscle causes Type V glycogenosis (McArdle's syndrome), in which the impaired mobilization of glycogen is accompanied by intolerance for exercise, which readiy causes muscular pain and cramps, stiffness, and weakness. The disease is not clearly manifest before early adulthood. Heterozygotes may have a lesser degree of symptoms. Other enzymes (α-amylase and α-glucosidase) can provide some breakdown of glycogen to mitigate the loss or deficiency of phosphorylase. Cardiac muscle is not significantly involved; however, smooth muscle (e.g., uterus) may be affected. There has been found an elevation of red-cell glycogen. Upon exercise there is failure of the normal rise of plasma lactate or pyruvate, which is a useful, but not pathognomonic, test. Patients with Type V have only half the maximal oxygen uptake of normal and can accumulate only one-fifth the normal maximum oxygen debt. An increased incidence among males indicates a sex-linked mode of inheritance and not a simple recessive trait, although siblings are frequently affected.

Type VI Glycogenosis

In this glycogenosis, known originally as Her's disease, the deficiency occurs in hepatic phosphorylase, either directly or via a defect in the activating system. The enzyme is not essentially missing, as in other glycogenoses. In leukocytes, which also display low phosphorylase activity in Type VI, the enzyme can be stimulated to normal by 1 mM AMP, suggesting a defect in cofactor activation for a certain type (Type VIa). With another subgroup (VIb, sometimes designated Type IX) the deficiency appears to reside with phosphorylase b kinase, which exhibits a greatly increased K_m for its substrate. Whatever the mechanism of deficiency, there is hepatomegaly of great extent from an early age with findings of intermittent hypoglycemia, moderate ketosis and lactic acidosis, and mild hypercholesterolemia, much as in Type III and less severe than in Type I. Functional (hyperglycemic) response to glucagon is low, but usually normal after epinephrine. However, distinction from Type I or subgroups of Type III may be difficult on this basis. This type of glycogenosis, like Type V, seems to have sex-linked heritable characteristics with transmission through the female and expression primarily in the male.

Type VII Glycogenosis

An enzymatic defect of glycogen synthetase occurs in this type, which is apparently rare. However, as with other recessive disorders, the extent of heterogeneity is not known. These patients have shown reduced carbohydrate tolerance as well as hypoglycemia upon fasting. There is a paucity of glycogen in the liver.

Other Glycogenoses

Some rare cases of glycogen storage disease have been recognized to be associated with deficiencies of one or another enzyme of the glycolytic pathway, such as phosphofructokinase or hexose phosphate isomerase. In such cases the accumulation of Gl-6-P disposes to conversion of the latter to glycogen. In a muscle glycogenosis due to deficiency of PFKase there have also been found increases in activities of UDPG-pyrophosphorylase and of a glycogen synthetase, which helps account for the accumulation of glycogen.

PROTEOGLYCANS

This group of complex carbohydrates, commonly called mucopolysaccharides (MPS), is composed of a few kinds with typical disaccharide repeating units in long chains. Their chemistry is summarized in Table

METABOLISM OF POLYSACCHARIDES

Table 6–3. *Major Types and Constituents of Proteoglycans (Mucopolysaccharides)*

	MAIN MONOSACCHARIDE CONSTITUENTS		PRESENCE OF:		
POLYSACCHARIDE	Amino Sugar	Uronic Acid or Other	N-acetyl Groups	O-sulfate Groups	Sulfamate Groups
hyaluronic acid (HA)	D-glucosamine	D-glucuronic acid	+	–	–
chondroitin	D-galactosamine	D-glucuronic acid	+	–	–
chondroitin-4-sulfate (C-4-S)	D-galactosamine	D-glucuronic acid	+	+	–
chondroitin-6-sulfate (C-6-S)	D-galactosamine	D-glucuronic acid	+	+	–
dermatan sulfate (DS)	D-galactosamine	L-iduronic acid	+	+	
keratan sulfate (KS)	D-glucosamine	D-galactose	+	+	–
heparin	D-glucosamine	D-glucuronic acid	–	+	+
heparan sulfates (HS)	D-glucosamine	D-glucuronic acid	+	+	+

6–3 and discussed in detail in Chapter 1. In contrast to glycoproteins, the protein in proteoglycans is a minor fraction. Because of the charactereistic content of hexosamines the name glycosaminoglycans, is sometimes used to define better chemically the MPS. The glycosaminoglycans are highly acidic owing to the presence of carboxyl groups of uronic acids and sulfyl groups (only sulfyl for keratan sulfate).

Chemical Reactions

SYNTHESIS. Proteoglycans are manufactured by appropriate connective tissue cells and ordinarily extruded from them. Other more specialized cells, e.g., platelets, leukocytes, and mast cells, can also form and store proteoglycans. All the enzymes involved in forming both the region of linkage with protein (the "core" of the proteoglycan containing typically the xylosyl-serine linkage) and the main carbohydrate chain are found in the microsomal part of the cell. The rough microsomal fraction appears to contain the enzymes for the initial steps, i.e., formation of the linkage region. Those enzymes forming the main chain of chondroitin sulfate are present more evenly in rough and smooth membranes, while sulfotransferases occur mainly in the smooth microsomal fraction. A continuous or assembly line of enzymes, from rough to smooth microsomes, finally reaching the Golgi apparatus, is thus suggested. Isotopic and other techniques indicate that the protein moiety of proteoglycans is also synthesized in or near the Golgi apparatus and at roughly the same time as the carbohydrate moiety. After or during biosynthesis the proteoglycans are transported through the cytoplasm to the periphery of the cell in vesicle-like compartments, which appear to fuse with the cell membrane and thus allow discharge of their contents into the extracellular space.

The linkage region containing the structure Gal-Gal-Xyl-serine (or threonine) is a common one for several MPS. With UDP-xylose as donor a specific transferase attaches xylose to the hydroxyl of serine in a polypeptide. The successive transfer of the two galactosyl units from UDP-galactose in the further formation of the carbohydrate-protein linkage region is catalyzed either by two different enzymes or by different loci on the same enzyme. The second of the galactosyl transferases is the more specific; only galactosyl-β-1$\rightarrow$4-xylose or a corresponding serine derivative will act as acceptor. This specificity prevents the attachment of a third galactosyl unit. The transferase which links the first glucuronic acid (from UDP-GA) to Gal-Gal-Xyl is different from that which adds glucuronic acid to NAcGal to lengthen the chain in chondroitin sulfate (CS).

A theory for the subsequent alternate addition of hexosamine and uronic acid from their UDP derivatives to the growing polysaccharide chain postulates the existence of one transferase enzyme with multiple active sites. Alternating addition of the two kinds of monosaccharides would occur simply by alternate occupation of binding sites by the terminal residues of the chain and the next UDP-monosaccharide to enter into attachment. Present evidence fits better with this theory than with the idea that the disaccharide unit, such as hyalobiuronic acid, is initially formed and then attached to the end of the chain.

The sulfate donor for biosynthesis of various sulfate esters of MPS (and of the sulfatide type of glycolipid: see Chap. 7) is phosphoadenosine-5-phosphosulfate (PAPS). The latter is formed from inorganic sulfate and ATP by a two-stage process. Tissues which can synthesize sulfated polysaccharides can generally synthesize PAPS. There appears to be a multiplicity of transferases which are specific for transferring sulfate from PAPS to individual glycosaminoglycans. Probably also the sulfation of amine groups is enzymatically distinct from that of hydroxyl groups.

The various components in sulfated MPS of connective tissue turn over at approximately the same rate, which suggests a close coordination of biosynthesis of the fully sulfated molecule de novo from all components. Most evidence suggests sulfation of a monosaccharide residue immediately or very soon following its attachment to the non-reducing end of the chain. However, already sulfated residues will accept another sulfate. Possibly a chondroitin sulfate molecule contains a mixture of non-sulfated, monosulfated and disulfated disaccharide units. Prior conjugation with a protein moiety is not essential before sulfation of acceptor polysaccharide units in vitro.

The reasons for O-sulfation of chondroitin sometimes at C-4 of NAcGal to form chondroitin-4-sulfate (C-4-S) and other times at C-6 to form chondroitin-6-sulfate (C-6-S) may be various. Different transferases or the nature of the receptor (presence of protein moiety, fine structure of the glycan in the receptor region) have been variously hypothesized. How does the chain stop growing? Telser has suggested that sulfation of a terminal NAcGal unit is the cause, since such a terminal sulfated sugar is not an acceptor for UDP-glucuronic acid.

The formation of N-sulfyl groups in heparin and heparan sulfate (HS) occurs by displacement of N-acetyl by the stronger anionic N-sulfyl bond. The process seems to be much more complete in heparin than in heparan sulfate. In spite of the strong chemical resemblance, the quite different functional properties and organ distribution of heparin and heparan sulfate suggest independent origins and a lack of biochemical interconversion.

There is evidence that not all the chondroitin sulfate proteins in a cartilage cell have similar rates of growth, turnover, or affinity for precursors available. Chondroitin sulfate chains which are most intensively labeled by tracer sulfate, glucosamine, and so forth are generally the ones which are the most anionic or the largest or both.

DEGRADATION. Little information is available which would explain the turnover of acidic proteoglycans in terms of catabolic enzyme activity. The proteolytic enzymes (of lysosomes) may be predominant in determining the degradation of mucopolysaccharides by attacking the protein moieties before the carbohydrate chains are further split by enzymes of higher specificity. The protein moiety of chondroitin sulfate in cartilage is particularly susceptible to breakdown by proteases such as papain with release of glycosaminoglycans into the blood stream and subsequently the urine.

The degradative enzyme most widely studied so far is testicular hyaluronidase, which cleaves the $\beta\text{-}1 \rightarrow 4$ hexosaminide bond between N-acetyl-D-glucosamine and D-glucuronic acid in hyaluronic acid (HA). Chondroitin, C-4-S, and C-6-S are also substrates for hyaluronidase, but the sulfated substrates are degraded more slowly than the non-sulfated. The enzyme does not act on dermatan sulfate (DS), keratin sulfate (KS), heparin or heparan sulfate. Hyaluronidase can transglycosylate the glucosamine unit of HA to another receptor oligosaccharide chain. In vitro this can produce hybrids of HA and CS, and the natural occurrence of such mixtures is a possibility.

Other hyaluronidases, with different, generally lower pH optima, have been found in kidney, liver, spleen, aorta, plasma, synovial fluid, bone, and fibroblasts of healing wounds. There is some evidence for the existence of glycanohydrolases which would degrade DS, KS, HS, or heparin. Such hydrolases, including hyaluronidase, produce oligosaccharides as end products. Further degradation, i.e., cleavage of terminal monosaccharide units from the non-reducing end of the chain, may occur by the alternate action of β-glucuronidase and $\beta\text{-}N$-acetylhexosaminidase, present in liver, spleen, and other tissues, but possibly not in skin or bone. The sulfated NAcGal units at the end of odd-numbered oligosaccharides derived from C-4-S and C-6-S after action of hyaluronidase and B-glucuronidase seem not to be susceptible to N-acetylhexosaminidase.

Sulfatase enzymes have been little characterized in mammalian tissues, but studies with ^{35}S-labeled chondroitin sulfates and the kinds of polysaccharides excreted in the urine indicate that desulfation occurs in vivo. Enzymes which detach the glycosaminoglycan at or near the protein linkages, i.e., β-xylosidase and β-galactosidase, as well as a general glycosylaminase, are known to exist, but identification of their action on natural substrates has not yet been made.

The amount of MPS catabolized in the human body per day has been estimated indirectly by Neufeld and Fratantoni on the basis of the amount of L-xylulose excreted in the urine of patients with the rare condition of pentosuria (Chap. 5). Since the amount of urinary xylulose (1 to 4 g.) represents the rate of turnover of the uronic acid cycle, and the major substrate flow through the cycle is probably the products of breakdown of MPS, the corresponding figure for the latter is about 3 to 12 g. per day.

Activities and Functions in Various Organs and Species

CONNECTIVE TISSUE. The mucopolysaccharides are mainly produced by or found in the mesenchymal cells which exist as fibroblasts in loose areolar stroma, layers of the skin and subdermal tissue, elastic layers of vascular walls, linings of joint surfaces (synovial membranes), and the interstices of parenchymatous organs. These cells further differentiate to form dense fibrous ligaments, matrix of cartilage, and calcified collagenous skeleton. Characteristics of connective tissue in some of these particular forms are discussed below.

In different connective tissues there are differences in the types of glycosaminoglycans, which have different functions at various sites. The patterns vary further with species and with age. Generally with aging there is a gradual decline in over-all tissue content of MPS. On the other hand, stimuli of various sorts—wounding, introduction of foreign substances or irritants into the tissues, ionizing radiation—provoke an increase of MPS owing to increased biosynthesis accompanying a generally increased cellular activity in the area. To a variety of stimuli and for all species and tissues this general response of a fibroblastic nature appears quite similar. The increase of glycosaminoglycans, in which nonsulfated polymers (hyaluronic acid and chondroitin) usually precede the sulfated ones, coincides generally with the production of collagen by the cells. In rat granulation tissue (connective tissue induced to proliferate) the half-lives of various sulfated MPS, as studied with ^{35}S, are of the order of 0.5 to 1.0 days—relatively fast compared with half-lives in other connective tissues.

The majority of acidic mucopolysaccharides are important constituents of the amorphous, extracellular "ground substance" of connective tissue. Along with collagen and elastin they support important fibrous and cellular elements and contribute to the load-bearing characteristics of anatomical surfaces. The function of the highly viscous hyaluronic acid in such sites as vitreous humor and synovial membranes is evidently to serve as a shock absorber. In many other sites it facilitates the gliding of surfaces. A hypothesis of Meyer is that the nature of the MPS in the fibrous proteins determines the architecture of the collagen bundles. Further, they may act as adhesives in cellular aggregation.

Biosynthesis of connective tissue MPS is probably regulated in some way by tensorial forces acting on the cells, not only by unusual circumstances (e.g., injury) but by ordinary daily locomotion. Thus, unstressed cells secrete the types of proteins and proteoglycans which make up the usual gelatinous ground substance, but undue tension calls forth an excess of protocollagenous material which becomes organized as fibrils or cartilage, following an outpouring from the cells of newly synthesized chondroitin sulfate. The particular responses, e.g., in wound healing, are finely adjusted to the degree of distortion of the mechanical forces normally acting on these tissues. Translated into biochemical terms, such forces may determine the degree of hydration of structures and the secondary chemical bonding of their molecular components, which may therefore be in some way involved in control of biosynthesis. That such controls are serviceable not only for gross distortions (as in injury) but also within normal physiological ranges is suggested by the ready occurrence of thinning of bone structures (osteoporosis) in immobilized individuals, e.g., bed-ridden patients or astronauts.

In view of the intense polyanionic nature of acidic glycosaminoglycans it has been supposed that they may serve as a focus for distribution and concentration of inorganic cations within tissues. Physically their action could be that of cation exchangers or chelating agents for multivalent cations. Their highly ionic character could serve to immobilize or interact with relatively large numbers of water molecules. The resultant large three-dimensional "organization" would have considerable influence on the porosity of the tissue to other solutes (see Chap. 1).

CARTILAGE AND BONE. In young and developing cartilage the major glycosaminoglycans are HA (about 35 per cent), CS (40 per cent), and DS (20 per cent). In cartilage at older sites and in older individuals the total content decreases and the percentages of HA and CS decrease, while that of KS increases, so that this kind of glycosaminoglycan may rise to 50 per cent of the total by the fourth decade in human cartilage. Whereas the C-4-S/C-6-S ratio may be about 1:1 at an early age, this ratio falls to about 1:4 or less by the fourth decade and after.

These changes with age are observed in costal cartilage and bronchial cartilage. Normal articular cartilage of joints changes relatively little with age.

Sulfated protein-polysaccharides are synthesized intracellularly and then secreted into the cartilage matrix by chondrocytes. The abundant hyaluronic acid of synovial fluid is synthesized by cells of the synovial membrane. As cartilage is replaced by metaphysial bone, most of the MPS and protein are lost. Involvement of the MPS in calcification would be expected because of their strongly anionic character, but the manner and mechanism of involvement are still obscure. Some evidence suggests that the action of the lysosomal enzymes alters the MPS in ways favorable to calcification. On the other hand, the activity of calcification has been correlated with that of formation of MPS.

The hormone calcitonin (a product of the ultimobranchial tissue related to parathyroid) causes in vitro an increased incorporation of glucose-^{14}C into MPS of embryo calf bone cells in culture; it has been postulated that the inhibitory effect of this hormone on bone resorption may be consequent to an increased capacity of the polyanionic MPS for binding of calcium. Anabolic steroids stimulate incorporation of ^{35}S into the MPS of the organic matrix of healing bone, whereas cortisone (notorious for depleting bone of calcium) markedly depresses incorporation. These steroids may exert such effects locally by their actions on carbohydrate metabolism or else on synthesis of the protein moieties of the proteoglycans. The effect of pituitary growth hormone to stimulate incorporation of sulfate into MPS of the ground substance of cartilage is mediated by a serum factor distinct from growth hormone itself and called "sulfation factors." This factor is probably directed primarily toward protein synthesis, however, so that increased synthesis of protein-polysaccharides is but one aspect of its more general action.

The biosynthesis of all MPS involved in formation of cartilage and bone appears to depend on sufficient manganese in the diet; a general lowering of content of various MPS (particularly CS) along with skeletal defects is observed in Mn-deficient animals. Ascorbic acid deficiency also seems to limit the appearance of galactosamine in newly formed cartilage.

In the connective tissue of cartilage and nucleus pulposus the soluble proteoglycans incorporate labeled glucose into the hexosamine residues prior to the insoluble glycans. Incorporation into galactosamine (of chondroitin sulfate) is more rapid than into glucosamine (of keratan sulfate), which suggests that the synthesis of each of the two polysaccharides is somewhat independent of the other. Sulfation of these heterosaccharides with ^{35}SO$_4$ occurs rapidly in vivo, suggesting attachment to a preformed precursor. There is a very long half-life of this sulfate-^{35}S in ester linkage with the saccharides of cartilage; slow turnover would be expected of such a structural component. On the other hand, the turnover time of hyaluronic acid in canine periarticular connective tissue is only about 2 days.

The fraction of CS in pig articular cartilage which is linked to glutamic acid incorporates both sulfate-^{35}S and glucose-^{14}C faster than does the serine-linked moiety. Whereas the glucose-^{14}C is taken up rather uniformly by subfractions of the CS, that of ^{35}S is relatively heterogeneous, which reflects the fact that some of the CS are more fully sulfated than other fractions. Some inorganic sulfate may be held to the esterified sulfate by cations in a chelating role.

BLOOD VESSELS. Possible relationships between MPS in blood vessel walls and the major disease conditions of atherosclerosis, hypertension, and vascular thrombosis call for careful consideration of normal type and functions and changes in MPS at this site. In spite of several theories there is no clear indication of the existence or mechanism of any etiologic relationship between the glycosaminoglycan pattern and atherosclerosis. Correlation of atheroma with changes in MPS are not definite and, moreover, evidence on the direction or nature of the changes is conflicting. Some general changes—not only with the occurrence of atheroma but with aging, with species, and with hormonal influence—can be discerned, however, and are worth noting, since they may eventually fit into more tenable theories.

In general, the fraction of hyaluronic acid in the MPS of the aorta decreases with age and also is relatively lower in atherosclerotic areas than in normal areas. The degeneration of elastic tissue in bovine coronary arteries is paralleled by a decrease of HA and an increase of highly sulfated MPS. Other changes with age are less definite. Some studies of aorta indicate a decrease in CS and an increase of DS and HS, with an increase in the amount of sulfation of the latter. However, other studies with peripheral arteries suggest decrease of both HA and HS with age and there is less HS in the atheroma-prone internal iliac artery than in the resistant external iliac. Certain other parts of the vascular tree (abdominal aorta, coronary, and intracerebral arteries) show a lower HA/CS ratio than other sites. The number of protein-polysaccharide linkages (or cross-linkages) seems to be increased for the glycosaminoglycans of old aortas.

The amount of MPS in aortas of different species seems to be proportional to the susceptibility to atherosclerosis. In early aortic lesions with fatty streaks the intimal layer shows accumulation of MPS (particularly DS) in human aorta, though in later stages with calcific plaques the MPS are very much reduced. The aortas of animal species susceptible to atherosclerosis (e.g., pig, guinea pig, chicken) contain less HS and more CS than the aortas of resistant species (hamster, rat). The aortic pattern of kinds of glycosaminoglycans in the susceptible species more nearly resembles that of embryonic arterial tissue and the intimal layer of the atherosclerotic plaque. As detected with sulfate-^{35}S, both synthesis and turnover of MPS are increased in the intimal layer of aortic atheroma of cholesterol-fed rabbits, but synthesis is more increased, which results in accumulation. Intermittent systemic hypoxia of rabbits causes increases in aortic content and rate of synthesis of sulfated MPS, which appears to be a result of repair processes. Importance of the latter in the production of atherosclerosis is therefore emphasized.

Since the HA/CS ratio is decreased in areas of atheromata and since sulfated MPS better retard movement of solutes than HA, one theory holds that a lower permeability of the aortic intima favors accumulation of lipids and other components of atheromatous plaques. It seems feasible to suggest that cholesterol in particular could become esterified to secondary hydroxyl groups of sulfate in CS or other sulfated MPS. The fact that atherosclerotic plaques show a decrease in HS could have some bearing on the availability of heparin, if the latter can possibly be derived from heparan sulfates. The plasma heparin level helps determine the activity of lipoprotein lipase. Klynstra further points out that both heparin and HS promote endogenous fibrinolysis, which tends to protect the vascular wall from fibrin accumulation. Some attempt has been made to compare the MPS of the aorta with those of platelets, which are so abundant in thrombi. Results suggest that the MPS of atherosclerotic areas resemble those of platelets more than do the MPS of unaffected areas.

When isolated segments of arterial wall are depleted of MPS by testicular hyaluronidase, sodium ion is also lost and the segments show less response to constrictor agents. Increase of Na^+ content of the wall is associated with increased constrictor response. Such findings suggest that the binding of cations by MPS is correlated with contractility, and the arterial walls of hypertensive rats do contain an excess of MPS.

Some hormonal effects on vascular MPS are at least partially delineated, while others are quite obscure because of conflicting evidence. The effect of chronic administration of glucocorticoids seems generally to be a decrease in aortic content of MPS, as well as a decrease in synthesis as indicated by decreased ^{35}S uptake and by relatively low sulfate content of the MPS. Other evidence suggests an inhibition of degradation of CS. Although glucocorticoids elevate blood lipids in both normal and cholesterol-fed rabbits, the latter seem to be protected by glucocorticoids against the aortic atherosclerosis usually induced by cholesterol feeding. In view of the positive correlation of atheromata with increased content, synthesis, and turnover of MPS, there is a strong implication that glucocorticoids protect via their inhibition of MPS metabolism. On the other hand, the changes in MPS may relate also to the tendency of glucocorticoid-treated animals, or of patients with an excess of these hormones, to general vascular fragility.

Thyroxine has long been known to protect experimental animals against atherosclerosis, but evidence on whether thyroxine increases or decreases content or metabolism of sulfated MPS is presently conflicting. Growth hormone specifically increases CS in the aortas of hypophysectomized rats, which is in accord with its anabolic effect on MPS recognized elsewhere.

The way in which insulin may be involved in regulating the sulfation of MPS in vascular connective tissue is obscure. When insulin is not sufficiently available there may be an excessive sulfation, and this depends upon the presence of growth hormone. However, in some special circumstances insulin itself conduces to excessive uptake of ^{35}S. Changes due to diabetes are also rather confusing. During an early phase following pancreatectomy of rats the sulfated MPS of the aorta are markedly increased in concentration with isotopic evidence of increased synthesis. In a later, more severely diabetic phase the concentration of sulfated MPS (but not HA) is decreased and synthetic activity is low. Findings in the early phase resemble those found with atherosclerotic vessels. As stated by the investigators Cohen and Foglia, these results imply that early diabetes might be a period of greater hazard for atherogenesis than fully developed insulin-deficient diabetes.

Vascular connective tissues of hamsters, monkeys, and humans contains increased amounts of MPS after treatment with estrogens, but this is not so for rabbits. The content of MPS of coronary arteries (cows) or aorta (rabbits) is higher during the estrogenic phase of the sexual cycle. Changes in vascular MPS as a result of giving different sex steroids are not easily correlated with atherogenic tendency, nor do different investigators agree on the extent or kind of change. It appears questionable that any change in MPS can be related to protection of premenopausal women from atherosclerosis, but this possibility should remain open to investigation.

SKIN. The majority of MPS in skin is HA with lesser amounts of DS and CS; traces of heparin and HS are also present. There is about 1 mg. glycosaminoglycan/g. of dry skin. The concentration of DS in the skin (of pigs) rises with age, while that of HA declines. The half-times for HA in the skin (rat, rabbit) are on the order of 2 to 5 days, whereas the CS turns over more slowly with half-times of 5 to 10 days. Hypophysectomy can result in doubling of the half-times for both these components. However, after hypophysectomy (rats) concentration of HA in the skin tends to increase while that of CS decreases. The effects on the sulfated MPS are in accord with the effects of growth hormone previously mentioned. Testosterone increases the synthesis and concentration of HA in the skin without any effect on the sulfated MPS. The skin of diabetic rats synthesizes various glycosaminoglycans at only about one-third the rate of normal. The HA and C-4-S are most significantly decreased with lesser changes in C-6-S, DS, and HS. Heparin, oddly, is increased. Insulin restores the pattern to normal, except that heparin remains increased.

Pituitary thyrotropin increases uptake of ^{35}S into MPS of various organs. This effect may help account for the gross accumulation of MPS in certain organs, including skin, in the course of thyroid disease. The condition is known as myxedema. A more common occurrence, which possibly has a similar etiology and mechanism with respect to abnormal synthesis of glycosaminoglycans in the retrobulbar connective tissue, is that of exophthalmos in hyperthyroidism.

With cultured skin fibroblasts about three-quarters of newly synthesized sulfated MPS are secreted into the medium and one-fourth are stored and later degraded intracellularly. In these fibroblasts there appear to be two mechanisms of degradation of MPS, as indicated by two distinct metabolic pools or types of MPS with half-lives of about 8 hours and of 3 days, respectively, according to data with ^{35}S.

LIVER. The small amount of sulfated MPS in liver is mostly chondroitin sulfate and heparan sulfate. This organ does not notably synthesize MPS, but it has a latent capacity which seems to be stimulated by injury. Thus the disease of cirrhosis and the well-known liver toxin carbon tetrachloride both promote the synthesis and accumulation of sulfated MPS. Liver lysosomes contain β-glucuronidase, β-N-acetylhexosaminedase and a hyaluronidase.

KIDNEY. The MPS are not prominent in kidneys, but are of interest in relation to some physiological and pathological changes. Because blood vessel disease in the kidney is very significant in diabetes, changes in MPS in the diabetic kidney have been explored. Advanced nodular lesions show particular increases of heparan sulfate, whereas those kidneys of diabetics with diffuse intercapillary glomerulosclerosis are typified by increase of hyaluronic acid. Under the influence of the antidiuretic hormone the interstitial cells of the kidney papillae take up much more labeled glucose into MPS. This and other evidence for increased MPS in the concentrating kidney suggest that the mucopolysaccharides could play a functional role in antidiuresis, such as trapping of cations along with water.

BRAIN. In the brain of several mammalian species are found HA, HS, CS, and DS. There is no significant heparin or keratan sulfate. Hyaluronic acid and chondroitin sulfate are the major glycosaminoglycans in brain. The various sulfated MPS incorporate ^{35}S at about the same rate in rat brain. In peripheral nerve and spinal cord the percentage of HA is higher than in brain, where sulfated MPS are more prominent than in peripheral nerve.

REPRODUCTIVE ORGANS. Estrogens induce accumulation of MPS in the endometrium and uptake of ^{35}S into the mucous membranes of the uterus, Fallopian tubes, and vaginal walls of the rabbit. Treatment with a progestagen and periods of hormone withdrawal lead to depletion of MPS. In estrogen-treated male rabbits the normal urinary bladder wall shows an increase of sulfated MPS but not in the area of a healing wound.

LYMPHATIC ORGANS. Formation of sulfated MPS by spleen, lymph nodes, and thymus is most particularly associated with the production of mast cells with their high content of heparin. Although lymph nodes and spleen take up sulfate-^{35}S more actively than thymus, the latter shows a more vigorous and prolonged increase in uptake following stimulation by glucocorticoids (in contrast to the general effect of these hormones on biosynthesis of sulfated MPS). This response of the thymus is said to be associated with its particularly high capacity to produce mast cells. Heparin in mast cells may act to bind histamine and 5-hydroxy-tryptamine (serotonin); extracellularly its functions appear to include clearance of lipemia and prevention of coagulation. Recent indications are that a heparin-like glycosaminoglycan may be naturally complexed with lipoprotein lipase.

BLOOD CELLS. Both leukocytes and platelets synthesize and contain MPS, which in each type of cell is mainly C-4-S with some HA. About 25 per cent of the hexosamine content of WBC and platelets is contained in glycosaminoglycans; the rest is in glycoproteins. The MPS of leukocytes show more polydispersion, indicating a greater variety of molecular weights and charge densities than the MPS of platelets. The HA of connective tissue has effects on platelets which include a reduction of glycolysis and a promotion of aggregation. It is believed that binding of divalent cations (e.g., Mg^{++}) by the carboxyl groups of HA could be involved in the antiglycolytic effect. These effects may be significant in relation to a common factor leading to aggregation, which is a leakage of ADP from the platelets.

LUNG. Little is known about MPS in this organ, but changes in MPS of the connective tissue components are of interest in relation to the development of pulmonary fibrosis and possibly emphysema. Experimental reduction of the normal blood supply to the lung stimulates the incorporation of ^{35}S. Since there is no increase in hexosamine content of the lung, the change seems to be an increased turnover rate for MPS. (Similar effects for aorta and liver are noted above.) These findings bear on the development of pulmonary fibrosis in various lung diseases. Possibly the changes reflect invasion of the area by WBC and mast cells as well as by fibroblasts.

URINE. In addition to study of the MPS in organs and tissues the MPS in the urinary secretion has received much attention, because the nature of MPS there tends to reveal the kinds and amount of breakdown products in normal turnover of tissue MPS, as well as the characteristics of MPS metabolism in pathological conditions, e.g., the mucopolysaccharidoses, the arthritides, and exophthalmos.

Normally most of the urinary MPS are chondroitin sulfates or chondroitin, with C-6-S being the major sulfated glycosaminoglycan. About 5 to 10 per cent of normal urinary MPS is HS and there are trace amounts of KS, DS, and HA. There is considerable variability in the rate of excretion of MPS from day to day and even more at different times of the day. About three-quarters of urinary MPS in humans are bound to polypeptides (containing predominantly serine, glycine, and glutamic acid), with one-quarter in free form. Each of the polymers has generally undergone depolymerization or desulfation or both. Heparin is excreted as a modified "uroheparin," which differs from heparin owing to the cleavage of about half of its N-sulfate groups.

Young children show a very high ratio of MPS/creatinine in the urine, which decreases until the late teens, after which the ratio stays constant. A progressive slow decline of the chondroitin sulfate/MPS ratio in urine occurs until the middle years; there is less of the non-sulfated glycosoaminoglycans (chondroitin) in the urine of children than in that of adults. Hypophysectomy in rats reduces significantly the level of MPS in the urine to about two-thirds of control level. This suggests that the higher MPS in the urine of children may be a function of growth hormone.

Special Pathological Conditions

Several pathological, as well as physiological, changes in the mucopolysaccharides have been discussed in the preceding section. There are some diseases with more obvious and particular changes in MPS which deserve special attention.

ARTHRITIDES. In osteoarthritis the affected bone and cartilage show a decrease in content of CS with some decrease in KS in advanced lesions. Yet uptake of ^{35}S into the CS of osteoarthritic cartilage is generally increased. Articular cartilage of osteoarthritic human femur displays markedly increased synthesis of not only MPS but also DNA and protein, which suggests reversion of the chondrocyte to a chondroblastic state. In rheumatoid arthritis, also, there is an increased production of hyaluronic acid per cellular unit. The increased unit turnover of MPS may be a cellular attempt to compensate for excessive breakdown of MPS due to increased activity of lysosomal enzymes. These may be derived from leukocytes called forth into the joint areas by stress, injury, or infection. The lysosomal enzymes include hyaluronidase and proteases. Hyaluronidase from leukocytes or plasma further has excessive activity on the synovial fluid of arthritic joints owing to a lowering of pH. The enzyme has a low pH optimum and is essentially inactive above pH 5. Again, the leukocytes contribute to the catabolism of MPS because their high glycolytic rate produces excessive lactic acid, which lowers the pH. As a result of the high hyaluronidase activity the chain length of CS is only about 60 per cent of that in normal areas. Possibly the excessive breakdown of protein-polysaccharides causes accumulation of products which not only favor calcification but act as irritants to the tissues. It is apparent that agents which are anti-inflammatory, e.g., glucocorticoids, would have a favorable effect on the pathologic disturbance of MPS metabolism in arthritic joints.

Aging itself does not seem to be an integral component of the arthritic process. Although costal cartilage shows changes (decrease in CS and the C-4-S/C-6-S ratio) due to aging, these changes are not seen in normal articular cartilage. Loss of vigor of repair processes due to aging could be responsible, though, for failure to compensate for breakdown of MPS promoted by stress or injury.

In rheumatoid arthritis and lupus erythematosus (a disease of joints and other tissues) there may be moderate increases in all fractions of urinary glycosoaminoglycans; the increases are noted particularly in active stages of the diseases.

MUCOPOLYSACCHARIDOSES. Some relatively rare genetic defects appear as gross distortions of normal amounts and kinds of MPS with marked clinical features. The different mucopolysaccharidoses (MPSS), like the glycogenoses, bear names of their discoverers and are commonly numbered by type as in Table 6-4. They are related in nature to the gangliosidoses (Chap. 7); some genetic conditions (lipomucopolysaccharidoses) contain features of both.

The MPSS are lysosomal diseases, in which the normal function of the lysosomes to degrade MPS is somehow deranged. Presumably this is because of the genetic loss of one or another of the glycosidases in the lysosomes, but a primary or accompanying abnormality in proteolysis of the proteoglycans is not excluded. The defect is widespread throughout the tissues, with little organ or even molecular specificity except for the particular sugar hydrolyzed by the affected glycosidase. Accompanying the loss of any one glycosidase are large increases in others, perhaps in response to the massive accumulation of MPS in the lysosomes. The accumulation of certain MPS (DS and HS) may interfere with the degradation of HA or of CS owing to inhibition of hyaluronidase. Sulfatase also

Table 6-4. *Urinary Glycosaminoglycans in Different Types of Mucopolysaccharidosis*

TYPE OF MUCOPOLY-SACCHARIDOSIS	KINDS OF GLYCOSAMINOGLYCANS IN EXCESSIVE AMOUNT IN URINE	OTHER CHARACTERISTICS
I	Dermatan sulfate, heparan sulfate (ca. 2-4:1)	Multiple physical and mental defects—severe
II	Dermatan sulfate, heparan sulfate (ca. 1:1)	Multiple physical defects—less severe
III	Heparan sulfate	Mental defects—severe
IV	Keratan sulfate, dermatan sulfate, and heparan sulfate	Multiple physical defects
V	Dermatan sulfate, heparan sulfate	Multiple physical defects
VI	Dermatan sulfate	Multiple physical defects

tends to be low in activity. Thus a multiple mucopolysaccharidosis can stem from a primary defect in degradation of certain MPS. The large excess of both HS and DS has not yet been explained in terms of a single glycosidase deficiency. Whereas β-galactosidase is commonly low in the MPSS, this does not explain selective accumulation of certain MPS or other distinctive traits of the different types, since galactoside bonds in the protein linkage region are common to all of the MPS except keratan sulfate. In any case, galactosidase is not decreased to the same extent as in generalized gangliosidosis (Chap. 7). Although β-galactosidase may be low in liver and spleen of patients with MPSS (Types I and III), skin fibroblasts and leukocytes of such patients show no deficiency of galactosidase.

Cultured fibroblasts from skin of patients with MPSS take up isotopic precursors of MPS faster than normal. An increased rate of synthesis may therefore help account for accumulation of MPS. This is not held to be a primary or only cause, however, in view of the observed changes in the hydrolases and the abnormal structure of the polysaccharides. The polysaccharides have, if any, only remnants of attached protein; the carbohydrate polymers are themselves smaller than normal.

Six main types (I to VI) of MPSS are now recognized and differentiated on clinical, genetic, and biochemical grounds. All are characterized by excessive secretion of one or more MPS in the urine (Table 6-4). The different types show various combinations of some of the following clinical abnormalities: mental retardation, dwarfism, bone and joint changes, skin changes, hepatomegaly and splenomegaly, aortic regurgitation, and corneal clouding. Type I (Hurler's syndrome) is more clinically severe than Type II (Hunter's syndrome). Whereas the proportion of DS/HS in the urine in Type I is 2-4:1, in Type II it is about 1:1. In Type III, in which physical defects are mild but mental retardation is severe, the urinary glycosaminoglycan is almost entirely HS. Types I and III are transmitted as classical autosomal recessive traits; Type II is sex-linked, in the manner of hemophilia.

In the liver of patients with MPSS there is more typically an accumulation of HS, whereas spleen shows more DS. Cultured skin fibroblasts often show large intracellular accumulations of DS or total MPS, depending on type, but the rates of excretion into the medium continue to be normal. An abnormal storage pool or form of MPS is therefore postulated to be separated from a smaller and normal secretory pool. However, another theory holds that a surplus of protein-denuded DS from fibroblasts is eventually transported to liver or spleen, where further action of hyaluronidase and β-glucuronidase produces the relatively shortened carbohydrate polymers found there and in the urine. The accumulation of HS and DS in bone, which does not normally contain these glycosaminoglycans, is supposed to be due to transfer from other tissues.

The white blood cells of patients with MPSS as well as individuals heterozygous for these inherited diseases display abnormal metachromasia in tissue culture, which is indicative of excessive occurrence of MPS. Homozygous and heterozygous individuals can be distinguished by the rate of disappearance of such metachromasia. These phenomena of metachromasia are also found in cultured fibroblasts.

A curious finding is that cells (or culture medium therefrom) of a normal genotypic line can correct in vitro the defect of an abnormal type. Two abnormal cell lines (e.g., Types I and II) can also correct each other's defect. The corrective factor is heat labile and macromolecular. The necessary degradative enzymes which are possibly lacking may be induced or prevented from loss to the cell by the corrective factor. In the sex-linked Type II the mothers of homozygotes are not only genetic carriers but examples of "autotherapy" in which the normal half of the somatic cell population provides to the abnormal half the corrective factor which operates in the case of mixed cultured fibroblasts.

Accurate classification based on these in vitro tests is useful for genetic counseling of heterozygotes. Fibroblasts from patients with Types I and V do not correct each other's metabolism of MPS, when mixed together, so evidently they share some common defect. In addition, they excrete similar proportions of DS and HS (Table 6-4).

The abnormality of quantity and type of MPS in the amniotic fluid may be used to diagnose certain MPSS prenatally. In Type I, for example, the amniotic MPS may be largely HS, which is undetectable in normal amniotic fluid, in which HA predominates.

OTHER DISORDERS. Cystic fibrosis, or mucoviscidosis, is an uncommon recessive disease in full homozygous expression but heterozygotes may include 2 to 5 per cent of the population. Abnormality of secretions of many exocrine glands includes changes in content of mucopolysaccharides and glycoproteins. Fibroblasts from such patients may show normal or else markedly increased MPS, suggesting two classes of cystic fibrosis. When MPS are increased the relative proportions of HA, DS, and CS are about normal. Unlike the MPSS, in cystic fibrosis the uptake of ^{35}S into cultured skin fibroblasts is normal, and, also unlike the MPSS, there is excessive secretion of MPS into the medium. In Marfan's syndrome, a rare disorder of generalized connective tissue, accumulation of HA in cultured skin fibroblasts has been noted.

Vitamin A has effects on metabolism of MPS. Although not well understood, one effect seems to be a decrease in membrane stability with release of proteolytic enzymes from lysosomes. Deficiency or excess of Vitamin A may decrease or accelerate uptake of ^{35}S, but the effect on catabolism is more definite.

GLYCOPROTEINS

As described extensively in Chapter 1, the glycoproteins are different from proteoglycans (i.e., MPS) in having multiple branched, relatively short oligosaccharide chains with a greater variety of monosaccharides and without repeating disaccharide units. The protein moiety is generally more prominent than in proteoglycans. An outer sequence, sialic acid (or fucose) → galactose → N-acetylglucosamine, linked to an inner core of about 3 mannose and 2 NAcGlu residues, is a general structural pattern common to many glycoproteins, e.g., those of serum and of basement membranes. Blood-group–active glycoproteins contain Gal, NAcGal, NAcGlu, N-acetyl- or N-glycolylneuraminic acid and fucose but no mannose.

Chemical Reactions

SYNTHESIS. The intracellular anatomical locale for progression of a stepwise, membrane-bound course of biosynthesis of carbohydrate components of glycoproteins is much like that for carbohydrates in proteoglycans. The polypeptide synthesis is relatively rapid and seems clearly to precede that of the polysaccharide portion, which takes about 15 to 20 minutes as determined for secreted glycoproteins with labeled precursors. The initial attachment of NAcGlu to an asparagine residue of the polypeptide (for some glycoproteins) or to serine or threonine residues (for others) may take place prior to or just following detachment of the polypeptide from the ribosomes. Most of the monosaccharides are added by a series of glycosyltransferases on the microsomal rough and smooth membranes comprising the endoplasmic reticulum (Fig. 6-2). These enzymes transfer sugars from their nucleotide derivatives to appropriate acceptors. Specificity is directed toward both the base and the sugar moieties of the sugar nucleotide and to the acceptor molecule in various ways. Not only the terminal sugar (site of direct attachment) but the penultimate sugar of the acceptor as well as the polypeptide portion — particularly the sequence of amino acids surrounding the linkage site — are significant in determining affinity of the enzyme for the acceptor.

As depicted in Figure 6-2, the Golgi region seems to be an arrival point for glycoproteins (GP) after their synthesis or passage through the endoplasmic reticulum; some transferase activity also may occur in the Golgi region. The completeness of action of the glycosyltransferases may not be absolute, since this process is not directly template-controlled as is that of polypeptides by RNA. This would account for some molecular variations in chain length or numbers of chains — the microheterogeneity which is recognized to characterize glycoproteins as well as proteoglycans. Positional isomerism could be due to lack of specificity of transferases or to groups of transferases with different isomeric specificity. The extent to which partial degradation coincident with biosynthesis may also add to heterogeneity is not known.

Separate glycosyltransferases control assembly of the common tri-

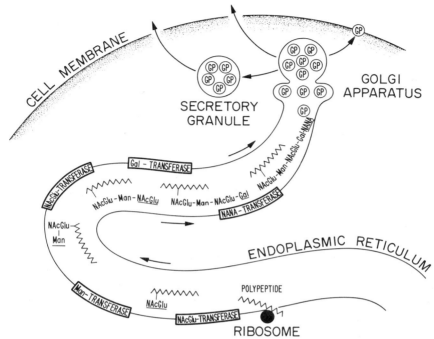

Figure 6-2. Stylized mode of intracellular attachment of sugars in biosynthesis of a typical glycoprotein. (Modified from Spiro, R.: New Eng. J. Med. *281*:991, 1969.)

saccharide sequence, NANA → Gal → NAcGlu. These enzymes are found in liver (for synthesis of fetuin, α_1-acid glycoprotein, and so forth), thyroid (thyroglobulin) and mammary gland (milk oligosaccharides as well as glycoproteins). Another series of 5 glycosyltransferases found in porcine submaxillary gland is presumably responsible for sequential formation of the branched oligosaccharide of the blood-group-specific glycoprotein in this organ.

The properties of some other individual transferases have been investigated. The polypeptide; UDP-NAcGal transferase for bovine submaxillary (BSM) gland glycoprotein requires Mn^{++} or Co^{++}. The K_m values for Mn^{++}, UDP-NAcGal, and receptor have been measured. The transferase for attaching galactose to hydroxylysine residues of collagen also requires Mn^{++} or Co^{++}. Similar transfer of Gal from UDP-Gal to acceptor protein of bovine retina requires Mn^{++}, but that of NANA does not. As expected for the final stages of the oligosaccharide synthesis, various sialyl transferases are found in the smooth membranes and the Golgi apparatus and tend to be more soluble enzymes than other transferases. Transferases for NAcGal or Gal found in gastric mucosa or milk of women of blood group A or B, respectively (both transferases if the individual is type AB), are active only on those receptor oligosaccharides which contain the 2'-fucosyl-galactose structure at the non-reducing end of the chain. Although glycosyltransferases for synthesis of glycoproteins usually prefer or require acceptors of high molecular weight, some CMP-sialyl transferases from rat mammary gland and from colostrum will utilize lactose as acceptor. Possibly a resultant nucleotide-trisaccharide complex can be utilized in the synthesis of glycoproteins. Mammals and birds are known to secrete UDP-disaccharides and UDP-trisaccharides, e.g., in goat colostrum.

There is evidence that during the process of incorporation of mannose from GDP-mannose into secreted glycoproteins (e.g., of liver) an intermediate mannose-lipid compound may be formed. There is no mannose in glycolipids (Chap. 7), the properties of which are quite different from that of the mannolipid. Two mannosyl transferases appear to be involved—one enzyme synthesizing the mannolipid and another transferring mannose from lipid to glycoproteins. Some preliminary evidence suggests that vitamin A may function to carry monosaccharides in a sugar-lipid complex which is intermediate in the formation of glycoproteins. In vitamin A deficiency there is a marked decrease in synthesis of glycoproteins at various body sites. Changes in metabolism of glycosaminoglycans with either deficiency or excess of vitamin A, as previously noted, may relate to this finding.

What determines that an oligosaccharide will not be further enlarged is not entirely clear. Evidently the occurrence of either sialic acid or fucose as the terminal unit concludes the biosynthesis of some glycoproteins. For the blood-group substances and other glycoproteins, in which sialic acid or fucose is attached as a branch to the main chain, a signal for conclusion of synthesis may be the anomeric configuration of the final residue—which is uniquely α-D for blood-group substances A and B.

Very little is known about cellular regulation of the rate or ordering of accumulation of the oligosaccharide chains in glycoproteins. Whether substrate or cofactor availability, enzyme synthesis, or other factors govern the production of glycoproteins is still mostly speculative. In the mammary gland there is the presently unique example of a shift from synthesis of glycoproteins to that of lactose by the influence of α-lactalbumin on the acceptor specificity of galactosyltransferase. The regulator protein, α-lactalbumin, appears during lactation.

Although some glycoproteins are contained in mitochondria and in fact can be synthesized on the inner mitochondrial membrane, most develop via the route of Figure 6-2 to become situated on the plasma membrane or else secreted. The secreted glycoproteins are generally quite different from those of the plasma membrane, which contain little, if any, mannose, have a high NAcGal/NAcGlu ratio and typically are linked to the polypeptide via serine and threonine rather than asparagine.

Most extracellular proteins are glycoproteins, while most within the cell are not. This has led to the theory that the carbohydrate of secreted glycoproteins may be added for the purpose of facilitating their excretion from the cell. If so, there could be a specific interaction of the carbohydrate moiety with a membrane receptor analogous to the transport into cells of monosaccharides via specific interaction with a membrane carrier (Chaps. 2 and 3). Eylar supposes that glycoprotein secretion could be a primitive evolutionary device which in certain cells and for some secreted proteins has been superseded by a mechanism involving formation of globules or zymogen granules. Such proteins (e.g., insulin, pancreatic, and salivary enzymes) are not carbohydrated, but the membrane of the globule or sac which transports the protein destined for secretion is itself a glycoprotein. The granular membrane interacts with the plasma membrane in the process of discharging its contents to the extracellular environment. Ribonuclease B and a minor fraction of serum albumin, which contain a small amount of hexose, may represent transition forms or vestiges of more primitive molecular types. Eylar postulates that some other noncarbohydrated extracellular proteins, like glucagon, ACTH, growth hormone, and serum albumin, may be secreted via a granular mechanism and that such a mechanism may provide for more rapid secretion in response to physiologic needs.

DEGRADATION. As with the proteoglycans, the lysosomes in liver, kidney, and several other organs are the sites of intracellular catabolism

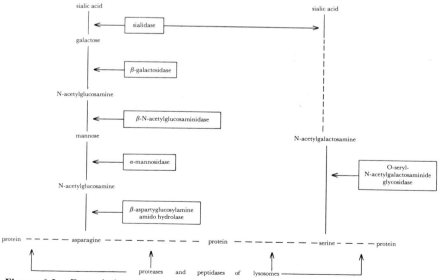

Figure 6-3. Degradation of hypothetical glycoprotein by lysosomal enzymes. (From Mahadevon et al.: Archiv. Biochem. and Biophys. *129*:529, 1969.)

of glycoproteins by an array of glycosidases, which can sequentially break down the carbohydrate chains. The pattern of glycosidase activity on a hypothetical glycoprotein resembling fetuin and some other glycoproteins is indicated in Figure 6-3. Proteases in the lysosomes hydrolyze the polypeptide bonds and thereby aid in the degradation of the carbohydrate chain by glycosides. The enzyme which splits the NAcGlu-asparagine linkage is an amidase rather than an N-glycosidase. Glycosylamine may be formed, followed by deamination to free ammonia and NAcGlu. Other evidence suggests that the peptide portion of the glycoprotein is entirely degraded before the carbohydrate is released from this attachment. In contrast to the order of degradation for the type of glycoprotein with the asparagine linkage, the N-acetylgalactosaminidase which splits NAcGal from serine or threonine works optimally when the peptide chain or part of it is still present.

Activities and Functions in Various Organs and Species

The biological functions of glycoproteins, including enzymes, hormones, antibodies, lubricants, metal transporters, clotting agents, and so forth, are so diverse that a unifying concept of the role of the carbohydrate moiety is an unlikely prospect. A common function as a means for cellular extrusion has been mentioned for secreted glycoproteins. Determination of molecular tertiary structure and degree of hydration would seem to be possible common effects. In layered membranes the carbohydrate portions of glycoproteins probably determine density of packing and porosity as well as active transport of some electrolytes. Sialic acid in mucous glycoproteins obviously helps confer useful viscosity to the secretions of the respiratory and digestive tracts and other mucosal surfaces. Moreover, sialic acid seems essential to the antigenic properties of the M and N blood-group substances and appears to be involved in the function of the glycoprotein hormones, such as human chorionic gonadotropin. However, loss of the sialyl residues may just allow rapid degradation in vivo, so that hormonal activity is effectively lost. Similarly, the loss of sialic acid from fibrinogen may permit more rapid clotting by change in stability of the molecule. Loss of sialic acid is known to make some glycoproteins more susceptible to proteolytic digestion.

LIVER. Glycoprotein synthesis in liver is particularly important because it is the site of synthesis of many serum glycoproteins. Liver microsomes, mainly of the reticuloendothelial cells, contain incomplete glycoproteins which lack sialic acid but are otherwise immunologically like the serum acid glycoproteins. When ^{14}C-labeled fucose is injected intraperitoneally into rats the glycoproteins of liver, small intestine, and serum become most highly labeled. Those of serum are much less labeled if the animal is hepatectomized, but some degree of synthesis of plasma glycoproteins appears to occur in sites other than liver.

Various changes in serum glycoproteins (discussed below) should be viewed as possible changes occurring in hepatic biosynthesis of glycoproteins. One important group of these consists of the plasma low-density lipoproteins (LDLP), which are both glyco- and lipoproteins. To what extent general changes in carbohydrate metabolism in liver can influence the synthesis, structure, and metabolism of these glycolipoproteins and their accumulation and turnover in liver and plasma is a subject for future investigation. Some influences are presently known. Experimental choline deficiency decreases incorporation of glucosamine-^{14}C into hepatic glycoproteins of the smooth microsomes and also into plasma LDLP. It is the postribosomal attachment of the sugar residues which is affected, since choline deficiency does not decrease the incorporation into the protein of a labeled amino acid (leucine). Conversely, phosphorylcholine has a strong stimulating effect on glycoprotein biosynthesis in liver slices. Conceivably phosphorylcholine could affect the phospholipids of the microsomal smooth membrane, but the total amount of these membrane phospholipids remains the same.

Another influence is that of ethanol; in rats excess ethanol decreases synthesis of hepatic glycoproteins and plasma LDLP. Again, the incorporation of labeled glucosamine, and apparently the synthesis of the intermediate UDP-N-acetyl-hexosamine, is impaired without interference with general synthesis of liver or plasma proteins. Probably relevant to these influences is the occurrence of fatty liver with either choline deficiency or chronic ethanol administration.

PLASMA. The various plasma glycoproteins have been enumerated and chemically described in Chapter 1. Although most of the serum glycoproteins, and fibrinogen in plasma, originate in the liver, those of the gamma-globulin fraction come largely from lymphoid cells. The turnover of plasma glycoproteins has been considerably studied. Homologous plasma proteins, labeled with ^{14}C in NAcGlu and sialic acid moieties, disappear from the circulation of rats at a rate of about one-thirtieth of the total plasma amino sugar per hour. Slow and fast components of turnover of glucosamine-labeled serum glycoproteins have been defined in sheep. The fast component has a half-life of about 12 hours and the slower component $t_{1/2}$ of about 135 hours. These components may represent two main groups of glycoproteins. Half-times of various serum glycoproteins have been found in certain species to be within a range of 5 to 10 days and generally shorter than that of serum albumin.

The half-time of orosomucoid is approximately 5 days in normal man and does not change with chronic inflammatory condition or parenchymatous hepatic degeneration. Nevertheless, in certain chronic diseases (e.g., diabetes mellitus, arthritis, neoplasms), during inflammatory reactions and following injury (traumatic or surgical) there are considerable increases in serum concentration of several serum glycoproteins, particularly those in the α electrophoretic component. Moreover, there are some indications that increased biosynthesis (probably hepatic) is responsible. Following injury in man the half-time of serum haptoglobin, as studied with glucosamine-^{14}C, is decreased from the normal value of 4 or 5 days to about half of that and the absolute turnover increased several-fold. A concomitant study of incorporation of methionine-^{35}S into the same glycoproteins has indicated that the carbohydrate moiety may be turned over faster than the polypeptide portion after injury.

The general rise in plasma glycoproteins in response to various stress situations appears to be mediated by adrenal glucocorticoids. Administration of cortisol causes elevations of glycoproteins like those found after trauma. It is said that the plasma accumulation of α-2-glycoprotein is so specific, sensitive, and linear a function of the amount of anti-inflammatory glucocorticoid present that it may serve as a highly valid measure of glucocorticoid activity. The biochemical mechanism by which the steroid or any other factor, perhaps from damaged tissue, operates to stimulate glycoprotein synthesis is quite obscure. There is some evidence that in serum of injured rats there is a specific humoral factor (sometimes called "mucopoietin") which increases the solubility state of serum glycoproteins from non-injured rats.

After various kinds of renal damage there is an enhancement of incorporation of labeled glucosamine into plasma glycoproteins. The mechanism is not yet known, but it is postulated that normally an inhibitor of synthesis of such glycoproteins is elaborated by the kidney and damage to the organ reduces production of the inhibitor.

The pattern of responses of the glycoproteins may be fairly specific to the type of disease or stress. For instance, the α_1-acid glycoprotein during certain chronic diseases typically lacks some of its sialic acid residues. After surgical injury there are rises in the α_1-acid glycoprotein, haptoglobin, fibrinogen "C-reactive" protein, and ceruloplasmin, but not α_2-macroglobulin. After physical exercise all the above, including α_2-macroglobulin, and also transferrin, α_1-antitrypsin, and prealbumin increase. In pregnancy and with the use of contraceptive steroids there are found increases in fibrinogen, ceruloplasmin, transferrin, and some α-globulins, but a decrease in α_1-acid glycoprotein. The latter glycoprotein has the capacity for binding progesterone and other steroid hormones; biological activity of the hormones is reduced by this binding. The interaction of progesterone and α_1-acid glycoprotein has been used as a model of binding of cortisone to cortisone-binding globulin.

In the alloxan-diabetic rat the incorporation of labeled glucose into hexosamine of serum glycoprotein proceeds at a normal rate, even though glycogen synthesis is virtually abolished. This does not bear out the postulation that inhibited utilization of glucose by the insulin-dependent glycogenic and other pathways would promote excessive utilization by insulin-independent routes, including glycoprotein synthesis (see

Chap. 5). However, the biosynthesis of glycoproteins other than those in serum, e.g., in retina and kidney, may be affected, and the increase of serum globulins (particularly α-globulins) commonly observed in human diabetes suggests the need for more such study.

BLOOD CELLS AND BLOOD-GROUP SUBSTANCES. Red cell membranes are rich in glycoproteins, which are no doubt important in transport function, in surface interaction between cells, and in antigenic properties. The identification of certain glycoproteins on red cell membranes as the "blood-group substances" (BGS) acting as agglutinogens in blood type agglutinations is well recognized (Chap. 1). It is the inheritance of glycosyltransferases which determines the formation of the particular BGS in individuals. Analogous enzymes determine glycolipids in red cell membranes, the carbohydrate units of the soluble glycoproteins in mucous secretions, and the oligosaccharides of milk and urine.

The H gene for determining a certain type of BGS (e.g., Type O) is expressed in the form of a fucosyl transferase which adds the non-reducing terminal fucose residue to the precursor substance. The resultant H-substance may serve in turn as substrate to produce either A or B substance (or both) by transfer of NAcGal, or Gal, respectively, to the end of the chain (Fig. 1-9). Sometimes A and B activities in an AB blood-group individual seem to be inseparable, owing probably to the occurrence of both types of oligosaccharide attached to the same polypeptide core. Le^b-activity may also be attached to the same molecule.

Because of the genetic independence of the four major blood-group systems (ABO, Hh, LeLe, and SeSe) there is much heterogeneity of these glycoproteins; part of this is associated with variability in sialic acid residues. Because of the interactions of the various gene types for blood groups individuals of certain blood types may be different from either parent, having inherited determinants of two enzymes, one from each parent, which combine to form a "gene interaction product." Type Le^b, characterized by the occurrence of two specific L-fucosyltransferases, is an example.

The density of sialic acid residues on the surface of red cells may determine their surface charge, consequently their adhesiveness to other cells. This may determine their survival time; in fact, as red cells age the number of sialic acid units per surface area declines.

Human peripheral blood lymphocytes in culture synthesize glycoproteins at a low rate. The basal rate can be increased up to 10- or 20-fold by an antigenic type of stimulus. This recent observation suggests interesting possibilities for in vitro definition of allergic phenomena and other immunologic relationships. The phenomenon of intercell recognition clearly depends on the sugar moieties; if certain sugars (e.g., fucose, NAcGal) are removed from lymphocyte membranes by glycosidases their general function remains unimpaired but they do not "find their way" as easily from the blood to the lymphoid tissue. Tumor cells show changes in normal proportions of neutral and amino sugars and sialic acid in their membrane glycoproteins; this has prompted the hypothesis that such cells therefore have diminished capacity for contact inhibition and intercell affinity, which could relate to their tendency to metastasis (wandering).

Glycoproteins of platelet cell membranes are known to be distinctive from those of the erythrocyte and to contain a variety of complex heterosaccharide units. A glycoprotein isolated from porcine blood platelets is said to have a heterosaccharide structure similar to that of Le^a blood-group substance as found in ovarian cyst fluid. The blood platelet membranes may differ from other cellular outer (plasma) membranes by being derived from the inner, smooth membrane of its parent cell, the megakaryocyte of the bone marrow. The platelet membranes possess influenza virus receptivity, which generally means the occurrence of NANA as available end-group. Such residues are very likely responsible, at least in part, for the surface charge of platelets, which is a fundamental factor in determining the processes of aggregation and subsequent blood coagulation. In analogy to the interaction of serotonin with NANA residues on the surface of smooth muscle and nerve cells, it is highly probable that sialic acid residues on platelet membranes are involved in the uptake of serotonin by platelets, in which this amine is highly concentrated. Also in analogy to smooth muscle this may relate to the contractile characteristics of platelets.

CONNECTIVE TISSUE. The ground substance of connective tissue contains numerous glycoproteins as well as proteoglycans. Keratan sulfate in cornea and cartilage is perhaps better considered a sulfated glycoprotein than a mucopolysaccharide in view of the kinds of its sugars and their branching structure. Some connective tissue proteins in skin and tendon closely resemble serum proteins, e.g., albumin or gamma globulin, but contain more carbohydrate. One inference is that these proteins, in fact, do derive from serum proteins, to which carbohydrate has been added in order to facilitate their transport into tissues.

Specific glycosyltransferases are responsible for the attachment of galactose or glucosyl-galactose disaccharide moieties to a variable percentage of the hydroxylysine units in the peptide chain of collagen. The special collagen of "basement membranes," which line capillaries, glomeruli and tubules of kidneys, alveoli of the lung, follicles of the thyroid gland, components of the eye, and so forth, contain essentially only the disaccharide units. Another series of glycosyltransferases link to asparagine various other sugars in a heterosaccharide chain found in some collagens, particularly that of glomerular basement membrane.

In diabetes mellitus there is a thickening of the glomerular basement membrane which is accompanied also by relative increases in both the amount of hydroxylysine and the numbers of disaccharide units linked to these amino acids. This may relate to the increased porosity of these abnormal membranes, which allows albumin and other proteins to escape in the urine. Presumably there is a change in the structural ordering of the peptide chains. Whether the synthesis of the disaccharide chains plays any role in this pathology is quite unknown. Alloxan-diabetic rats have increased activity of the glucosyltransferase which attaches glucose to the galactosyl-hydroxylysine unit. Whether this transferase is induced to higher activity by elevated levels of glucose is an important question.

In wound areas there is an accumulation of plasma-derived fucoglycoproteins. Such macromolecules appear to accumulate as the result of thrombic blockage of the circulation. There are several glycoproteins in the aortic wall; one of these is closely associated with elastin. In bovine heart valves the glycoproteins appear to be more rapidly synthesized than the proteoglycans. Biosynthetic activity of the aortic valve is higher than that of the pulmonary valve — possibly because of greater stress put on the aortic valve for the left ventricle.

SALIVARY GLANDS. Glycoproteins provide needed viscosity to the saliva, which must intermix, penetrate, and provide adhesive medium for ingested food during mastication and digestion. Different intensities of salivary secretion (in the dog) can produce variations in the content of fucose and NANA in the glycoproteins secreted from the submaxillary or parotid glands. With lower intensity the submaxillary glycoproteins contain generally more NANA and less fucose than those secreted at higher rates of salivation. However, the parotid glycoproteins have a higher concentration of NANA when secreted at higher rates. The parotid glycoproteins contain a distinctly higher mannose/galactose ratio than submaxillary glycoproteins.

Salivary glycoproteins in incubated oral saliva undergo fairly rapid glycosidic cleavages with total loss of fucose and sialic acid and loss of about half of the hexosamine. This may be due to oral bacteria or to other conditions in the mouth — no loss occurs if the saliva is collected directly from the secretory ducts. The low gastric pH further predisposes to loss of sialic acid from swallowed saliva as well as to slow loss of fucose and galactose. The occurrence of glycosidases in the intestinal tract is speculative. Possibly they may occur in the brush border of the cells in analogy to disaccharidases (Chap. 2), or bacterial flora may be responsible for degradation of the oligosaccharide chains of glycoproteins ingested or secreted into the gastrointestinal tract.

GASTROINTESTINAL MUCOSA. Intestinal mucins are largely glycoproteins rather than proteoglycans. Precursor glucosamine in vivo in the rat appears progressively in the course of 2 or 3 hours in the microsome, mitochondria, brush border, and luminal glycoprotein. Besides a component of rapidly labeled and secreted glycoprotein, there is another more stable intracellular glycoprotein in the mucosal cells, which is slowly labeled and not apparently secreted. Secretion of mucosubstances into the stomach antrum by the mucosa is inhibited by glucocorticoids.

This may be relevant to the tendency for excessive glucocorticoids to predispose to the occurrence of gastric or duodenal ulcer.

BRAIN. About 5 to 10 per cent of the total proteins of brain are glycoproteins. These glycoproteins contain about half of the sialic acid in brain (the rest in glycolipids). During the first few days after birth the brain (of the mouse) incorporates ^{14}C-labeled fucose and glucosamines intensively into two main fractions of glycoproteins and after that time the incorporation is not prominent. The brain glycoproteins are distributed in the membranes of various subcellular structures, particularly the synaptic regions. Just as glycoproteins at the cell surface in many other cells determine interactions with the environment, so also do glycoproteins at the surface of nerve cells. Brunngraber has commented that the great heterogeneity plus the very high specificity of the carbohydrate components of such surface-active brain glycoproteins (and glycolipids) could determine vital functional as well as structural characteristics of the central nervous system, perhaps indeed lie at the biochemical basis for specificity of synaptic transmission, including such functions as memory.

URINE. A number of biologically active glycoproteins, e.g., gonadotropins, are found in normal urine. Also there are low-molecular-weight glycopeptides, some of which appear to be derived from blood-group substances and have characteristic serological activity. There are hydroxyproline-containing glycoproteins which may arise from collagen. A glycopeptide with novel carbohydrate-protein cross-linkage, digalactosylcysteine, has recently been identified in normal human urine.

Special Disorders

Pathological changes occurring in diabetes and other chronic diseases and in acute inflammation or injury have been mentioned. Some rare diseases, which show accumulation of particular glycoproteins as well as glycolipids, seem to have their genesis in deficient glycosidases of particular type, e.g., β-galactosidase or α-fucosidase. More specifically for glycoproteins there are two lysosomal deficiencies, that of mannosidase and aspartylglucosaminidase, which are tentatively designated as Types I and II glycoprotein storage diseases. Type I, with storage in liver and brain of mannose-containing residues of glycoprotein, and Type II, with aspartylglucosaminuria, are characterized clinically by progressive psychomotor retardation and severe mental retardation, respectively. However, some cases of glycoprotein storage disease may show minimal clinical evidence, e.g., only splenomegaly.

OLIGOSACCHARIDES

The chemistry of the oligosaccharides in milk has been described in Chapter 1. Lactose, the disaccharide abundant in milk, is formed from

UDP-galactose and glucose in the mammary gland. As previously mentioned in Chapter 5, the glucose moiety derives mainly from the blood glucose while the galactose originates de novo from small precursors in the mammary gland. Lactose synthetase catalyzes the transfer of galactose from UDP-Gal to glucose.

The enzyme is resolvable into two protein components, A and B, which are inactive for lactose synthesis individually. Either the A or the B proteins from milk of cow, sheep, goat, or human are interchangeable for purposes of enzyme activity, although the B proteins are antigenically different. The A protein alone is able to transfer galactose to N-acetylglucosamine instead of to glucose. A similar enzyme from rat liver forms the same disaccharide structure. When either the A protein or the liver enzyme acts in the presence of B protein (identical with α-lactalbumin), lactose is formed in preference to N-acetyllactosamine. Probably during lactation the function of α-lactalbumin or B protein is to switch synthesis of glycoprotein to that of lactose. In the mammary gland, which is the only organ that can synthesize α-lactalbumin, there is a rapid rise in the latter after parturition. In vitro the syntheses by mammary gland of A and B proteins (along with other proteins) are synergistically promoted by insulin, hydrocortisone, and prolactin.

Fucose and sialic acid contained in other oligosaccharides of milk appear to derive in parallel with galactose from common precursors (e.g., glycerol, fructose) (see Chap. 5). As in the synthesis of oligosaccharide units of glycoproteins, the nucleotide-linked precursors, GDP-fucose and CMP-NANA, are intermediates in the formation of milk oligosaccharides. The higher oligosaccharides of milk (Chap. 1, Fig. 1-6) are formed by sequential transfer of glycosyl units from sugar nucleotides. Nucleotide-linked oligosaccharides occur in milk but their role, if any, in the synthesis of free oligosaccharides is not known.

The presence in human milk of 2'-fucosyllactose is found only in secretors of soluble blood-group substances, regardless of blood type. This trisaccharide evidently is formed by a fucosyltransferase like that which forms A,B, or O(H)-type soluble BGS; this enzyme is absent or inactive in non-secretors of BGS, i.e., secretors of Le^a-type glycoprotein (Chap. 1, Fig. 1-9). The specific fucosyltransferase is, in fact, found in soluble form in human milk from BGS secretors but not the milk of non-secretors.

The serum contains virtually undetectable concentrations of oligosaccharides, but no doubt very small amounts are present to account for the appearance in urine of small but definite amounts. Normally, human urine contains about 60 mg. of dialyzable hexosamine per 24 hours. Generally this is more than in the non-dialyzable fraction, i.e., glycoproteins or proteoglycans.

The urine may contain three groups of oligosaccharides. One of these includes the neutral, glucosamine-containing oligosaccharides,

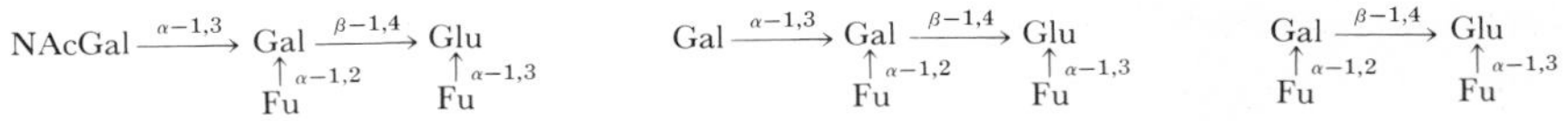

$$\text{NAcGal} \xrightarrow{\alpha-1,3} \text{Gal} \xrightarrow{\beta-1,4} \text{Glu} \qquad \text{Gal} \xrightarrow{\alpha-1,3} \text{Gal} \xrightarrow{\beta-1,4} \text{Glu} \qquad \text{Gal} \xrightarrow{\beta-1,4} \text{Glu}$$
$$\begin{array}{cc} \uparrow_{\alpha-1,2} & \uparrow_{\alpha-1,3} \\ \text{Fu} & \text{Fu} \end{array} \qquad \begin{array}{cc} \uparrow_{\alpha-1,2} & \uparrow_{\alpha-1,3} \\ \text{Fu} & \text{Fu} \end{array} \qquad \begin{array}{cc} \uparrow_{\alpha-1,2} & \uparrow_{\alpha-1,3} \\ \text{Fu} & \text{Fu} \end{array}$$

Type A_1 \qquad\qquad Type B \qquad\qquad Type O

Figure 6-4. Proposed structure of oligosaccharides in urine of blood group-characteristic secretors. (From Lundblad, A.: Biochim. Biophys. Acta *165*:202, 1968.)

which are identical with those in milk and which are present in the urine of pregnant and lactating women. Another is a series of fucosyl oligosaccharides which are related to blood-group substances and secretor status. Certain ones have been, at least tentatively, identified and, as seen in Figure 6-4, are typical for the blood-group type. These BGS-type oligosaccharides are distinct from milk oligosaccharides in the urine in that the latter contain NAcGlu as the only hexosamine, whereas the BGS type contain NAcGal. However, the origin of these oligosaccharides does not seem to be from the soluble BGS, because the latter do not contain glucose. They may be derived from or be related to the glucose-containing, blood-group-active glycolipids present on the surface of red blood cells. A larger amount of fucosyl oligosaccharides is secreted in the urine after a fucose-containing meal, particularly in BGS secretors. A third group of urinary oligosaccharides contains NANA. The latter may be attached to lactose (at C-3 or C-6 of Gal), *N*-acetyllactosamine (at C-6), or galactosyl-*N*-acetyl-galactosamine (at C-3 of one or both components). These NANA-containing oligosaccharides are fairly constant in human urine, not related to blood-group type and not influenced by dietary conditions except that total fasting for several days decreases the output.

Although hepatic oligosaccharides of glucose have been supposed to be breakdown products of glycogen, with glucose-^{14}C as precursor it appears that they may be formed from UDP-glucose prior to or simultaneously with formation of glycogen. The kinetics of ^{14}C incorporation suggest that the oligosaccharides are neither precursors nor breakdown products of glycogen. The physiological significance of these hepatic oligosaccharides, which are increased in the livers of starved animals, remains to be investigated.

REFERENCES

For Glycogen Metabolism

Darnton, S. J.: Glycogen metabolism in rabbit kidney under differing physiological states. Quart. J. Exptl. Physiol. *52*:392–400, 1967.

Goldberg, N. D., and O'Toole, A. G.: The properties of glycogen synthetase and regulation of glycogen biosynthesis in rat brain. J. Biol. Chem. *244*:3053–3061, 1969.

Hers, H. G., DeWulf, H., and Stalmans, W.: The control of glycogen metabolism in the liver. F.E.B.S. Letters *12*:73–82, 1970.

Hultman, E.: Studies on muscle metabolism of glycogen and active phosphate in man with special reference to exercise and diet. Scand. J. Clin. and Lab. Invest. *19*:Suppl. 94, 1–60, 1967.

Leloir, L.: Regulation of glycogen metabolism. National Cancer Institute Monograph 27:3–18, 1967.
Nelson, S. R., Schulz, D. W., Passoneau, J. V., and Lowry, O. H.: Control of glycogen levels in brain. J. Neurochem. 15:1271–1279, 1968.
Ryman, B. E., and Whelan, W. J.: Dual function and common identity of proteins in glycogen metabolism: a hypothesis. F.E.B.S. Letters 13:1–4, 1971.
Smith, E. E., Taylor, P. M., and Whelan, W. J.: Enzymic processes in glycogen metabolism. In: Carbohydrate Metabolism and its Disorders, Vol. I, pp. 89–138, ed. by F. Dickens, P. J. Randle, and W. J. Whelan, Academic Press, New York, 1968.
Villar-Palasi, C., and Larner, J.: The hormonal regulation of glycogen metabolism in muscle. Vitamins and Hormones 26:65–111, 1968.

For Glycogenoses

Fried, R. A.: The glycogenoses: von Gierke's disease, and maltase deficiency, and liver glycogen phosphorylase deficiency. Am. J. Clin. Path. 50:20–28, 1968.
Hsia, D. Y.: The diagnosis and management of the glycogen storage diseases. Am. J. Clin. Path. 50:44–51, 1968.
Pearson, C. M.: Glycogen metabolism and storage diseases of Types III, IV, and V. Am. J. Clin. Path. 50:29–43, 1968.
Steinitz, K.: Laboratory diagnosis of glycogen diseases. Adv. in Clin. Chem. 9:227–354, 1967.

For Proteoglycans

Bollet, A. J.: Connective tissue polysaccharide metabolism and the pathogenesis of osteoarthritis. Adv. Int. Med. 13:33–60, 1967.
Dodgson, K. S., and Lloyd, A. G.: Metabolism of acidic glycosaminoglycans (mucopolysaccharides). In: Carbohydrate Metabolism and Its Disorders, Vol. I, pp. 169–212, ed. by F. Dickens, P. J. Randle, and W. J. Whelan, Academic Press, New York, 1968.
Klynstra, F. B., Böttcher, C. J. F., Van Melsen, J. A., and Van der Laan, E. J.: Distribution and composition of acid mucopolysaccharides in normal and atherosclerotic human aortas. J. Atherosclerosis Research 7:301, 1967.
Meyer, K.: Biochemistry and biology of mucopolysaccharides. Am. J. Med. 47:664–672, 1969.
Muir, H. M.: The structure and metabolism of mucopolysaccharides (glycosaminoglycans) and the problem of the mucopolysaccharidoses. Am. J. Med. 47:673–690, 1969.
Neufeld, E. F., and Fratantoni, J. C.: Inborn errors of mucopolysaccharide metabolism. Science 169:141–146, 1970.

For Glycoproteins

Brunngraber, E. G.: The possible role of glycoproteins in neurofunction. Perspectives in Biology and Medicine, 12:467–470, 1969.
Eylar, E. H.: On the biological role of glycoproteins. J. Theoret. Biol. 10:89–113, 1965.
Ginsburg, V., and Neufeld, E. J.: Complex heterosaccharides of animals. Ann. Rev. Biochem. 38:371–383, 1969.
Gottschalk, A.: Biosynthesis of glycoproteins and its relationship to heterogeneity. Nature 222:452–454, 1969.
Marshall, R. D., and Neuberger, A.: The metabolism of glycoproteins and blood-group substances. In: Carbohydrate Metabolism and Its Disorders. Vol. I, pp. 213–258, ed. by F. Dickens, P. J. Randle, and W. J. Whelan, Academic Press, New York, 1968.
Spiro, R. G.: Glycoproteins: Their biochemistry, biology and role in human disease. New Eng. J. Med. 281:991–1001, 1043–1056, 1969.
Spiro, R. G.: Glycoproteins. Ann. Rev. of Biochem. 39:599–638, 1970.

For Oligosaccharides

Huttunen, J. K., and Mietinen, J. A.: Quantitative determination of urinary neuraminyl oligosaccharides by gas-liquid chromatography. Anal. Biochem. 29:441–458, 1969.
Lundblad, A.: Isolation and characterization of a urinary oligosaccharide characteristic of blood group-O(H)-secretor. Biochim. Biophys. Acta 165:202–207, 1968.
Sie, H. G., Das, I., and Fishman, W. H.: ^{14}C labeling of glucosyl oligosaccharides during starvation and during hydrocortisone-induced glycogenesis. Arch. Biochem. and Biophys. 138:679–683, 1970.

CHAPTER

7

CHEMISTRY AND METABOLISM OF CARBOHYDRATES IN GLYCOLIPIDS

CHEMISTRY

Although oil and water do not mix, living tissue accomplishes an astounding degree of integration, traffic and peaceful coexistence between the two. This occurs at a myriad of interfaces between polar and nonpolar loci at cell surfaces or within cells. Largely responsible for the existence and proper function of these interfaces is a class of compounds, the glycolipids, which combine carbohydrates and fats within the same small molecule. The main glycolipids are hybrid compounds of the 18-carbon aminoalcohol, sphingosine, a long-chain fatty acid, and one or more (up to seven) monosaccharides. A combination of sphingosine and fatty acid* is called a ceramide. The monosaccharides are usually neutral hexoses but may be hexosamine or one or more of the sialic acids. Sphingosine is the base contained also in the phospholipid sphingomyelin. This resembles glycolipids in having the highly hydrophobic region of the ceramide portion of the molecule and another region which is strongly hydrophilic. The glycolipids have regions of even more contrasting

---

*See Masoro, E. J.: Physiological Chemistry of Lipids in Mammals for description of this constituent of glycolipids.

polarity than sphingomyelin and a greater range of solubility in a variety of polar and non-polar solvents.

On the basis of their carbohydrate constituents, there are two main classes of sphingoglycolipids, the ceramide hexosides and gangliosides. The latter are characterized by the presence of sialic acid. Ceramide hexosides are largely neutral, except for those containing hexosamines (aminoglycolipids) or sulfates (sulfatides). Some less common glycolipids do not contain sphingosine.

Ceramide Hexosides

In ceramide hexosides (CH) or "cerebrosides"* the primary hydroxyl group of the sphingosine base is linked through a glycosidic bond (usually β) to C-1 of a monosaccharide (glucose or galactose). One of these, ceramide galactoside, is depicted in Figure 7-1. Besides ceramide monohexosides (CM), there are ceramide dihexosides (CD), trihexosides (CT_3), tetrahexosides (CT_4), and rarely pentahexosides. CD may contain lactose (with glucose linked to ceramide) or digalactose. CT_3 often contain N-acetylgalactosamine, as do CT_4 and gangliosides. CT_3 may also contain 1:2 or 2:1 mixtures of glucose and galactose. Two types of CT_4 have been defined. One is "globoside" of the erythrocyte stroma, which has the structure: ceramide-glucose (4←1) galactose (4←1) galactose (6 ←1) N-acetylgalactosamine. Another has been called "asialoganglioside," because it is the common structure remaining when gangliosides have been stripped of their sialic acid residues. This common CT_4 structure is ceramide-glucose (4←1) galactose (4←1) N-acetylgalactosamine (3←1) galactose. Human adenocarcinoma (gastric or bronchogenic) has an unusual CT_4 glycolipid, which contains fucose.

Gangliosides

In acidic glycolipids the acidic reaction is due to the presence of a sialic acid. These are commonly called gangliosides and are present mostly in brain. The four major gangliosides of human or beef brain have a common basic structure—the asialoganglioside or CT_4 mentioned above, to which are bound 1 to 3 molecules of sialic acid (Fig. 7-1). The sialic acid in brain gangliosides is N-acetylneuraminic acid or NANA, whereas in gangliosides of other tissues N-glycolylneuraminic acid is the predominant form of sialic acid. In the normal major monosialoganglioside (GM_1) the NANA is located at C-3 on the middle galactose molecule.

---

*The term cerebroside is sometimes used in a restricted sense to designate only ceramide galactoside, i.e., the major CH in brain, but often (as in the present text) refers to any neutral CH.

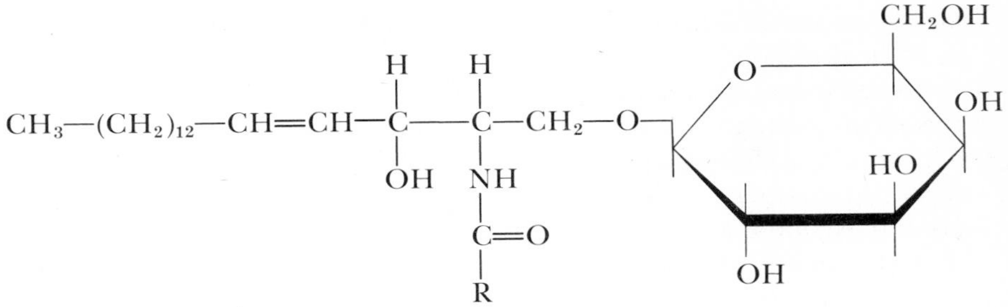

ceramide galactoside

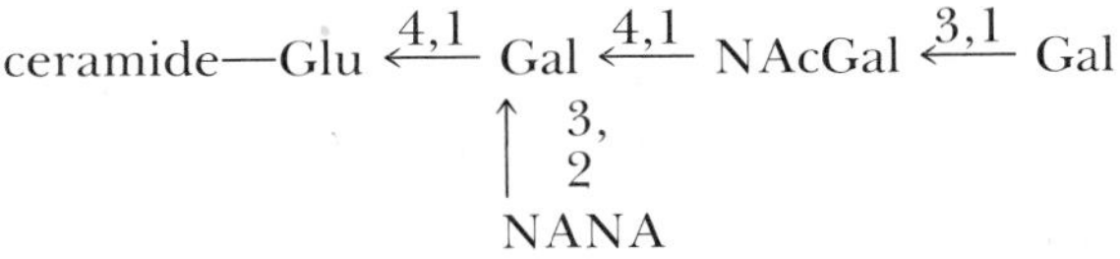

(structure shown: $CH_3-(CH_2)_{12}-CH=CH-\underset{OH}{\overset{H}{C}}-\underset{NH}{\overset{H}{C}}-CH_2-O-$ [galactose ring with CH_2OH, OH, HO, OH]; NH connected to $C=O$ bonded to R)

ceramide galactoside

$$\text{ceramide}-\text{Glu} \xleftarrow{4,1} \text{Gal} \xleftarrow{4,1} \text{NAcGal} \xleftarrow{3,1} \text{Gal}$$
$$\uparrow 3,2$$
$$\text{NANA}$$

monosialoganglioside, GM_1

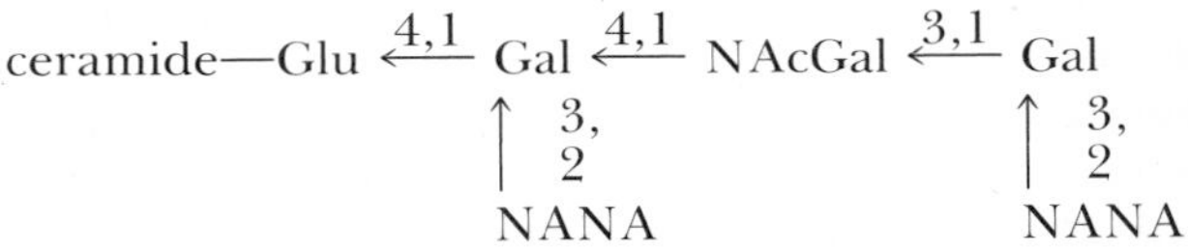

$$\text{ceramide}-\text{Glu} \xleftarrow{4,1} \text{Gal} \xleftarrow{4,1} \text{NAcGal} \xleftarrow{3,1} \text{Gal}$$
$$\uparrow 3,2 \qquad\qquad\qquad \uparrow 3,2$$
$$\text{NANA} \qquad\qquad\qquad \text{NANA}$$

disialoganglioside, GD_1

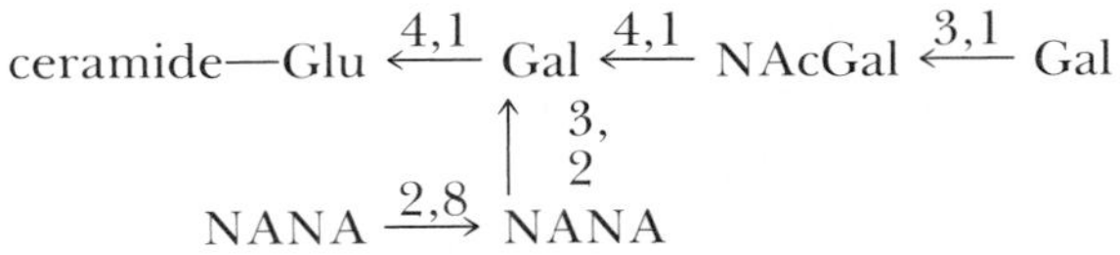

$$\text{ceramide}-\text{Glu} \xleftarrow{4,1} \text{Gal} \xleftarrow{4,1} \text{NAcGal} \xleftarrow{3,1} \text{Gal}$$
$$\uparrow 3,2$$
$$\text{NANA} \xrightarrow{2,8} \text{NANA}$$

disialoganglioside, GD_2

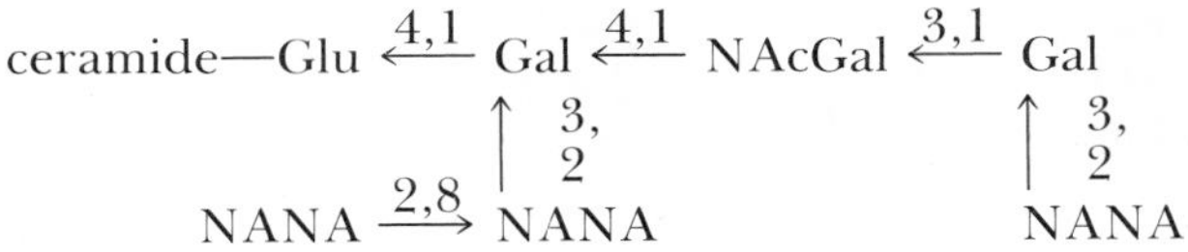

$$\text{ceramide}-\text{Glu} \xleftarrow{4,1} \text{Gal} \xleftarrow{4,1} \text{NAcGal} \xleftarrow{3,1} \text{Gal}$$
$$\uparrow 3,2 \qquad\qquad\qquad \uparrow 3,2$$
$$\text{NANA} \xrightarrow{2,8} \text{NANA} \qquad\qquad \text{NANA}$$

trisialoganglioside, GT_1

Figure 7-1. Structures of ceramide galactoside and four principal gangliosides.

In one of the disialogangliosides (GD_1) both galactose units are substituted at C-3 with NANA. In the other disialoganglioside (GD_2) the two sialic acids are bound to each other and to the middle galactose molecule. The two types of binding are combined to attach the three sialic acids to the galactose units in the trisialoganglioside (GT_1).* There are six or eight known additional but minor gangliosides which constitute about 10 per cent of the total in normal human brain. A more complex pattern of gangliosides is present in the cerebral cortex of higher mammals (e.g., sheep, monkey, man) than in lower species (e.g., rabbit, guinea pig, rat). In pig brain, three of the minor gangliosides have been identified; one of these contains 4 units of neutral hexose and is a disialoganglioside. Further developments from separation by thin-layer chromatography and identification of individual minor gangliosides can be expected in the future.

Sulfates

Some ceramide hexosides (mono-, di-, or tri-) have a sulfate attached to one of the hexoses, generally to C-3 of a terminal galactose unit. These sulfated cerebrosides are called sulfatides and are found in a variety of tissues (e.g., liver, lung, kidney, spleen, skeletal muscle, and heart) as well as in brain, where they are most abundant. In normal human brain, sulfatides are present in one-fourth the concentration of neutral CH. The addition of sulfate confers even higher polarity to the hydrophilic end of the ceramide hexoside and provides a negative charge similar to that possessed by gangliosides containing sialic acid.

Other Glycolipids

In the brain white matter of several mammalian species has been found a neurolipid of an unusual structure consisting of a diglyceride which incorporates D-galactose in β-glycosidic linkage at one of the primary hydroxyls of the glycerol moiety. This unusual glycolipid is not found in tissues other than brain or spinal cord and presumably its function is that of a structural component of myelin. The same glycolipid, except for different fatty acid components, is found in the chloroplasts of green plants. In sheep brain 0.4 per cent of total lipid is a glycerogalactolipid fraction with two components characterized as diacyl-glycerylgalactoside and monoalkyl-monoacyl-glyceryl-galactoside. Ether linkages have also been found in the glycerogalactolipids of bovine brain. Moreover, an extra aliphatic chain attached in ether linkage to the

---

*Some investigators claim that a NANA is linked to N-acetylgalactosamine rather than to the terminal galactose of the trisialoganglioside.

sphingosine moiety has been noted among the ceramide hexosides of bovine brain and lung.

OCCURRENCE AND FUNCTION IN VARIOUS ORGANS AND SUBCELLULAR STRUCTURES

The carbohydrate constituents of glycolipids are more characteristic for different organs than they are for species. Correspondingly, glycolipids, as haptens, show strong organ specificity but relatively little species specificity. Carbohydrate content and sequence in glycolipids, like those of glycoproteins, are important determinants of immunologic specificity of mammalian cell surfaces. For example, comparison of the immunologic properties of the two main CT_4 (see above) indicates that the arrangement of the carbohydrate portion confers immunospecificity. An example is the glycolipid in human adenocarcinoma, which contains the same carbohydrates (galactose, glucose, N-acetylglucosamine, and fucose) and in the same sequence as in some of the blood-group substances. This tumor glycolipid shows greater cross-reaction with group A substance and RBC of type A than with those of other blood types. Hakamori and co-workers suggest that the higher incidence of carcinoma of the stomach in individuals of blood-group type A may be due to greater "acceptance" by the host of the tumor cells with glycolipid similar to that of the host's own substance. Such distinctions afford interesting possibilities for the development of tumor "vaccines." Also, greater selectivity of participants in organ transplants may in the future be gained by consideration of the specificities of glycolipids as well as blood-group glycoproteins.

Nervous System

Mammalian glycolipids were primarily identified in brain, where their concentration is much higher than in other tissues. The ceramide hexosides constitute about 4 per cent of the fresh, wet weight of brain tissue. Most of the CH is in the white matter and about 70 per cent is in the myelin sheath. The monohexoside, ceramide-galactose, accounts for 95 per cent of the neutral ceramides of nervous tissue.

Gangliosides are present in much lower concentration than CH, i.e., about 0.1 to 0.3 per cent of the wet weight of brain. Also in contrast to CH, gangliosides are about 4 times more concentrated in gray matter than in white matter. High amounts are found in cerebral cortex, cerebellum, caudate nucleus, and thalamus, while low amounts are found in centrum semiovale corpus callosum, and optic tracts. Optic and sciatic nerves, pineal and adrenal glands, and sympathetic ganglia (in spite of their name) contain no gangliosides.

Gangliosides appear to be more concentrated in dendritic processes of nerves, i.e., those areas which contain terminal axons and synaptic endings. Among subcellular fractions they are found in the microsomes and the crude mitochondrial fraction. Further subfractionation has identified gangliosides particularly with the cholinergic type of synaptic membrane, where Na^+, K^+-activated ATPase, adenyl cyclase, and phosphodiesterase are also concentrated.

The common subcellular distribution of acetylcholine and gangliosides in brain tissue has suggested a participation of gangliosides in the transport of acetylcholine from synaptic vesicles through the presynaptic membrane. Various other amines, e.g., serotonin, epinephrine, norepinephrine, dopamine, histamine and gamma-amino butyric acid (GABA), are also found in ganglioside-rich nerve endings, so gangliosides may have receptor properties for various biogenic amines. This has been demonstrated particularly for serotonin (5-hydroxy-tryptamine). Interaction between gangliosides and serotonin was established by a series of studies linking the effects of neuraminidase, of diets high in galactose, and of restoration of serotonin response by addition of gangliosides. Woolley and Gommi suggested that an action of serotonin is to provide active transport of Ca^{++} through the cell membrane by a sequence of reversible and cyclic reactions with ganglioside—a sort of ion exchange mechanism. A ganglioside with particular activity as a serotonin receptor is probably a disialobiose-ganglioside.

Others implicate gangliosides in the membrane properties at inhibitory synapses, where GABA is also said to be functional in the process of electrical transmission. Synaptic inhibitory activity is associated with increased conductance of potassium, which can well be imagined to involve the acidic groups of sialic acid in gangliosides in a transport mechanism like that for Ca^{++}.

Different anatomical regions of ox brain show the same pattern of about eight gangliosides. However, retina shows a somewhat different pattern in both dog and ox. Human brain cortex shows no change in ganglioside pattern between infancy and 14 years of age except for a progressive increase in one component. The isolated myelin of rat brain contains a relatively constant amount of NANA at all ages, but the proportion among the individual major gangliosides changes owing to a progressive increase in the major monosialoganglioside, GM_1. Mature (adult) human brain contains more neutral CH and sulfatides than does immature (infant) brain. A general supposition is that in a primarily lipid membrane the sugar moieties of glycolipids in brain or elsewhere provide channels for the passage of water-soluble molecules, e.g., minerals and polyfunctional organic compounds and perhaps for water itself.

Kidney

In kidney, where membrane active transport is an important activity, a variety of glycolipids have been identified. The neutral (non-acidic) portion consists mainly of aminoglycolipids, which are CT_4 with the probable structure: ceramide-glucose (4←1) galactose (4←1) galactose (3←1) N-acetylgalactosamine. Then, in decreasing order of concentration, occur CT_3 (mainly ceramide-Glu (4←1) Gal (4←1) Gal), CD (a mixture of ceramide lactosides and digalactosides), and CM (containing either glucose or galactose, mostly the former). Thus, the predominance of oligosaccharides in CH of kidney is in contrast to the primarily monosaccharide content of brain cerebrosides. The human kidney contains altogether about 4 mg. of total neutral glycolipids per g. of dry weight and there is moderate decrease in concentration with age.

As found in guinea pig kidney, there are gangliosides which resemble the major ones found in brain plus other additional components. In human kidney are found sulfatides which are both monohexosulfatides and dihexosulfatides. The former have the same structure as brain sulfatides (i.e., ceramide- Gal-3-SO_4), while the latter have the structure: ceramide-Glu (4←1) Gal-3-SO_4.

Red Blood Cells

The most abundant glycolipid in human red cell stroma is the tetrahexoside, "globoside," the structure of which already has been mentioned. The oligosaccharide of globoside closely resembles that of the CT_4 in kidney with the difference that in globoside the terminal N-acetylgalactosamine is attached to C-6 of the adjacent galactose rather than C-3. As suggested before, globoside carries the capacity to inhibit the agglutination of erythrocytes of a given blood group by its corresponding isoagglutinen.

In bovine erythrocytes the hexosamine constituent is glucosamine, while in hog and guinea pig red cells the primary glycolipid contains galactosamine, like globoside of human RBC. Indeed, the glucosamine/galactosamine ratio of the lipid fraction of erythrocytes seems to be characteristic of animal species. The "Forsmann hapten"* of sheep red cells is a glycolipid similar to globoside. In general, the glycolipids in the RBC of man, sheep, guinea pig, and rabbit are high in hexosamine content, while those of horse and cat RBC are high in sialic acid. However, in human RBC there are acidic glycolipids which contain fucose in addition to glucosamine, glucose, galactose, and sialic acid, and some of these glycolipids have been associated with blood group A specificity (see above). The ganglioside type of lipid from equine RBC stroma has been

---

*Forsmann found that substances (some recognized as non-protein haptens) from organs of one species, when injected into another species, generated antibodies which were heterophilic, i.e., reacted with antigenically related substances of still a third species, e.g., sheep red cells.

termed "hematoside" and contains, besides galactose and glucose, N-glycolylneuraminic acid. Also in bovine erythrocytes there is a tetrahexosyl ganglioside containing glucosamine and N-glycolylneuraminic acid.

Spleen

The ceramide hexosides of spleen are primarily mono- and dihexosides. The latter are more prominent in human, equine, and bovine spleen and contain lactoside as the sugar moiety. A dihexoside with the same haptenic property was isolated from human dermoid cancer and called "cytolipin H." The monohexosides of equine, bovine, and human spleen contain glucose. It may be this same normal glucoside which accumulates so markedly in human spleen in Gaucher's disease (see below). There are small amounts of CT_3 found in bovine spleen. Glycolipids containing hexosamine and sialic acid are found in higher concentration in human and equine spleen than in bovine spleen.

Serum

CM (containing glucose) and CD (containing lactose) each comprise about 40 to 50 per cent of the total CH in serum. Among the CD is "cytolipin H." There is also a small amount of a CT_3, which contains two moles of galactose and one of glucose. Another CT_3 contains the sequence: glucose-galactose-N-acetylgalactosamine (this may be the same oligosaccharide as the Forsmann hapten). The serum, like several other tissue sites besides brain, contains small amounts of CM and CD sulfatides.

Milk

The glycolipids of bovine milk are principally ceramide glucoside and ceramide lactoside (CL). Thus, milk glycolipids are quite similar to those of blood serum. In milk, there are also lesser amounts of higher ceramide oligosaccharides.

Placenta

The placenta contains about equal amounts of neutral glycolipids and gangliosides. The four neutral glycolipids have the same composition as those of serum, but the most complex ceramide hexosides, i.e., tri- and tetrasaccharides, predominate in contrast to the high amount of CM and CD in serum, milk, or spleen. The low amount of CD and the relatively high proportion in the ganglioside fraction of a monosialolactose component (known as GM_3) suggest that in the placenta there is low neuraminidase activity, since the latter will readily split GM_3 to CL and NANA.

Placenta contains a higher concentration of gangliosides than kidney, liver, lymph node, lung, or muscle. Only the spleen, red cells, and neural tissues contain higher amounts. Since the placenta has an important membrane function the role of gangliosides in this organ may be the sort of cation transport which has been ascribed to these gangliosides in other sites.

White Blood Cells

From normal human leukocytes have been separated a fraction of CM containing galactose and a fraction of mixed CD containing galactose and glucose in the ratio of 2:1. The CD constitute the major glycolipid fraction and represent about 15 per cent of the total leukocyte lipids, which is relatively much higher than most tissues. This may reflect a special role of this glycolipid in structure of leukocytes and in their membrane function. More polar glycolipids are found in very minor concentration.

Intestine

The transport functions of intestinal mucosa would predict the significant occurrence of glycolipids. These have been found to be primarily of a complex type. A major component, as identified in the dog, is a novel ceramide pentaglycoside which has the apparent sequence: Glu-Gal-Gal-NAcGal-NAcGal. Another component is a ganglioside which contains two residues of galactose, one of glucose, and one of sialic acid. Gangliosides containing hexosamine are present in very minor amount.

METABOLISM

Synthesis

Details of the pathways, sequence of intermediates, enzymes involved, and other characteristics of synthesis of glycolipids are little understood for most tissues. Even for brain, knowledge is quite incomplete. Although certain evidence has suggested that some neutral glycolipids (mainly tri- and tetrasaccharides, as suggested in Fig. 7-3) are formed upon degradation of the acidic glycolipids, in general it appears that ceramides and gangliosides are formed independently by rather separate pathways. On the other hand, sulfatides seem to be a product by one route or another (Fig. 7-4) of the synthesis of cerebrosides (ceramide hexosides).

The differing time sequence of formation of cerebrosides and gangliosides of mouse brain, as indicated by analysis of labeling after in-

(Text continued on page 246)

CARBOHYDRATES IN GLYCOLIPIDS

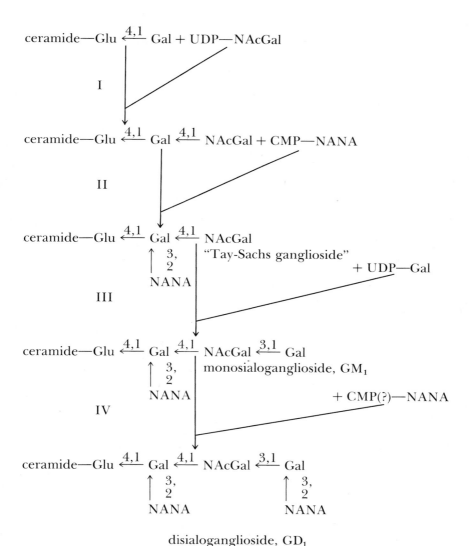

Figure 7-2. Probable sequence of reactions in biosynthesis of gangliosides.

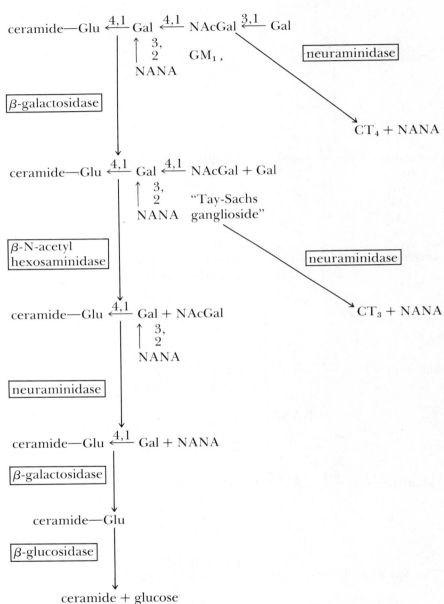

Figure 7-3. Probable sequence of reactions in catabolism of gangliosides.

CARBOHYDRATES IN GLYCOLIPIDS

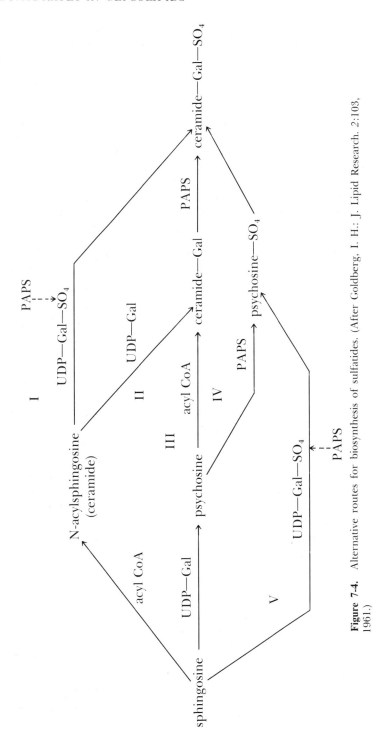

Figure 7-4. Alternative routes for biosynthesis of sulfatides. (After Goldberg, I. H.: J. Lipid Research. 2:103, 1961.)

traperitoneal injection of ^{14}C-labeled glucose or galactose, suggests that gangliosides are actively formed during the first 20 days of postnatal life and earlier than cerebrosides, which show a sharp rise of rate of incorporation of glucose-^{14}C into cerebroside galactose at 22 days of age. Qualitatively, there are similar findings of time difference for labeling of cerebrosides and gangliosides by glucose-U-^{14}C in rat brain slices. Both in vivo and in vitro studies suggest multiple peaks of labeling of gangliosides during the first few days of life. Of course, the different anatomical location (mainly myelin sheath for cerebrosides and neuronic dendrites for gangliosides) would in itself suggest that these two types of glycolipids do not have a close biosynthetic relationship.

In the in vivo study mentioned above (Moser and Karnofsky), there was more radioactivity incorporated into galactose of cerebrosides from glucose-U-^{14}C than from galactose-^{14}C. This suggested that, at least for the precursor pool in brain, glucose is more readily converted to UDP-galactose, the probable immediate hexose donor, than is galactose itself. On the other hand, gangliosides of mouse brain incorporated labeled galactose more extensively than labeled glucose. This is further evidence for a separation of biosynthetic pathways. Galactose (and galactosamine) of gangliosides also differed from galactose of cerebrosides in that some randomization of ^{14}C from C-6 of precursor glucose occurred in the former but almost none in the latter case.

Information on the initial steps in synthesis of ceramide hexosides is relatively lacking. According to study with isotopically labeled glucocerebroside, this, when administered to rats intracerebrally or intraperitoneally, does not serve as a precursor of the more highly glycosylated, non-acidic sphingoglycolipids of brain, kidney, liver, or spleen of rats. (The negative finding with ceramide glucoside in vivo could be due to inadequate penetration to the synthesizing site — yet the cerebroside was degraded to ceramide.) On the other hand, ceramide lactoside is formed from ceramide glucoside and UDP-galactose by homogenates of rat brain, kidney, liver, and spleen. Highest incorporation from UDP-galactose-1-^{3}H was found with spleen and lowest with liver.

Some studies have suggested that in brain sphingosine reacts with UDP-galactose to form sphingosylgalactose, called psychosine, and is subsequently acylated to become ceramide galactoside (Fig. 7-4). Sphingosylpolyhexosides may also be formed prior to acylation as a means of synthesis of the series of neutral glycolipids.

In the case of the gangliosides, there is more evidence that the synthesis proceeds through successive build-up of ceramide glycosides. The sequence shown in Figure 7-2 appears likely according to individual reactions demonstrated in brain tissue of chicks, pigs, and rats and in rat kidney. For embryonic chick brain, there is evidence that reactions I and II of Figure 7-2 may be reversed. If reaction III is deficient, there could be a piling-up of "Tay-Sachs ganglioside" as occurs in Tay-Sachs disease.

However, another defect in catabolism of gangliosides could also explain this accumulation (see below).

Some investigators suggest the occurrence of a pathway which does not include GM_1 as a precursor of polysialogangliosides. The latter interpretation has been inferred from the observation that the corresponding sugar moieties of all four major gangliosides in the brain of 7-day-old rats were labeled at about the same rate with very similar specific activities after subcutaneous injection of glucose-U-^{14}C. However, specific activities of the various sugars were quite different. Glucose, galactose, and neuraminidase-resistant NANA had similar and higher specific activities than galactosamine or neuraminidase-labile NANA. The latter are readily removed from polysialogangliosides, while NANA in monosialoganglioside is resistant to the enzyme.

As indicated in Figure 7-2, the active precursor for transfer of NANA to the glycosidic chain does not involve the usual UDP complex with the sugar but rather another nucleotide, cytidine monophosphate (CMP). There is some evidence that CMP-NANA donates NANA more extensively to the monosialoganglioside than to the di- or trisialogangliosides. This further suggests (in addition to the differing specific activities mentioned above) that there are different metabolic pools of NANA contributing to the neuraminidase-labile and the neuraminidase-resistant types, and that the former may involve a precursor other than CMP-NANA.

Other differences in biosynthesis among the sugars include the observation that radioactivity from labeled glucosamine appears in NANA and galactosamine but none in the galactose or glucose portions. Comparisons of specific activities of various carbohydrate moieties of gangliosides with those of various carbohydrate precursors suggest that the precursor pools for synthesis of NANA and galactosamine are very small compared with those for glucose or galactose of gangliosides.

As indicated with ^{14}C-labeled glucose or galactose, sulfatides of brain are formed more slowly than cerebrosides, which have higher specific activities and earlier incorporation of ^{14}C. This suggests the conversion of cerebrosides to sulfatides. However, there may be common precursors for both cerebrosides and sulfatides. Sulfate may be introduced at various points in the course of synthesis as indicated in Figure 7-4. In any case, the immediate donor of activated sulfate is the nucleotide, 3'-phosphoadenosine-5'-phosphosulfate (PAPS). In schemes I and II of Figure 7-4, the ceramide is the precursor of the more complex glycolipid, whereas in schemes III, IV, and V, psychosine is intermediate. There is evidence for various alternative synthetic routes, which could be different for different end products, e.g., cerebrosides or sulfatides. Evidence is lacking for the occurrence of UDP-Gal-SO_4, so synthesis by routes I or V is less likely.

An enzyme catalyzing the transfer of sulfate to ceramide galactoside (scheme II or III) has been isolated from rat kidney and brain. Its activity correlates with the extent of incorporation of sulfate-^{35}S into sulfatides of brain in vivo and with the rate of myelination. There is greatest activity around the twentieth postpartum day, with a subsequent rapid decline. With kidney, however, the activity of the sulfotransferase continues to increase up to the fiftieth day. The difference between brain and kidney may reflect the stability of the myelin of brain once formed, in contrast to rapid turnover in labile membranous components in kidney such as those of mitochondria and microsomes.

In malnourished animals (oversize litters), the time of onset of synthesis of sulfatides in myelin (7 to 8 days postnatal in the rat) is normal, but the maximal rate of synthesis is markedly decreased. This may be related to the observation that the biosynthesis of gangliosides seems to require concomitant protein synthesis, since puromycin inhibits the incorporation of labeled sugars. There was somewhat less inhibition of the synthesis of the neutral glycolipids than of gangliosides.

Turnover and Degradation

After labeling of the brains of 13-day-old rats by administration of galactose-1-^{14}C or glucosamine-1-^{14}C, the decline of activity in cerebrosides indicated a half-life of about 45 days. Gangliosides appeared to turn over faster — the half-life was 24 days with glucosamine-1-^{14}C and only 10 days with galactose-1-^{14}C. The higher value with glucosamine could be due to reutilization of this labeled carbohydrate.

After injection of sulfate-^{35}S into young (12-day-old) rats, there was little subsequent turnover of the radioactivity incorporated into the myelin fraction. Evidently, sulfatides formed at this time are extremely stable and long-lasting. However, if the sulfate-^{35}S is injected into adult rats, the sulfatides of the myelin sheath show two components of turnover — one with a half-life of 11 days and another with a half-life of about 100 days.

Cerebroside galactosidase is present in spleen, kidney, intestine, and lung as well as in brain, but is not found in liver or heart. In brain, glucosidase activity has clearly been separated from that of galactosidase. On the other hand, in rat intestine, and perhaps other tissues, there is an enzyme which cleaves either ceramide glucoside or ceramide galactoside to the constituent hexose and ceramide. An enzyme that cleaves the terminal galactose off the trihexoside, ceramide-Glu-Gal-Gal, is present in brain, liver, kidney, spleen, and small intestine of the rat.

Globoside, the prominent tetrahexoside of RBC stroma, may be initially degraded in the liver, since after injection of the tritiated aminoglycolipid a labeled ceramide trihexoside is found in the liver.

Studies of hydrolysis of various ceramide hexosides and gangliosides by brain, kidney, and liver enzymes have suggested a probable sequence of degradation of monosialogangliosides according to the pathway proposed by Gatt (Fig. 7–3). Several of the enzymes involved are found in the same subcellular fraction of brain and have similar physical properties; they may, therefore, be components of a membrane-bound multienzyme complex. A similar series of reactions provides a pathway for the complete, stepwise degradation of complex neutral (non-acidic) glycolipids. It is unlikely that sialidase (neuraminidase) normally attacks the tri- or tetrahexosylgangliosides. However, in one pathological condition, Tay-Sachs disease, not only does the "Tay-Sachs ganglioside" increase in brain tissue but the tri- and tetrahexosides shown in Figure 7-3 appear. Polysialogangliosides are converted to monosialogangliosides by a neuraminidase present in animal tissues.

PATHOLOGICAL CONDITIONS

Nervous System

With a fortunately rare incidence, there are specific enzyme deficiencies in brain, and in other tissues, which are characterized by bizarre accumulations of one or another glycolipid as well as by bizarre and generally devastating clinical effects. In one leukodystrophy (abnormality of the formation of white matter or myelin), there is a three- to four-fold increase in content of CM sulfate not only in white matter but also in gray matter of the brain. This is called metachromatic leukodystrophy. In this hereditary condition, there is a decrease in a sulfuric acid esterase—called arylsulfatase (A) because its activity is determined with synthetic arylsulfate esters. The enzyme is markedly deficient in kidney and liver as well as nervous tissue. In the peripheral nerves, the increase in ceramide-Gal-3'-SO_4 is associated with focal lamellar splitting and a segmental pattern of demyelination that has been attributed to alteration of the net charge in a lipoprotein matrix.

In another type of leukodystrophy (globoid), which occurs rarely in human infants, there is a widespread degeneration of white matter accompanied by accumulation of neutral cerebrosides and, in contrast to the metachromatic type, a depletion of the sulfatides. In this condition, deficiency of another enzyme, cerebroside sulfotransferase, interposes a block in formation of the sulfatides from neutral ceramides. In so-called "maple syrup disease" (in which the urine has the odor of maple syrup) there is a marked loss of both cerebrosides and sulfatides of cerebral white matter, but not of other glycolipids.

A similar change has been implicated in a neurological disease which is more common, has a much more protracted course, is not hereditary, and appears usually after childhood, i.e., multiple sclerosis. Abnormalities in the concentrations of cerebrosides and cerebroside sulfates in the white matter of the cerebrum have been noted in patients with multiple sclerosis, although "total lipids" and cholesterol are usually normal. Other, though lesser, changes occur in the concentration of neutral glycolipids of cerebral white matter in Alzheimer's disease. The idea that some degenerative neurological diseases (e.g., multiple sclerosis) may represent autoimmune phenomena involving ceramide hexosides has been suggested. In some patients with multiple sclerosis and others with amyotrophic lateral sclerosis, antibodies to monosialoganglioside have been found in the serum.

Disturbances in metabolism of gangliosides may give rise to various pathological and clinical phenomena of the nervous system. In Tay-Sachs disease, also known as familial amaurotic idiocy, the disastrous clinical effects, present from birth, are associated with massive accumulation of the "Tay-Sachs ganglioside" (Fig. 7-2). Whereas this ganglioside is normally present at a level of 3 to 6 per cent of the total gangliosides of brain, in this disease it constitutes more than 90 per cent. Nevertheless, there are still present normal amounts and patterns of other mono- and polysialogangliosides. For this reason, a defective anabolic reaction seems unlikely and a deficiency of some catabolic reaction is suspected. Since the corresponding ceramide trihexoside derived from Tay-Sachs ganglioside by removal of sialic acid is also present in excess, the hydrolysis of N-acetylgalactosamine from the ganglioside (Fig. 7-3) may be the site of deficiency, but evidence for this is lacking.

Another inborn error of metabolism analogous to Tay-Sachs disease and causing similar clinical effects is called generalized gangliosidosis. In this disease, the normal monosialoganglioside (GM_1) and the corresponding ceramide tetrahexoside are present in excess. The glycolipids accumulate not only in neurons but also in hepatic, splenic, and other histiocytes and in renal glomerular epithelium. Presumably, the initial step of ganglioside catabolism, i.e., the hydrolysis of the terminal galactose molecule, is at fault.

In "gargoylism" three of the normally minor gangliosides constitute 30 per cent of the total in brain. Two of these are defined as monosialogangliosides which are "fast-running" on thin-layer chromatograms. However, in this disease, extensive abnormalities in proteoglycans (Chap. 6) of various organs overshadow the changes in glycolipids of the brain.

Experimental galactosemia in rats, which is produced by excess galactose feeding, has resulted in a decrease of brain serotonin receptor activity to less than 10 per cent of control levels. A corresponding decrease or absence of the serotonin-specific ganglioside, without change in other major galactolipids, was demonstrated. By implication, the mental deficiency or aberrations of human galactosemia could be explained as being due to defective formation of the galactose-containing ganglioside which serves as the serotonin receptor in the brain.

Hypercapnia (increased CO_2 content in the body fluids) causes a change in ganglioside composition, i.e., less NANA and hexosamine, in the brain of humans or cats. This may explain, in part, the altered excitability of neurons associated with hypercapnia and respiratory acidosis.

Spleen

In a rare familial disorder known as Gaucher's disease, there is extensive accumulation of ceramide glucoside in the reticuloendothelial cells of various organs, particularly the spleen. An increase of 5 to 30 times the normal amount of this constituent causes enlargement of spleen, liver, and lymph nodes; there is also infiltration of other tissues, such as the bone marrow, which leads to ready bone fractures. In the infantile form, the brain is affected and there is an early high mortality. The adult form may extend over many years. The metabolic lesion seems to be in the catabolism of the more complex glycolipids, e.g., globoside of RBC stroma or gangliosides in brain. Glucocerebrosidase, the enzyme which cleaves ceramide glucoside to ceramide and glucose, shows a mean level of 15 per cent of normal in the spleens of patients with Gaucher's disease.

Kidney

Another sphingolipodystrophy, called Fabry's disease, may be analogous biochemically to Gaucher's disease since in Fabry's disease (a sex-linked condition appearing chiefly in males) there is accumulation of a CT_3, ceramide-Glu-Gal-Gal. This could be another degradation product of the CT_4, globoside. However, the pathology differs from that in Gaucher's disease, since there are lipid deposits occurring primarily in renal glomeruli and tubules and in blood vessels throughout the body. Many organs and tissues are involved. A deficiency of ceramide trihexosidase has been demonstrated in small intestine mucosa. Besides the CT_3, a dihexoside containing entirely galactose is also found in high concentration in kidney.

Blood

Some abnormalities of glycolipids in serum or red cells have been correlated with neurological or psychiatric conditions and may hold some clue to their pathology. There is a synergist for serotonin activity present in the serum of normal subjects but occurring in higher concentration in serum of schizophrenic and perhaps other psychiatric patients. This synergist has certain characteristics (e.g., it is destroyed by neuraminidase) which suggest that it is a ganglioside. In the glycolipid fraction from erythrocytes of patients with Huntington's chorea there is a marked increase of NANA. Future study may reveal whether this is accompanied by abnormal gangliosides in brain which would help explain this degenerative neurological condition.

In a rare genetic disease, Chediak-Higashi syndrome, characterized by granulocytopenia, there is an accelerated turnover of glucose in ceramide glucoside of leukocytes. The rate of degradation is increased more than that of biosynthesis. The phenomenon may reflect general increased turnover of leukocytes. A similarly increased rate of incorporation of labeled glucose into sphingolipids occurs in leukocytes from patients with myelogenous leukemia.

REFERENCES

Brady, R. O.: Immunochemical properties of glycolipids. J. Am. Oil. Chem. Soc. *43*:67–69, 1966.
Brady, R. O.: The sphingolipidoses. New Eng. J. Med. *275*:312–318, 1966.
Burton, R. M., Garcia-Bunel, L., Golden, M., and Balfour, Y. M.: Incorporation of radioactivity of D-glucosamine-1-C^{14}, D-glucose-1-C^{14}, D-galactose-1-C^{14} and DL-serine-3-C^{14} into rat brain glycolipids. Biochemistry *2*:580–585, 1963.
Burton, R. M., and Howard, R. E.: Gangliosides and acetylcholine of the central nervous system. VIII. Role of lipids in the binding and release of neurohormones by synaptic vesicles, Ann. N.Y. Acad. Sci. *144*:411–432, 1967.
Carter, H. E., Johnson, P., and Weber, E. J.: Glycolipids. Ann. Rev. Biochem. *34*:109–142, 1965.
Gatt, S.: Enzymatic hydrolysis of sphingolipids V. Hydrolysis of monosialoganglioside and hexosylceramides by rat brain β-galactosidase. Biochem. Biophys. Acta *137*:192–195, 1967.
Goldberg, I. H.: The sulfolipids. J. Lipid Research *2*:103–109, 1961.
Hakamori, S.-I., Koscielak, J., Bloch, K. J., and Jeanloz, R. W.: Immunologic relationship between blood group substances and a fucose-containing glycolipid of human adenocarcinoma. J. Immunology *98*:31–38, 1967.
Lapetina, E. G., Soto, E. F., and De Robertis, E.: Gangliosides and acetylcholinesterase in isolated membranes of the rat-brain cortex. Biochem. Biophys. Acta *135*:33–43, 1967.
Lowden, J. A., and Wolfe, L. S.: Studies on brain gangliosides. III. Evidence for the location of gangliosides specifically in neurones. Canad. J. Biochem. *42*:1587–1594, 1964.
Maker, H. S., and Hauser, G.: Incorporation of glucose carbon into gangliosides and cerebrosides by slices of developing rat brain. J. Neurochem. *14*:457–464, 1967.
Makita, A. and Yamakawa, T.: Biochemistry of organ glycolipid. I. Ceramide-oligohexosides of human, equine and bovine spleen. J. Biochem. *51*:124–133, 1962.
Martensson, E.: Neutral glycolipids of human kidney. Isolation, identification and fatty acid composition. Biochim. Biophys. Acta *116*:296–308, 1966.
Moser, H. W., and Karnofsky, M. L.: Studies on the biosynthesis of glycolipids and other lipids of the brain. J. Biol. Chem. *234*:1990–1997, 1959.
Shapiro, B.: Lipid metabolism. Ann. Rev. Biochem. *36*:247–270, 1967.
Suzuki, K., and Korey, S. R.: Study on ganglioside metabolism. I. Incorporation of D-(U-^{14}C) glucose into individual gangliosides. J. Neurochem. *11*:647–653, 1964.
Svennerholm, L.: Gangliosides and other glycolipids of human placenta. Acta Chem. Scand. *19*:1506–1507, 1965.
Wild, G., Woolley, D. W., and Gommi, B. W.: Effects of experimental galactosemia on the measured serotonin receptor activity of rat brain. Biochemistry *6*:1671–1675, 1967.
Woolley, D. W., and Gommi, B. W.: Serotonin receptors. V. Selective destruction by neuraminidase plus EDTA and reactivation with tissue lipids. Nature *202*:1074–1075, 1964.
Woolley, D. W., and Gommi, B. W.: Serotonin receptors. VII. Activities of various pure gangliosides as the receptors. Proc. Nat. Acad. Sci. *53*:959–963, 1965.
Yamakawa, T., Irie, R., and Iwanaga, M.: The chemistry of lipid of posthemolytic residue or stroma of erythrocytes. J. Biochem. *48*:490–507, 1960.

CHAPTER 8

TURNOVER AND FATE OF CARBOHYDRATES IN THE CIRCULATION

The various organs and tissues of the body are knit together functionally by the circulation and so is their carbohydrate metabolism. The circulation becomes the medium not only for body homeostasis but for support of special needs in critical situations by surging shifts of metabolic interchanges. No metabolic processes are more important than those which maintain an adequate level of blood glucose for the functions of the central nervous system, which depends so strongly upon glucose supply via the circulation. The means by which the body adapts to the historically erratic and uncertain exogenous supply of glucose (the "feast or famine" of man's primitive ancestors and most carnivores) have been partly discussed in other chapters, but should be considered from the integral standpoint of circulatory turnover and fate and what governs these factors. Linked to the characteristics of turnover of glucose are those of other major circulating carbohydrates, i.e., lactate, pyruvate, and glycerol. Interacting with these, and with special characteristics of their own, are two monosaccharides in the diet, galactose and fructose. The characteristics in the circulation of each of these carbohydrates will be discussed.

GLUCOSE

Methods of Evaluation of Distribution and Turnover

Prior to the advent of isotopic tracers it was possible to evaluate rates of utilization and production of blood glucose only by such methods as measuring the rate of removal of an intravenous load of glucose, measuring the arteriovenous difference of blood glucose concentration in combination with the blood flow through a certain organ or area, or observing the results of extirpation of an organ. It was evident that after removal of the liver in a dog or like-sized animal the blood glucose concentration was reduced to about half normal in 30 to 60 minutes. It was further evident from arteriovenous studies that the liver produced at least 85 per cent of the blood glucose and the kidney essentially accounted for the remainder in the early postabsorptive state.

About 25 to 30 years ago the availability of carbon isotopes made possible the addition of trace amounts of labeled glucose to the circulation in studies which could then monitor the rate of disappearance of glucose by the rate of disappearance of the tracer, e.g., glucose-^{14}C. In the "steady state," i.e., when the concentration of blood glucose remains essentially the same, the rate of disappearance represents also the rate of appearance or the turnover rate. The latter can be stated as an "absolute" rate (quantity/unit of time) or fractional rate (per cent of total/unit of time). There have been two general ways of assessing glucose turnover by means of glucose-^{14}C. One is by single, rapid injection of the tracer followed by successive sampling of the blood or plasma to measure the rate of decrease with time of the specific activity of the glucose in the blood compartment or "pool." The labeled blood glucose rapidly equilibrates (in about 5 minutes in the rat and about 15 minutes in an animal the size of sheep or man) with the glucose in the extracellular fluid, so that the dynamics of an extracellular pool of glucose is mainly measured. Repetitions of the injection, i.e., successive measured injections of tracer (SMIT), as developed by the group of Wrenshall, Hetenyi and coworkers, have permitted observations of changes in the dynamics of glucose turnover consequent to a metabolic perturbation, e.g., injection of a hormone. The size of the glucose mass being measured for turnover rate can be estimated by extrapolation from the semilogarithmic plot of the line of decreasing specific activity of blood glucose back to the "time zero" or time of injection. This point on the ordinate represents the specific activity (Sp. A.) of the miscible glucose mass had it actually mixed instantly (Fig. 8-1). The measured specific activity usually displays an early curve, then a straight line, which can be described by a series of exponential functions.

Curve-stripping by subtraction starting with the slowest component of turnover allows calculation of the separate exponentials (Fig. 8-1). It is also possible to construct models of multiple glucose pools (usually one to three) and by a computer devise sizes of pools and rates of interchange, inflow, and outflow which best fit a theoretical curve to the actual curve of specific activity.

A simplified method, as used by Shipley and co-workers with rats, employs the Stewart-Hamilton dilution principle, whereby $K = \dfrac{d}{A}$ and K

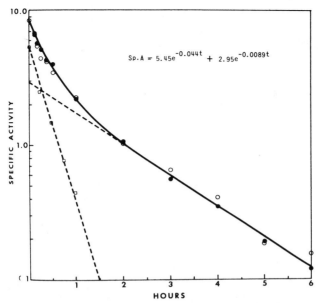

Figure 8-1. Specific activity of blood and body glucose-^{14}C after intravenous injection in rats. Specific activity on the ordinate is % dose/mg. of glucose carbon corrected for a rat weighing 100 g. Dots are blood glucose and open circles are whole rat glucose. Calculated point at zero time is 100 per cent dose divided by total glucose carbon in the rat. Squares define subtractions of successive points on the extrapolated dashed line (second slope, 0.0089) from the observed smoothed curve in order to derive the first slope (0.044). Time in the equation is min. (From Shipley et al.: Am. J. Physiol. *213*:1149, 1967.)

= rate of disposal, d = dose of tracer, and A = area under the curve of specific activity over a period of time representing the main period of total loss of activity. This method of analysis is not dependent upon a kinetic model of glucose mixing in various pools, although analysis of the specific activity curve into components facilitates calculation of the area. All glucose mass must enter or pass through the sampled pool (blood glucose). Rates of glucose turnover thus measured are quite close to those obtained from kinetic analysis.

A reason for the gradual decrease in slope of specific activity of blood glucose is a recycling of glucose-^{14}C after metabolism to lactate by various tissues. Lactate returns through the circulation to the liver, where it is resynthesized to glucose (Chap. 5). This well-known "Cori cycle" is supplemented by an analogous alanine cycle (Fig. 8-2). The alanine cycle serves a role in the transport of ammonia from peripheral tissues to the liver for production and excretion of urea. The two cycles may vary in relative activity, depending on circumstances, but these cycles plus some early intermixing of glucose with glycogen, glycerol, and perhaps some other glucogenic amino acids contribute to the need for postulating more than one "pool" of "glucose" to explain the nature

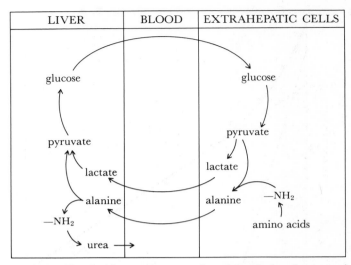

Figure 8-2. The Cori lactate-glucose cycle and the alanine-glucose cycle.

of the specific activity curve of blood glucose after single injection. A guide to the extent of this recycling has been the rate at which carbon-14 appears in other blood glucose carbon positions after glucose-1-^{14}C or glucose-6-^{14}C is injected. Either the Cori lactate cycle or the alanine cycle can be responsible for this. Still another way in which the extent of recycling is evaluated is by comparison of the rate of turnover of glucose-^{14}C with that of glucose-^{3}H after single injection. Unlike ^{14}C, the tritium is detached* from the intermediate molecules in the course of recycling and therefore measures a "purer" and lesser mass of glucose with a higher turnover rate.

In the other general method of measurement of glucose turnover rate with ^{14}C (introduced by Searle and co-workers and by Steele and co-workers with dogs) there is a continuous intravenous infusion of trace amounts of glucose-^{14}C, usually after a single, priming injection to bring the specific activity at an early time to a level which can be conveniently sustained. An equilibrium level is asymptotically approached, so that the specific activity becomes fairly constant. From the specific activity of the repeatedly sampled glucose and the rate of tracer glucose-^{14}C injected, the rate of appearance of the unlabeled component (glucose-^{12}C) can be derived. The method does not depend upon pool analysis to calculate inflow rate after "equilibrium" is reached, but the size of the "total" glucose pool or mass can be estimated (using also the blood glucose con-

---

*Glucose labeled at the 1, 2 or 6 positions has been used, but glucose-2-^{3}H is most rapidly and surely removed in metabolism at the hexose isomerase stage.

centration) by extrapolation of the line of equilibrium specific activity to zero time. This method permits calculation of separate and different inflow and outflow rates when the specific activity and concentration of glucose are changing over short intervals.

When carefully compared in dogs, sheep, or humans the two methods have given very similar or identical results when the same tracer, e.g., glucose-U-^{14}C, is employed. Results with both methods are influenced by the recycling effect. This effect is less readily evaluated by the priming-infusion method, as can be done in the case of the single injection method by factorial analysis of the specific activity curve. With both methods, the recycling phenomenon, if not taken into acccount, can cause the calculation of the glucose mass by the extrapolation technique to be spuriously high. With the method of single injection there is some hazard of utilizing significant amounts of tracer prior to initial mixing, but by methods of independent validation this has appeared to be negligible or minimal.

Some tracer studies of glucose turnover (either single injection or priming-infusion method) have been accompanied by measurement of carbon dioxide-^{14}C in the breath which has enabled the calculation of rates of oxidation of glucose to CO_2. In the priming-infusion method this is represented by the difference in specific activity of the CO_2 (after an equilibrium is attained) and that of the blood glucose. In both methods it is necessary to introduce estimates or data on the size and turnover of the CO_2 pool (one or more) and make simplifying assumptions concerning precursor-product relationships between glucose and CO_2.

Distribution and Turnover in the Early Postabsorptive State

As a base line for comparison with other nutritional states or for study of hormonal or drug effects, it has been customary to investigate the above parameters in the early postabsorptive condition when there is a presumed steady state, as indicated by a generally constant "fasting" blood glucose concentration. In humans this is usually early in the morning after an overnight fast, i.e., about 12 to 14 hours postprandial. At this time the blood glucose concentration is usually steady, although varying among individuals normally from about 60 to 90 mg./dl. by the enzymatic methods and about 70 to 100 mg./dl. of blood by the less specific methods for total reducing compounds. Actually this period of "steady state" is a time of gradual transition from postfed to a more distinctly fasting state. It is a time when glycogen stores are dwindling and providing less and less to hepatic glucose output and gluconeogenesis is providing increasingly more. In the dog after an overnight fast glycogenolysis accounts for only about a third, while the remainder is derived from gluconeogenesis. It is probable that even during the usual span between meals for humans eating three meals daily, the gluconeogenic forces, and those factors controlling them, must come into play (Chap. 5). The relative contributions of major substrates (e.g., lactate, alanine, glycerol) to gluconeogenesis have already been discussed in Chapter 5 and will be given further attention below.

From arteriovenous glucose balance studies the estimates of splanchnic glucose production for man in the postabsorptive state range from 120 to 420 g./day. Such variability of estimates reflects the variability of this metabolic function as measured by different investigators with patients under different dietary and other conditions as well as the factor of transitional change into a true fasting state, as mentioned above. Considering various tracer studies by single injection or priming infusion in humans, the range is similar (about 100 to 300 g. glucose/day), but most findings are in the range of about 130 to 200 g./day or about 80 to 120 mg./kg. body wt./hr. By compilation of studies in species of widely differing sizes, from rats to cattle and including man (Fig. 8-3), the glucose turnover rate (like other metabolic activities) is seen to be an exponential function of the body size with a power exponent of about 0.80. In terms of fractional turnover rate, the findings with glucose-U-^{14}C indicate in man a value of about two-thirds of 1 per cent/min., in the dog about 1.5 per cent/min., and in the rat about 2 per cent/min.

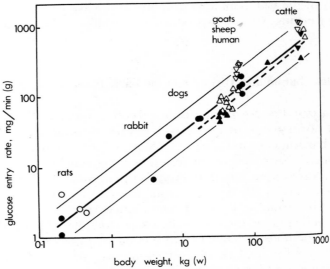

Figure 8-3. Glucose entry rate vs. body weight of several species. Thick continuous line is regression for data on rats, rabbits, dogs and humans (non-ruminants) with ±2 standard deviations as thin lines. Dashed line is regression for non-lactating sheep and cattle. Equations for regression include $G = 4.60\ W^{0.804}$ and $G = 2.86\ W^{0.80}$ for postabsorptive ruminants and non-ruminants, respectively. (Reprinted from Federation Proceedings 28:224, 1969.)

The range of present data for non-ruminants is so large that any difference between ruminants and non-ruminants is of questionable significance (Fig. 8-3). As described in Chapter 5, the ruminant depends entirely on gluconeogenesis for glucose turnover, rather than on dietary glucose, and uses to a large extent the ruminal substrate, propionate.

The extent of recycling in the early postabsorptive state has been measured by the methods mentioned earlier and also by comparison of glucose turnover with the turnover of lactate-^{14}C and its conversion to glucose. There seems to be a difference in recycling dependent on species and possibly on method used. In dogs the recycling (from lactate) is only 3 to 8 per cent, as measured by comparing glucose and lactate turnover. In sheep the recycling is 10 to 20 per cent, in cows it is about 20 per cent, and in man it has appeared to be about 15 to 20 per cent whether judged by randomization of ^{14}C between blood glucose carbon positions C-1 and C-6 or by comparison between glucose and lactate turnovers. In the rat is found the highest degree of recycling, about 20 to 30 per cent or more, according to the difference in slope of decline of specific activity of ^{3}H-labeled vs ^{14}C-labeled in the blood. Thus with glucose-2-^{3}H the fractional turnover rate of glucose in the rat is 3 per cent/min. instead of 2 per cent/min. as with glucose-U-^{14}C. In normal human subjects the difference between the rates of turnover of glucose-^{3}H (1-labeled) and glucose-^{14}C (1-labeled) is less than half as much as in the case of rats.

In single injection studies with glucose-^{14}C in man the apparent quantity of the total body "glucose" mass, as derived by extrapolation of the major component of specific activity slope back to the time of injection, has ranged from about 75 to 150 mmoles of "glucose" (including some other readily equilibrated metabolites, as mentioned above) normalized for a standard weight of 70 kg. Assuming that this total pool of "glucose" is distributed in a hypothetical space at the same concentration as that measured in the blood in each case, such a space has appeared to occupy 20 to 30 per cent of the body volume (1./kg.), which is somewhat higher than the extracellular fluid volume. Like values for "glucose" distribution are found for rats by single injection and for dogs by single injection and priming-infusion techniques. Measurements in sheep and cows have provided similar estimates, although the "glucose" mass per kg. is slightly lower owing to lower glucose concentration in ruminants.

Fate of Circulating Glucose in the Early Postabsorptive State

Since, under ordinary nutritional circumstances, about 120 to 130 g. of carbohydrate per day are required to satisfy the energy requirements of the human brain, according to the amount of oxygen consumed, and since the turnover of blood glucose during the postabsorptive state occurs at a rate not often more than 200 g./day, it follows that at least during such periods of the day the brain is using more than half of the blood glucose turnover. Of that glucose utilized by brain about half is converted to lactic acid, a quarter to carbon dioxide, and most of the rest to alpha amino acids. According to forearm arteriovenous differences in man, the quantity of glucose being used by skeletal muscle in the resting, postabsorptive state is about 30 g./24 hrs., i.e., about 15 per cent of the total glucose turnover. Measurements in the dog with skeletal muscle studied in situ provide a considerably higher estimate. In either case part of the measured glucose utilization may be due to glycolysis by red cells in the muscle mass. Whatever glucose is utilized by resting skeletal muscle can almost entirely be accounted for as lactate released by the muscle; aerobic utilization is almost nil. The splanchnic area drained by the portal vein, i.e., gastrointestinal tract, pancreas, and spleen, is found in sheep to utilize about 20 per cent of the total glucose turnover. Both kidney and heart muscle are poor utilizers of blood glucose. In the resting state, kidney (cortex) may use glucose for only 10 per cent of its oxidative metabolism. Both of these organs extract more lactate and pyruvate from the blood than glucose and use fatty acids in general in preference to carbohydrate for oxidative metabolism. Resting skeletal muscle also oxidizes fatty acids as a major energy source. Liver utilizes insignificant amounts of glucose in the postabsorptive state; instead, production of glucose by the liver is active at this time. According to the amount of glucose consumed by red blood cells in vitro, it has been estimated that the total RBC mass may use 25 g. glucose/24 hrs., i.e., over 10 per cent of the total turnover. The white blood cells have a much higher rate of uptake of glucose per cell than the RBC, but the total number of WBC is much less, so that the white cells in the marrow and peripheral blood are estimated to glycolyze an additional 4 g. glucose/day. Whereas platelets are much less active in metabolism, their greater abundance than that of WBC results in a similar value of about 4 g. glucose/day utilized for glycolysis by platelets. The renal medulla is dependent on glycolysis from glucose for energy, but because of its small total mass the amount of blood glucose utilized by renal medulla seems to be on the order of 2 g./day for man.

Those tissues which use glucose primarily or exclusively by glycolysis, i.e., skeletal muscle, blood cells, and renal medulla, are therefore disposing back to the circulation per day about 50 g. or more of the blood glucose turnover, and to this is to be added a similar amount glycolyzed by brain. It is therefore evident that a large percentage of utilized glucose is returned to the blood as lactate (or alanine). Gluconeogenesis from part of this lactate and alanine accounts largely for the recycling phenomenon for ^{14}C described above.

Correlations of turnover of glucose-^{14}C (after single injection) and formation of $^{14}CO_2$ have indicated that amounts from 50 to 70 per cent

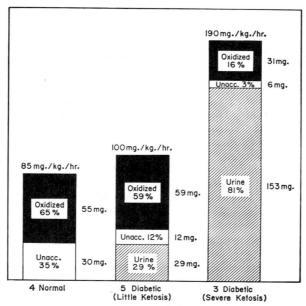

Figure 8-4. Turnover and disposal of body "glucose" by analysis of ^{14}C in blood glucose and breath carbon dioxide after single intravenous injection in human subjects. (From Shreeve: Amer. J. Med. *40*:724, 1966.)

of utilized glucose are oxidized fairly directly to carbon dioxide (Fig. 8-4), although studies by the priming-infusion technique have given somewhat lower values. In terms of another parameter 10 to 30 per cent of the total carbon dioxide derives from glucose normally in the postabsorptive, resting state, but, again, values are lower by the infusion than by the single injection technique. About 35 per cent of glucose-^{14}C injected intravenously in trace amount into normal humans has been converted to CO_2 (total amount in breath and body) in 3 hours; in rats about 50 per cent has already been expired by 4 hours, and in mice 60 to 70 per cent by 2 hours after intraperitoneal injection. Other studies on disposal of glucose-^{14}C in rats and mice have indicated that within 1 to 2 hours after parenteral injection about 4 to 6 per cent is in lipids (half in phospholipids), 5 to 10 per cent in protein and 7 to 11 per cent in glycogen. Less than 1 per cent is found in nucleic acids (mice). In the rat less than 5 per cent of glucose-^{14}C appears in the blood as glucose metabolites, mostly as lactate, 15 min. after intravenous injection.

Physiological Effects

Acute Effects of Feeding

To the introduction of exogenous glucose into the circulation there are certain responses designed to accommodate the imposed load both by diminution of the usual rate of endogenous glucose production and by acceleration of the rate of disposal by various routes. This is accomplished to a considerable extent by hormonal changes and particularly that of insulin. The capacity for adaptation to the acute load is known as glucose tolerance and is a common way of defining and evaluating a diabetic state. For such purposes there have been two main types of tolerance tests: by rapid intravenous injection of a load of glucose (usually from 25 to 75 g. in an adult human) or by oral administration of a single load of glucose (usually 50 to 100 g.). In man when there is a normal response of insulin output from the pancreas stimulated by the elevated blood glucose concentration, the intravenous load disappears at a logarithmic (constant fractional) rate of about 2 to 4 per cent/min. until the previous blood level is nearly restored. After an oral load the range of response accepted as "normal" is also somewhat variable. Some individuals tend to show a rise during the first 30 to 60 minutes up to a peak blood glucose concentration of about 160 mg./dl. with a subsequent decline to values usually less than about 120 mg./dl. at 2 hours after ingestion (or within 20 mg./dl. of the fasting concentration). Other individuals normally have a quite flat tolerance "curve," i.e., virtually no rise in blood glucose. Once it was supposed that this represented slow, and possibly impaired, absorption, but this has been refuted. It is furthermore now evident that variation in oral glucose tolerance among repeated tests in the same subjects can be as great as variation among different normal subjects.

It has become increasingly clear that normal glucose tolerance is achieved largely by the hepatic responses of increased glycogenesis, decreased gluconeogenesis, decreased glycogenolysis, and possibly increased glycolysis. For these hepatic changes to occur adequately there must be in response to the glucose load an output of pancreatic insulin, which then acts within the liver in ways described in Chapters 5 and 6. By the kinds of studies with carbon-14 described earlier, there is found in dogs a 50 per cent decrease in rate of hepatic glucose production after an oral load of glucose and in humans (studied also by glucose infusion) a decrease of 50 to 75 per cent. In pigs the decrease is even greater by ^{14}C analysis. The decrease may be 85 per cent in man according to splanchnic arteriovenous glucose balance. Although extra insulin is generally necessary for the normal degree of hepatic response, there is some evidence for regulation of glucose metabolism in the liver merely by the level of circulating glucose. Raising the blood glucose decreases hepatic glucose production and increases glucose utilization by the liver.*

---

*Even in the fasting animal, when the adaptive glucokinase has fallen to low levels, the correlation between hepatic glucose utilization and circulating glucose concentration is evident at high ranges of glucose concentration (550 mg./dl.) which are far above the K_m of hexokinase. This suggests the operation of some other mechanism, e.g., pyrophosphate; glucose phosphotransferase, to account for this hepatic glucose uptake (Chap. 4).

After an acute intravenous load the liver may absorb up to 50 per cent of this load, converting most of it to glycogen, as demonstrated in the dog. However, with an ordinary oral load of glucose, even though it passes first through the liver via the portal vein, the liver takes up only a minor amount, measured as about 15 per cent in the dog. After an oral glucose load in dogs the general utilization rate increases about 50 per cent. In normal humans infusion of glucose to produce elevations of about 20 to 35 mg./dl. is accompanied by slight (10 per cent) or no increase in absolute turnover rate. However, oral glucose loading (to produce slightly higher elevations) has caused a doubling of the disappearance rate for glucose-^{14}C as well as a similar increase in the extent of conversion of labeled glucose to CO_2. Particularly was there an increase in a component of CO_2 derived more slowly from glucose. This could be due to an increased rate of synthesis of intermediates such as glycogen or fat. The normal disposition of glucose into various organs or kinds of compounds under conditions of glucose load is little known and needs further study. Surprisingly little (about 5 per cent) is converted to carcass fat and less than 1 per cent to liver fat in the rat. The maximal disposal rates for glucose at very high concentrations in the blood, artificially imposed by infusion in normal rat, dog, or man, can be 5 to 10 times the rate in the fasting, postabsorptive state. The characteristics of the disposal rate-concentration relationship (in man) are those of a Michaelis-Menten equation, which suggests a saturable cell-membrane glucose transport system (Chap. 3). In the normal subject the K_m (for half-maximal velocity) has appeared to be 62 mg./dl., which is in the range of normal postabsorptive glucose concentration. The V_{max} (maximal transport velocity) is approximately 60 g. glucose/hr./m^2 body area.

The kinetics of circulating glucose after load depend upon the nature and degree of pancreatic insulin response and upon the response of the tissues to the elevated blood insulin. In general, the pancreas is able to respond to glucose with previously stored insulin and with insulin newly synthesized. Within these categories there may be other multiple phases of insulin response. Probably related to this is the observation that after intravenous injection of glucose in the dog there are harmonic oscillations of the rates of appearance and disappearance of glucose which are only partially damped and have periods of 1 to 2 hours. In normal humans the small storage pool of insulin seems to be an important determinant of the intravenous glucose tolerance.

Multiple-humped curves of blood glucose concentration after oral administration of glucose are occasionally observed and periods of 1 to 3 hours are normally found. Gatewood and co-workers ascribe this phenomenon partly to the division of the glucose pool into hepatic-portal glucose and systemic glucose when the load is introduced primarily into the former compartment (oral administration). Such computer-based analysis may also require consideration of multiple components of insulin response, not to mention other hormones involved acutely in blood glucose regulation. Besides the multiple pancreatic response to blood glucose elevation, there is after oral administration of glucose the release of insulin-stimulating "intestinal glucagon" (plus secretin and perhaps other factors), which adds to the intensity and complexity of pancreatic insulin production.

The oral glucose tolerance is said not to be significantly affected by differing rates of intestinal absorption, provided an early maximal absorption is assumed. However, the latter is questionable in view of evidence for prolonged gastric retention of hypertonic glucose solutions, which can be quite variable. Since there is some evidence for increased rates of intestinal glucose transport in diabetic subjects, this component plus the question of difference in stomach emptying are factors which deserve further consideration in the evaluation of oral glucose tolerance tests.

One aspect of the oscillatory characteristic of the blood glucose curve in tolerance tests is a phase (or phases) of hypoglycemia, which in an oral load test usually occurs from 2 to 5 hours after ingestion and may be accompanied by symptoms. There has been increasing recognition that an exaggerated extent of this phenomenon, known as "reactive hypoglycemia," occurs rather commonly in otherwise normal subjects, perhaps up to one-quarter of the population. The condition may be an early manifestation of a derangement of insulin production which can precede a diabetic state (see below). There is in normal subjects (and to a lesser extent in obese or diabetic patients) a diurnal variation in glucose tolerance with higher blood glucose values following an oral load in the afternoon than in the morning. In the afternoon about one-quarter of normal subjects have a tolerance curve which would be considered in the diabetic range by usual criteria. The prognostic significance of this "afternoon diabetes" is not presently known.

As tested by intravenous glucose tolerance, the occurrence of alimentary hyperglyceridemia, produced by a meal of butterfat, is accompanied by increased rate of glucose removal. Elevation of free fatty acids in the blood does not influence the intravenous glucose tolerance, but elevation of glycerol enhances it. One theory for the effect of ingestion of triglycerides is a stimulation of release of insulinogenic enteric hormones. Another theory relates to the possible stimulation of insulin secretion by ketone bodies (observed in dogs); fat ingestion may be accompanied by hyperketonemia.

Certain amino acids (leucine, arginine) can stimulate insulin secretion and leucine may independently decrease gluconeogenesis. Tryptophan and its metabolites may in various ways affect carbohydrate metabolism, including either production or utilization (Chap. 5). Other amino acids can also interact with carbohydrate metabolism.

It is obvious that glucose tolerance can be a resultant of many genetic as well as dietary and other environmental factors. Glucose intolerance is not a single homogeneous disorder which expresses a simple autosomal recessive trait. At least 16 distinct genetic disorders, with mutations at presumably different loci, may be associated with abnormal glucose tolerance. Therefore, the occurrence of glucose intolerance in an individual is not easily predictable from the antecedent family history. It will be affected by antecedent diet, by age, and by other factors as discussed below.

Chronic Dietary State

In normal subjects or patients with very mild diabetes, the oral glucose tolerance is improved by diets high in carbohydrate (dextrose or polymers of dextrose). This effect is not associated with increased insulin levels and has therefore been attributed to increased sensitivity of peripheral tissues to insulin. Adaptive changes in enzymes of glucose utilization is another explanation. This kind of effect of high-carbohydrate diet does not hold for severe insulinopenic diabetics, whose glucose tolerance, on the contrary, is worsened by high carbohydrate feeding. In the opposite direction a low-carbohydrate diet in normal man decreases both the intravenous and the oral glucose tolerance. In clinical testing it is customary practice to assure that candidates for glucose tolerance tests consume at least 150 g. of carbohydrate per day for about 3 days prior to testing, so that "false positive" results will not be encountered. The effect of low- or high-carbohydrate diet on the fasting (postabsorptive) blood glucose level seems to be variable.

Meal-fed rats have a 40 per cent greater absorptive rate for glucose than do nibbling rats, which is correlated to an increase in intestinal weight. However, in the meal-fed rats there is a slower rise of blood glucose; evidently this is due to a greater capacity for disposal of the glucose load. A greater capacity for fatty acid synthesis from glucose carbon in adipose tissue and greater glycogenesis by muscle (and perhaps liver) are characteristic of meal-fed rats and help to explain the greater disposal capacity. It can be surmised that insulin mediates these changes by both chronic and acute effects.

There can be special changes in glucose tolerance and other parameters following a diet high in sucrose. These will be discussed in the section on fructose, below.

The effect of chronic dietary factors other than carbohydrate on glucose tolerance or glucose turnover is little understood. A minimum level of protein in the diet appears to be necessary for normal glucose tolerance. Human subjects with prolonged protein deficiency (as in kwashiorkor) commonly have glucose intolerance. It is not known whether this is due to the effect of particular amino acids, as suggested above, or to a more general effect of protein nutrition on maintenance of essential enzymes of carbohydrate metabolism. Some minimal amount of potassium intake may be a significant factor in glucose tolerance; K^+ supplementation helps to sustain normal glucose tolerance in fasting obese subjects. A deficiency of chromium (Cr^{+++}) leads to impaired glucose tolerance in animals and humans. Some evidence suggests that administration of Cr^{+++} helps to restore glucose tolerance in patients with kwashiorkor or in normal elderly subjects. Others have observed improvement of adult human diabetics after giving chromium. Adequate levels of manganese in the diet seem to be needed for adequate glucose tolerance. Manganese-deficient guinea pigs show diabetic-like glucose tolerance curves. Clinical deficiencies or any relationship to diabetes mellitus are not clearly identified. The mean manganese content of blood of diabetics is approximately half that of normal, but administration of Mn^{++} does not ordinarily lower blood glucose levels in normal or diabetic subjects.

Prolonged Fasting

When normal physical activity is maintained during continued fasting in humans, the blood glucose declines slowly during the first few days to a minimum of about 40 to 50 mg./dl. at about 5 days. Usually the blood glucose concentration then returns to normal during the second week. Even at the time of nadir in blood sugar there are generally no hypoglycemic symptoms. The central nervous system is probably already shifting over from a reliance on blood glucose to the utilization of ketone bodies for energy. The latter change is a distinctive feature of starvation. After a week or more of fasting by humans the cerebral glucose consumption may decrease by 75 per cent and ketones (beta-oxybutyric acids produced by the liver during excessive lipid mobilization and hepatic oxidation) become the predominant fuel for the CNS. This allows the hepatic glucose production and the early postabsorptive "swell" in gluconeogenesis (Chap. 5) gradually to decrease during prolonged fasting.

In dogs after 2 or 3 days of fasting the rate of turnover of blood glucose has decreased to about two-thirds of the value in the early postabsorptive state. In sheep the decrease in turnover by this stage of fasting is only 10 per cent, but in cows fasted for 2 to 5 days there is a decrease to about 50 per cent of normal in flow of glucose through the system. In such fasted cows there was a decrease in the mass of the primary glucose pools but an increase in a third larger pool of inter-mixing non-glucose substances. After a week of starvation by human obese subjects there was a 30 per cent reduction in glucose pool size, turnover, and oxidation according to the priming-infusion technique with glucose-1-^{14}C, including $^{14}CO_2$ analysis. After several weeks of starvation by such subjects the total glucose output, according to hepatic and renal arteriovenous measurements, has fallen to one-third of the initial value. By this time the liver is producing only about one-half of the total glucose production and the kidney the other half instead of the normal ratio of about 5 or 10:1 for liver:kidney. Likewise in dogs after 30 days of fasting the glucose turnover rate has declined to about one-half of the value in the early postabsorptive state.

Blood alanine is ordinarily pre-eminent quantitatively as a precursor for hepatic gluconeogenesis (Chap. 5). It accounts for about 50 per cent of the total amino acid consumption. The decline in hepatic gluconeogenesis appears to be consequent to a decline in arterial levels of alanine. This decline in turn depends upon a marked diminution of release of alanine from the muscle tissues. If the all-important protein reserves were allowed to deplete at the initial rate in early fasting, the long-fasting individual would much sooner become moribund. If stress accompanies the fasting the mechanism for protein conservation may be significantly compromised (see below). The uptake of glycine by liver in prolonged fasting also diminishes markedly, even though this amino acid does not fall in arterial concentration but instead tends to rise. In place of liver the kidney takes up glycine increasingly (fivefold) as the arterial glycine concentration rises. This helps to account for the expanded role of the kidney in gluconeogenesis during a prolonged fast.

During a fast of one week in obese humans the amount of blood lactate converted to glucose increases about 50 per cent and the fraction of glucose turnover derived from lactate increases about twofold. The gluconeogenesis from lactate may be sustained or increased by the increased influx and metabolism of free fatty acids by the liver (Chap. 5). Also the increased concentration of FFA, as starvation progresses, may act as a brake to utilization of glucose by various peripheral tissues (the Randle hypothesis of the glucose–fatty acid cycle). The utilization rate decreases, of course, just as do the rates of appearance and turnover, since the blood glucose concentration is somehow sustained, as already mentioned. In 4-day fasted rats the oxidation of an intraperitoneal load of glucose-^{14}C is only one-third the value in the fed rat, and the extent of syntheses of fatty acids, glyceride-glycerol, and non-saponifiable fats from glucose-^{14}C by isolated adipose tissue and liver from such rats is greatly decreased. The reduction of glucose tolerance even by a low-carbohydrate diet has been mentioned above—during extensive fasting in normal human subjects the glucose intolerance is even greater and constitutes a diabetic-like state. The main mechanism of this glucose intolerance appears to be a decline in capacity of the pancreas to respond to glucose challenge with a prompt and adequate output of insulin.

Hibernation among certain animals is a special case of prolonged fasting. The blood glucose concentration declines moderately (about 20 per cent) during hibernation by ground squirrels. The capacity for renal gluconeogenesis increases during hibernation and probably for both hepatic and renal gluconeogenesis during arousal from hibernation. Formation of CO_2 and fatty acids from glucose-^{14}C is very low in hibernating animals relative to formation of protein, glyceride-glycerol and glycogen. Evidently the energy supply during hibernation comes predominantly from non-carbohydrate sources and glucose turnover is considerably reduced in comparison with the aroused, active state. Whereas in the active state only 9 per cent of injected glucose-^{14}C remains after 1 hour in the extracellular space (of ground squirrels), in the hibernating state 63 per cent still remains after 24 hours. The rate of utilization is thus sixtyfold less. The metabolic rate itself is decreased about forty-fivefold.

Hormonal Effects

Insulin

The biochemical effects of insulin on gluconeogenesis, glycogen metabolism, glucose transport, and glucose utilization, all of which are components of glucose turnover and tolerance, have been discussed in previous chapters. Insulin acts on both production and disposal of blood glucose in ways which cooperatively tend to lower the blood glucose concentration. In addition insulin seems to increase the space available to the mass of circulating glucose, which further augments the hypoglycemia. As tested with glucose-^{14}C in cows or dogs, the acute administration of insulin has appeared to increase the size of the glucose space on the order of 50 per cent or more. Some studies suggest increases in intracellular hepatic concentration of glucose following insulin, in spite of contrary findings with analogues of D-glucose in vitro (Chap. 3). The concentration of glucose in other organs, e.g., muscle, may also be increased by insulin. Analysis of dynamics of glucose pools (in the cow) indicates that insulin restricts the backflow of glucose from a secondary to the primary pool.

Insulin modulates the blood glucose concentration and turnover probably more by a regulation of glucose production than by a difference in utilization during the usual daily shifts in nutritional state. The increased output of insulin in response to elevated glucose goes directly via the portal vein to the liver. There it acts locally to limit the hepatic glucose output. After peripheral infusion of insulin (in sheep), there is a delay of about 80 minutes before the entry rate of glucose into the circulation declines, whereas after portal infusion there is a prompt reduction in entry rate which drops to about half of the initial value by 30 minutes after start of infusion. Evidence and mechanisms for rapid, direct effect of insulin on hepatic gluconeogenesis and glycogenesis have been discussed in Chapters 5 and 6. In addition, with insulin availability there is a diminished uptake by the liver of alanine, the prime substrate of gluconeogenesis. This can occur despite unchanged arterial blood levels of alanine.

Some credence may be given to the concept that insulin depresses gluconeogenesis (and thereby hepatic glucose production) secondarily and indirectly by limitation of flow of FFA to the liver (Chap. 5). However, at present this mechanism appears to be more influential in pathophysiological states (including unusual fasting) than in normal daily regulation, which depends upon direct action on the liver by small changes in circulating insulin.

Insulin lowers the concentration of blood glucose at which peak rates of utilization may occur but does not increase these saturation rates. This argues for a supposition (mentioned in Chap. 3) that insulin may lower the K_m for components of transport or utilization, thereby changing the efficiency of utilization at different blood glucose levels. In regard to the effect of insulin on the general fate of blood glucose, it is well established that glycogen and lipids show most pronounced quantitative changes, i.e., increases in content after the action of insulin, although the general anabolic effect extends to proteins and possibly to nucleic acids. Upon acute administration of insulin in vivo (in the mouse) there are two- to threefold increases in the conversion to lipids of glucose-^{14}C injected subsequently. These increases are mainly in the class of non-polar lipids (fatty acid moieties) and to a lesser extent in polar lipids (glycerol moiety). Clinically these relationships are difficult to verify, although there are inferences in the observations that obesity in both man and animals is somehow correlated with hyperinsulinism.

The multiple effects of insulin on carbohydrate metabolism are difficult, and perhaps impossible, to comprehend by attributing these effects to a common biochemical mechanism of hormonal action. Rather, the hormone appears to exert multiple mechanisms, all of which are directed toward positive adaptation to the incursion of food (mainly carbohydrates, but also other kinds). Insulin is the hormone of "good times," i.e., the feasting state. From an evolutionary point of view the multiple adaptation mechanisms appear to have become properties of a single protein molecule rather than to have been developed by the more cumbersome means of multiplication of protein or other molecular species of hormones. This principle seems to be extended to other hormones, to substrates which act as cofactors in positive "feed-forward" or negative "feed-back" mechanisms. The plentipotentiality of insulin (or at least some of it) may be linked to that of the hormonal "second messenger," cyclic AMP, which has multiple intracellular effects.

Adrenal Steroids

The influence of the "glucocorticoids" (11-oxysteroids) on carbohydrate metabolism has also been discussed in other chapters. In general, the effects are quite opposite to those of insulin. There is a major tendency to an increase in gluconeogenesis and a quantitatively minor, but perhaps qualitatively important, effect of inhibition of glucose utilization. Inhibitory effects are identified for certain peripheral tissues, e.g., lymphoid tissue, skin, and adipose tissue. The catabolic effect of deficient glucose metabolism in these tissues may be one of the factors which initiate the flow of amino acids from such sites to the liver, where they provide the substrate for increased transamination and gluconeogenesis.

The full effect of excessive adrenal steroids on the hepatic glucose production usually includes an increased blood glucose concentration and an increased glucose mass. The turnover has not been evaluated, but would probably be increased on an absolute basis in view of increased production. In the converse situation, i.e., adrenalectomy, in rats there is a diminution in plasma glucose concentration, in glucose mass, and in absolute rate of glucose turnover. The half-lives of neither glucose-^{3}H nor glucose-^{14}C are changed in these adrenalectomized rats, indicating unaltered fractional turnover rates and no change in the extent of recycling via lactate plus alanine. Since the conversion of alanine-^{14}C to glucose is much reduced in such rats, the proportional contribution from lactate to recycling may be increased.

The adrenal glucocorticoids have a fundamental role in the general "alarm reaction" to stress, as elucidated by Selye. Insofar as carbohydrate metabolism is concerned, this role seems to be one of protection and support of the vital organs, mainly the brain, by assurance of glucose supply in times of acute stress which historically in evolution would be typically accompanied by partial or complete interruption of feeding. The carbohydrate changes are part of a larger set of changes which are essentially adaptive to exigencies of the environment. The glucocorticoids "permit" the operation of other adaptive mechanisms in the body. However, the glucocorticoids also function in habituation, i.e., they set a limit to bodily responses which could otherwise be overreactive (e.g., by excessive inflammatory response) to noxious, unusual stimuli. Curtailment of carbohydrate utilization by some tissues (e.g., leukocytes in the case of inflammation) is one likely mechanism by which the glucocorticoids exert this "go-easy" function. In any case, clinical and experimental evidence indicates that adjustments, including those of carbohydrate metabolism, to even minor environmental disturbances are dependent upon the modulating effects of the adrenal glucocorticoids.

Growth Hormone

This polypeptide from the posterior pituitary has significant effects on carbohydrate and fat metabolism which are not properly recognized by the name which it gained mainly on the basis of its most obvious clinical effects. The blood levels of this hormone are in essentially opposite phase to that of insulin. That is, the level decreases in the postprandial period and rises during the postabsorptive period. At the latter time a major effect seems to be the facilitation of lipid mobilization which is accompanied by a rise in FFA. This rise can explain partly the inhibitory effect of growth hormone on glucose uptake from the blood, but the full effect of this inhibition follows some time after the FFA rise and includes some more directly inhibitory action on glucose uptake by peripheral tissues. One facet of the complex functions of this hormone is an early transient increase in glucose uptake from the blood (in dogs). This may be consequent to stimulation by growth hormone of insulin secretion. The latter effect is a direct one on the pancreas (possibly assisted by FFA stimulation) and is evidenced by exceedingly high blood insulin levels in patients with acromegaly, the clinical state of excessive growth hormone due to pituitary tumor. Yet, having once provoked this insulin output, growth hormone by its powerful inhibitory effect on glucose uptake seemingly "turns around" and nullifies or limits the action of the excessive insulin. The net result is a degree of glucose intolerance which is equivalent to diabetes in as many as one-third of patients with acromegaly. In dogs growth hormone appears to increase the output of glucose by the liver and increase glucose turnover, but in humans the evidence is negative in this regard and, in fact, a decreased glucose turnover after growth hormone suggests predominance of the inhibitory effect on glucose uptake. Perhaps it is best to view the function of this hormone also as protective and supportive to the organism in terms of conservation of the vital resources of carbohydrate and protein, which function it exerts partly by mobilizing lipid as the most "expendable" fuel. By stimulating insulin output and also by increasing amino acid uptake by some tissues the hormone serves anabolic functions and thus deserves its name of growth hormone.

Gonadal Steroids

The effect of ovarian steroids, primarily estrogens, on carbohydrate metabolism is seen most clearly in pregnancy. Pregnant women have an increased incidence of glucose intolerance which ranges from about 10 per cent by the standard oral test to about 30 per cent by the cortisone-sensitized oral test. Results with the intravenous test have been variable, but there is also higher than normal incidence by this test. So far a finding of oral glucose intolerance in pregnancy, except in frankly diabetic women, has not been correlated with obstetric complications in the mother or child. However, a positive test during pregnancy is more likely in women with an underlying diabetic trait and may foretell diabetes in the mother occurring years later even though the test reverts to normal after pregnancy.

The high incidence of glucose intolerance by the cortisone glucose tolerance test is thought to be due to potentiation of action of cortisol by estrogens. The latter steroids increase the amount of cortisol-binding globulin (CBG) in plasma. Combination of the glucocorticoid with CBG may provide a depot of slowly metabolized corticoid, which becomes available particularly in the liver where the steroid permits excessive gluconeogenesis. Protein mobilization and gluconeogenesis are both activated more extensively in early fasting of the pregnant compared with the non-pregnant rat. Pregnant sheep have a higher glucose turnover rate than non-pregnant (but normal glucose mass and space). The increased gluconeogenesis and turnover, particularly during early fasting, perhaps is a protective mechanism to provide a continuously feeding fetus nourished by an intermittently eating mother.

Some effects of estrogen on blood glucose characteristics can be noted in the non-pregnant woman. Glucose tolerance is greatest just at the beginning of the menstrual cycle and appears to bear an inverse correlation with the level of endogenous estrogens. Curiously the effect of estrogen in diabetics seems to be paradoxical in that improvement in glucose tolerance may occur upon administration of the steroids. In diabetic females control of the diabetes may be more difficult about the time of menstruation. Providing estrogens to female acromegalic patients has appeared to oppose the action of growth hormone on carbohydrate metabolism by decreasing the fasting plasma glucose concentration and improving the oral glucose tolerance. The effects, however, have been transient.

Prolactin

The effect of this pituitary polypeptide hormone on carbohydrate metabolism has been little studied. In the hypophysectomized dog administration of prolactin for several days elevates the blood glucose concentration, decreases sensitivity to insulin, and ameliorates secondary hypoglycemia in the glucose tolerance test. In the normal dog also fasting glucose concentration is increased, as is the glucose turnover. Whether by its effect on lactose synthesis by mammary gland (Chap. 6) or by some extramammary effect of prolactin, during lactation in sheep there is a threefold higher rate of turnover of blood glucose, while the glucose mass and space of distribution remain the same as in non-lactating sheep. The rate of lactose formation in lactating sheep indicates that up to 60 per cent of the total glucose turnover could be contributing to lactose formation (which is known to utilize blood glucose carbon almost entirely). In fed lactating goats also the removal of glucose from the blood traversing the mammary gland accounts for 60 to 85 per cent of the total glucose turnover. In fasting lactating goats the blood glucose entry rate is only one-third that of fed lactating goats, and of the circulating glucose only about 40 per cent is utilized by the udders. These large decreases in glucose uptake during fasting are accompanied by falls in mammary blood flow and O_2 uptake to half the values in the fed state. The fall in mammary uptake of glucose is the main reason for the fall in milk secretion. The importance of adequate nutrition to maintain accelerated rates of hepatic glucose production (or external provision in the case of non-ruminants) and thus meet the large needs of the mammary gland for blood glucose during lactation is evident from these studies.

Epinephrine and Glucagon

In dogs peripheral intravenous injections of epinephrine or of glucagon produce about the same degree of hyperglycemia, and in each case it is due to a rapid, several-fold increase in rate of appearance of glucose into the circulation, as indicated by the glucose-^{14}C priming-infusion technique. The appearance is more rapid and intense with epinephrine than with glucagon. With epinephrine the increase is about sixfold during the first 4 minutes. With either hormone the rate of glucose disappearance increases about one and one-half to twofold. The absolute rate of blood glucose turnover is therefore markedly accelerated. In the case of epinephrine, as the hormone of emergency reaction, this is evidently a part of the reaction which would permit and sustain extraordinary, prompt, and extreme muscular activity as well as other functions. Epinephrine is called forth when the blood sugar concentration falls to hypoglycemic levels (about 40 mg./dl.); the effect is more pronounced the faster the fall. The relatively slow and weak effects of epinephrine directly on the liver suggest that the rapid response to in vivo injection is an indirect effect on hepatic glucose production. The epinephrine could allow increased hepatic output by transiently inhibiting insulin release, which is a well-established effect of epinephrine on the pancreas. Another possible mechanism is via the central nervous system, since there is known to be a "hyperglycemic center" in the floor of the fourth ventricle which is responsive to elevated blood levels of epinephrine. The further mechanism of hyperglycemia (whether hormonal or neuronal) after activation of this center is not known.

Glucagon through its gluconeogenic action (and also by glycogenolysis) may play a significant role in physiological control of glucose turnover. Glucagon lowers blood levels of amino acids and increases urea excretion; presumably it facilitates hepatic uptake of amino acids and their conversion to glucose. In depancreatized dogs there is a prompt and large increase in glucose production upon intravenous infusion of glucagon with prompt hyperglycemia unless amounts of insulin at least 10 times the basal rate are administered. Such insulin does not counteract the effect of glucagon on hepatic production, but maintains the glucose concentration by increasing the rate of disappearance of glucose. Glucagon itself stimulates insulin release which in normal dogs is sufficient to prevent more than slight increase in glucose concentration following the administration of glucagon. The secretion of glucagon appears to have a prandial rhythm. The plasma level falls during glucose infusion and rises during starvation.

Thyroid

Although there is no prominent effect of thyroid hormone on control of glucose metabolism, patients with hyperthyroidism sometimes exhibit mild glucose intolerance. Frank diabetes is more common in patients with thyrotoxicosis than in the general population. The high level of plasma FFA in hyperthyroidism might be responsible for either increased gluconeogenesis or decreased glucose utilization or both. Another possible interacting mechanism is via epinephrine, the action of which is augmented by thyroid hormone and could include an inhibitory effect on insulin release by the pancreas. Nevertheless, studies with ^{14}C have indicated that the turnover rate of glucose in hyperthyroid patients in the postabsorptive state is within normal limits. Analyses with position-labeled glucose have suggested some abnormalities particularly in the pentose cyle pathway, but the results so far are contradictory.

In hypothyroid patients studied with glucose-^{14}C the turnover of glucose has appeared to be less than normal and the glucose mass reduced in size. The latter finding relates partly to low levels of blood glucose concentration in the postabsorptive state. Abnormalities of recycling of glucose-^{14}C have not been investigated in thyroid patients but might help explain other abnormal phenomena. An increased incidence of glucose intolerance in hypothyroid patients and a higher than normal rate of hypothyroidism in diabetics are both reported. The seeming paradox of a diabetic trait or tendency with either hyper- or hypothyroidism may relate to the autoimmune kind of thyroiditis which is frequently part of the pathogenesis of hypothyroidism. There is a high incidence of thyroid antibodies in the serum of diabetics.

Hypothalamus

There is a role for the hypothalamus in promoting hepatic glucose production in that either electrical stimulation of the hypothalamus or glucopenia of the area will activate the production of adrenal glucocorticoids, evidently via the ACTH-releasing factor from the hypothalamus and ACTH from the pituitary. This introduces the possibility, increasingly recognized and given supportive evidence in recent years, that emotional factors can affect such metabolic functions as the production and turnover of glucose in the blood. Pathological disturbances of the hypothalamus may give rise to hyperglycemia or be associated with other evidence of diabetes.

Exercise

During moderate exercise the rate of utilization and oxidation of total extracellular glucose, as determined by glucose-^{14}C kinetics in man and dog, increases about two- to threefold. The utilization by the exercising muscle itself may increase ten- to thirtyfold. During this kind of exercise the blood glucose concentration declines slightly. However, this means that the glucose production rate has increased almost to the same extent as utilization in order to avoid hypoglycemia. Even though blood lactate concentration may not increase appreciably during such exercise, increased gluconeogenesis from lactate seems to be a central factor in the implementation of increased glucose production. Plasma glycerol available from increased peripheral lipolysis during exercise is another source of increased gluconeogenesis. As exercise becomes more intense, the plasma glucose concentration does not decline further but in fact may show an increase. This probably depends upon the increasing role of rising blood lactate concentration, which in turn leads to greater gluconeogenesis. That is, the Cori cycle comes increasingly into play. Glycogenolysis in both muscle and liver, perhaps evoked by more epinephrine, also becomes a more significant factor during intense exercise.

In spite of the increased turnover and utilization of carbohydrate during exercise, this accounts for only a small part of the increased oxidative rate. In fact, turnover and utilization of FFA are increased even more than that of carbohydrate. The relative contribution of plasma glucose to total CO_2 production actually declines. Of the required energy during exercise in dogs, 70 to 80 per cent is furnished by FFA and only 10 to 15 per cent by glucose. This is true even if exogenous glucose is supplied to the exercising dog. The large dependence on FFA for fuel is characteristic of chronic exercise rather than acute, intense exercise. The use of FFA and the conservation of glucose for expenditure of energy by working muscles are another example of the way in which the organism protects its vital central nervous system by avoiding the extravagant use in other tissues of glucose, which is needed to supply more critically the CNS. When extra glucose is continually supplied to the exercising dog, the additional glucose utilization appears to go into storage substance, i.e., glycogen and fat. The rate of this storage function under conditions of hyperglycemia in the normal dog can be just as great during exercise as at rest. Indeed, exercise (unlike insulin) in dogs may increase the saturation rate (V_{max}) for glucose utilization by 50 g./m.^{2}/hr. In the absence of exogenous glucose it is the prior store of muscle and liver glycogen which ultimately determines the point of exhaustion in severe exercise. Pretreatment of dogs with methylprednisolone exaggerates the increase in turnover of glucose produced by exercise and allows dogs to run approximately twice as long. This increase in carbohydrate turnover and increased exercise tolerance is attributable to a larger store of hepatic glycogen in the dogs treated with the glucocorticoid.

The amount of physical activity before and during a glucose tolerance test very significantly affects the degree of tolerance. It is well known that normal persons tend to have a diabetic response to a glucose tolerance test when bedridden and an increased tolerance after performing muscular work. The increased glucose tolerance is related to the intensity of exercise rather than the total amount of work. The exercise effect seems not to involve a difference in insulin response to glucose, which is consistent with other findings about the "exercise factor" that promotes glucose utilization (Chap. 3). The exercise effect seems to depend upon the presence of insulin, however, and at least in part it may be attributed to increased utilization of insulin consequent to greater blood flow. Some other humoral factor, as claimed by some investigators, is not precluded.

Age

Hypoglycemia of the newborn and to a lesser extent older infants and young children is a common phenomenon and has been investigated extensively. In brief, there may be an imbalance of production of the various hormones governing glucose metabolism or a deficiency of responsiveness of the tissues to some hormones. For instance, in young puppies or human infants the liver may not respond to glucagon or epinephrine with adequate increases in glucose production at needed times. Nevertheless, the basic turnover rate of glucose in newborn puppies is two to four times that of adult dogs (per kg. body wt.). In some infants and young children there may be inappropriate secretion of insulin; this may be due to unusual sensitivity to stimulation by some amino acids (e.g., leucine) or to an improperly operating "glucostat" for release of pancreatic insulin.

With advancing age (middle age and elderly), there is a well-known gradual decrease in glucose tolerance, though the mechanism of this effect is not clear. Increasing blood glucose levels in the glucose tolerance test with advancing age seem not to be associated with any substantial decreases in blood insulin levels, yet some alteration of the release mechanism may occur so that the higher blood glucose level is needed to provoke the usual insulin response. There may be a diminishing sensitivity of peripheral tissues to insulin with age, since insulin levels during glucose tolerance tests tend to increase in the absence of notable change in glucose tolerance. In ambulatory persons over 60 years of age, about 15 to 20 per cent of an unselected group of subjects are likely to have glucose tolerance in the diabetic range and after 65 the incidence rises to 25 per cent or more.

Cold Exposure

Although the effect on glucose turnover of either acute or chronic cold exposure has not been measured, indirect evidence suggests that it is markedly increased. In the cold-stressed rat which is shivering (in the acute stage), gluconeogenesis is found to be increased as much as fivefold. This extra new glucose may be utilized by skeletal muscle with a comparable increase in lactate formation and the Cori cycle, which may supply extra ATP to the muscle for shivering activity. The increase in gluconeogenesis found in cold-exposed rats is also the resultant of substantial protein and fat catabolism, which are dependent upon hormonal (glucocorticoid and thyroid) effects. Besides supplying further substrate (glycerol) for gluconeogenesis, the excessive breakdown of triglycerides and flow of fatty acids to the liver stimulates gluconeogenesis (Chap. 5). Cold exposure increases markedly the capacity of rats to oxidize an intraperitoneal load of glucose. Over a 4-hour period the amount of glucose-^{14}C converted to $^{14}CO_2$ may be twice as great as in cold-exposed rats. Hepatic lipid formation is less in such rats.

Other Environmental Stress

Exposure of young men to a hypobaric-hyperoxic environment for 30 days adversely affects oral glucose tolerance in a way which may persist for at least two weeks of recovery period. Regular exercise during the exposure and recovery period does not alter this decreased glucose tolerance. Under high-altitude (hypoxic) conditions mice oxidize less blood glucose to CO_2 and convert more to glycogen of liver and skeletal muscle. The over-all oxidation of glucose intermediates to CO_2 does not appear to be abnormal, but there is a retention of unidentified glucose metabolites in certain tissues, i.e., liver, muscle, and heart. Although no data are yet available on the effect of either deceleration or weightlessness on glucose turnover in mammals, chickens exposed to 3 G for a prolonged period (24 weeks) have increased flux and utilization rates, presumably to help meet increased energy requirements.

Pathological Changes

Diabetes Mellitus

The present discussion can only briefly review the effects of diabetes on aspects of glucose turnover, utilization, and fate, which have been investigated extensively. The early concept derived from the glucose balance studies by Soskin and Levine—that at an elevated blood glucose concentration the diabetic animal or man utilizes glucose at an essentially normal rate—during the past 20 years has been verified and given more factual support by dynamic techniques using ^{14}C-labeled glucose. Thus, in maturity-onset, non-ketotic human diabetics with postabsorptive blood glucose concentrations generally above 200 mg./dl., the technique of single injection of tracer glucose-^{14}C has repeatedly demonstrated that the absolute rate of glucose turnover is equal to or greater than that of non-diabetic subjects (Fig. 8-4). Even when allowance is made for increased recycling of glucose through the Cori cycle, which has sometimes appeared to be 50 to 100 per cent higher than normal in diabetics, the irreversible loss rate has been as high in maturity-onset diabetics as in normal subjects. Because of a considerably increased glucose mass the fractional rate of removal or turnover of blood glucose in such diabetics is diminished. In juvenile-type, ketotic diabetics with more profound insulin lack the total turnover rate of glucose can be much higher than normal, but much of this turnover at higher blood glucose levels is lost in the urine, so that the absolute amount utilized by the tissues may be only about 50 per cent of normal (Fig. 8-4).

An increased turnover rate implies an increased production rate and, as discussed before, in the diabetic this is achieved largely via increased gluconeogenesis from the major substrates, i.e., alanine, lactate, glycerol, and others. Hormonal control within the liver—by a lack or ineffectiveness of insulin and/or by an excess of glucogenic hormones—is responsible for this increased gluconeogenesis and production rate in diabetes. Relatively high postprandial levels of glucagon in human diabetics (either adult or juvenile-type) suggest a failure of the normal physiological suppression of secretion after carbohydrate intake. The mechanism may be a failure to suppress levels of blood amino acids, e.g., alanine, which stimulate a release of glucagon. Hyperglucagonemia in diabetes emphasizes the fact that glucose overproduction rather than underutilization is a predominant characteristic of the disease. This has been further emphasized by the excessive extent of conversion of ^{14}C-labeled precursors to blood glucose in diabetics.

Development and use of radioimmunoassay in the past 15 years has revealed that, whereas juvenile-type diabetics clearly have a profound deficiency of insulin production by the pancreas, the type of diabetic who gets the disease in a rather mild form usually in the middle adult years is often able to respond to exogenous glucose with virtually normal amounts of insulin, although somewhat slowly. Theories about this type of diabetes include disturbances of amount, timing, and biological quality of the insulin released by the pancreas in response to stimulation by glucose or other substrates (amino acids, fatty acids), which in turn may be modulated by other hormonal influences on the pancreatic insulin production. Another concept is some insensitivity or resistance to the action of insulin, whether it is an innate phenomenon of the glucose-utilizing cells or, again, a conditioning of the cells and tissues by the influence of hormones (e.g., glucagon, growth hormone, glucocorticoids, insulin "fragments") or substrates (e.g., FFA) which oppose insulin at the cellular level. In any case, diabetics of the maturity-onset type have resistance not only to their own insulin but to exogenous insulin. There is an impedance to the utilization of glucose which seemingly does not allow a normal rate of utilization at a normal blood glucose level, even when the same amount of insulin is present.

Some evidence suggests that in maturity-onset diabetics the glucose homeostasis after perturbation occurs relatively normally with respect to restoration of the elevated postabsorptive level. Thus, an additional increase in blood glucose above the abnormal fasting level calls forth insulin in amounts adequate to return the glucose to the original elevated level. It is postulated that in such diabetics other regulatory mechanisms for insulin secretions, e.g., FFA concentrations, may be functioning in ways which do not allow a normal fasting level of glucose to be maintained. From another viewpoint, the pancreas may share the diabetic state and can provide a normal rate of insulin output only in response to elevated levels of blood glucose. The concept of a fundamental change in the glucose transport kinetics in diabetes receives further support from the work of Moorhouse and co-workers, who calculated that the K_m of blood glucose concentration, i.e., for half-maximal transport velocity, was about 140 mg./dl. for either non-ketotic or ketosis-prone diabetics, compared with about half this value for healthy subjects.

Thus the preservation of fasting euglycemia seems to be of relatively low priority in glucose homeostasis. The diabetic organism permits the blood glucose to rise in order to achieve an adequate rate of glucose utilization. This mechanism seems to be operable in preference to an increase in plasma insulin, which might achieve the same purpose at normal blood glucose levels. The therapeutic moral seems to be one of complying with nature to the extent of allowing moderate hyperglycemia in diabetics and not overadministering insulin in attempts to achieve euglycemia. Such attempts in the past have contributed to the "brittleness" of juvenile diabetics by provoking the "Somogyi effect," which is a rebound of hyperglycemia and insulin resistance after a period of hypoglycemia (absolute or relative). It is now known that even in normal subjects hypoglycemia provokes high blood levels of growth hormone, which increases insulin resistance by various mechanisms.

Nevertheless, the proper level of blood glucose for diabetics cannot be ultimately judged until more is known about the relationship of the blood glucose concentration (and its lability) to the slow but seemingly inexorable development in diabetics of thickening of basement membranes of blood vessel walls. Glycoproteins in these membranes are formed partly from glucose. In labile (juvenile-type) diabetics very high blood glucose levels can cause extreme glycosuria with accompanying dangerous urinary loss of water and electrolytes. Another factor is that with prolonged excessive hyperglycemia there may be significant conversion of glucose to sorbitol and fructose within cells of peripheral nerve and of the lens, thus leading by osmotic effects to neuropathy and to cataracts.

The definition of a diabetic glucose tolerance curve is somewhat arbitrary and controversial. Although most "normal" individuals show a return of blood glucose concentration to 120 mg./dl. or less by 2 hours after ingestion of a glucose load, a substantial number (up to one-third) of adults of all ages may have 2-hour postprandial glucose levels above 140 mg./dl. of blood. Small differences in minimal glucose intolerance, if reproducible (which is uncertain), may have diagnostic, or at least prognostic, significance. On the other hand, Whichelow and Butterfield have suggested that a 2-hour postprandial level above 200 mg./dl. must be shown to justify a diagnosis of true diabetes. In their studies of deep forearm arteriovenous glucose differences, "borderline" diabetics (those with 2-hour concentrations of 120 to 200 mg. glucose/dl. blood) have as good muscle utilization of glucose as normal subjects.

Nevertheless, there is evidence that maturity-onset diabetics with elevated fasting blood glucose utilize glucose for general anabolic synthesis at lower rates than normal. Thus, Figure 8-4 indicates that non-ketotic diabetics, as well as the ketotic, juvenile type, have a much smaller than normal proportion of unaccounted-for carbon within the total glucose utilization. This implies a decrease in anabolic activities (synthesis of glycerides, glycogen, protein); other more direct evidence from tracer conversions supports this interpretation. The oxidation of trace amounts of glucose-^{14}C to CO_2 in mild diabetics has seemed to occur at rates within normal range, although not so for severe diabetics (Fig. 8-4). Oxidation of labeled glucose in glucose loads is under further investigation. There may eventually be some application of labeled glucose or intermediate carbohydrate in a kind of tolerance test by rate of appearance of labeled CO_2 in the breath.

There are mathematical analyses of both the glucose and insulin curves during the glucose tolerance test which attempt to define diabetes systematically. There are also claims that simply the careful discrimination of the fasting (postabsorptive) concentration within the presumably normal range (60 to 90 mg. of "true" glucose/dl. of blood) provides correlation with glucose intolerance after glucose load and can be used in itself as a diagnostic parameter of "chemical" diabetes.

Obesity

The occurrences of a mild decrease in glucose intolerance and a lower respiratory quotient (CO_2 evolved/O_2 consumed) in a large fraction of obese people have long been known. There are genetic strains of animals exhibiting both diabetes and obesity. Investigations with ^{14}C-labeled fats and carbohydrates have further demonstrated that obese humans have a rapid turnover of plasma triglycerides and fatty acids, that the oxidation of fatty acids is not depressed by carbohydrate intake as it is in normal subjects, and that the turnover and oxidation of glucose are somewhat decreased. According to single-injection or priming-infusion studies with glucose-^{14}C, the turnover rate is about 20 to 30 per cent less than in non-obese subjects.* Studies of recycling by the use of tracer amounts of both lactate-^{14}C and glucose-^{14}C in the postabsorptive state have not suggested that obese subjects have a lesser per cent of glucose oxidized or a greater per cent of glucose recycled through the Cori cycle in spite of other evidence for abnormal lactate metabolism (see below). When glucose tolerance tests are conducted in conjunction with ^{3}H- and ^{14}C-labeled glucose in obese mice or humans, it is found that rates of formation of tritiated water are essentially normal, whereas rates of formation of $^{14}CO_2$ are decreased. This implies an alteration during the postprandial state in the intermediary metabolism of glucose metabolites, i.e., an abnormal diversion to products other than CO_2.

In comparison with the non-obese, non-diabetic (or borderline diabetic, as mentioned above) the obese person, even though non-diabetic, shows a decreased uptake of glucose by muscles as determined by arteriovenous forearm differences. The uptake after a glucose load is reduced to about one-third of normal. Instead of using about three-quarters of the glucose load (oral or intravenous) as in normal subjects, the muscles of obese persons consume only about one-fifth of the load. Whether increased uptake by adipose tissue or liver or other tissues serves to compensate for depressed muscle consumption is not directly known. There is evidence that enlarged fat cells of obese people take up less glucose per cell than normal, but if the total number of fat cells is increased (as it is likely to be if obesity starts early in life) then the total consumption by adipose tissue, mainly for glycerol formation, may be greater than normal. As indicated with ^{14}C-labeled glucose in genetically obese mice, the formation of hepatic lipids from glucose may be considerably increased, though this does not appear to account for a major quantitative outlet for glucose utilization. Formation of adipose tissue glycerol is probably much greater quantitatively.

Much evidence has indicated that obese human subjects, if non-diabetic, have a markedly increased pancreatic response to glucose and possibly to other stimuli. This is well established also for rodents, in some cases even if diabetic as well as obese. There is hypersecretory activity of the pancreatic beta cells by various criteria. Whether this is genetically

---

*Yet these are estimates on the basis of mg. glucose/kg. body wt. and the difference would be less, and questionably existent, if expressed on the basis of body surface area.

determined or substrate-induced (e.g., by hyperglycemia) or both has been difficult to delineate. However, the evidence from certain kinds of obese mice suggests a fundamental derangement of the pancreas, possibly of the alpha as well as the beta cells. In any case, the effect of hyperinsulinism to promote hyperlipogenesis from glucose and thereby foster obesity is a rather likely factor in the etiology or at least perpetuation of the condition. There is also resistance to the action of insulin on glucose utilization by various tissues (particularly muscle) in obese subjects, but this may be secondary to obesity, since the resistance to glucose utilization consistently tends to diminish or disappear if the obesity is controlled and ameliorated by reduced food intake.

Besides the common occurrence of glucose intolerance in obesity, the converse is clinically true, i.e., a majority of maturity-onset diabetics are obese, particularly in the early stages of the disease. Thus, there is a strong interrelationship of diabetes and obesity, which has received much study, but the etiologic or other connections are not yet clear. Another significant clinical factor in the interrelationship is the high incidence of diabetes in close family relatives of obese subjects and vice versa.

Injury and Stress

After severe injury or other acute stress, there is often hyperglycemia. Kinetic studies with glucose-^{14}C indicate a doubling of the glucose turnover rate and oxidation rate in man after injury. These findings are consistent with other signs, e.g., increased urea production and excessive hyperglycemia, which indicate a large increase in rate of gluconeogenesis after injury or major surgery. This change is dependent, of course, upon a large protein catabolism to supply amino acids as a major source of gluconeogenesis. Efforts to stem this catabolic tide (e.g., by administration of insulin and glucose) have recently engaged clinical attention. Hypernatremia, which occurs in various states of disease and acute injury, may be associated with carbohydrate intolerance. This effect of hypernatremia is presently unexplained.

Excess Ethanol

A condition of alcohol-induced hypoglycemia is sometimes evident in individuals who have consumed much alcohol during or after prolonged fasting. In human subjects after a short fast (overnight), ethanol administration does not provoke hypoglycemia or change the glucose turnover rate. However, recycling of glucose via lactate is diminished. After a 3-day fast hypoglycemia is readily induced by ethanol in normal humans. The high $NADH/NAD^+$ ratio in the liver induced by ethanol converts the essential gluconeogenic substrates, pyruvate and oxaloacetate, to the reduced states (lactate and malate, respectively) and thus prevents their conversion to glucose, particularly if the substrates are in short supply, as in fasting (Chap. 5). Besides a decrease in glucose production rate there is some increase in rate of glucose removal from the circulation after ethanol intake, which also helps account for the hypoglycemia. Interestingly, hypoglycemia in long-fasting subjects cannot be induced by ethanol if they have been treated with glucocorticoids or if they are obese. This has been explained on the basis of resistance to decrease in gluconeogenesis; recent evidence indicates, in addition, a decline in peripheral glucose utilization in obese subjects after ethanol intake. In dogs ethanol not only reduces hepatic glucose output but inhibits glucose utilization, so that hypoglycemia is produced only when the former effect exceeds the latter. The acute effect of a moderate amount of alcohol on oral glucose tolerance in humans (when the two substances are taken together) is a decrease in the tolerance. However, if ethanol is given in more chronic dosage for some hours prior to a test, there is a potentiation of insulin response to intravenous glucose with more rapid disappearance of glucose from the blood. These effects are little understood, but may relate to the effect of ethanol on production of catecholamines by the adrenal medulla.

Other Pathological Changes

Glucose tolerance is decreased and insulin responses to glucose are increased in patients with known atherosclerosis. The intravenous glucose tolerance test conducted in patients with history of myocardial infarction and who are back at work has not appeared to be abnormal. However, in another study, the oral glucose tolerance was considered abnormal in about 40 per cent of men over forty with history of coronary artery disease compared with about 20 per cent of controls. The peripheral (forearm) uptake of glucose is not decreased in young men with history of myocardial infarction in spite of decreased oral glucose tolerance, which may imply that such patients have a tendency to the "overproduction" kind of diabetic diathesis.

Conditions which result in depletion of potassium in the body (primary aldosteronism, chronic uremia, certain diuretics) are known to be characterized by deterioration of glucose tolerance. The effect of K^+ may be on a capacity for insulin production or more generally as a cofactor for several enzymatic reactions needed for carbohydrate metabolism.

Hemodialysis of patients with uremia improves their intravenous glucose tolerance and also seems to increase their insulin responses to intravenous glucose—to values even greater than normal. A peripheral tissue insensitivity to insulin is somehow improved by hemodialysis. These are not likely to be effects on the balance of electrolytes mentioned above, since uremic patients tend to have hyperkalemia and hyponatremia.

In children who had kwashiorkor in infancy (1 to 4 years of age) and were then treated, a defect in glucose tolerance is commonly observed even several years later. The intravenous glucose tolerance may be only 60 per cent of that of a control group. Whether pancreatic damage with islet cell deficiency underlies this effect of early malnutrition is not yet known.

By some tests, e.g., intravenous glucose tolerance, cortisone-modified oral glucose tolerance and respiration of glucose-^{14}C (either 1- or 6-labeled) to $^{14}CO_2$, persons with deficiency of glucose-6-phosphate dehydrogenase (G6PD) in their red blood cells (Chap. 4) can be demonstrated to have minor abnormalities of the total glucose metabolism. However, G6PD deficiency does not appear to occur more frequently in diabetics than in non-diabetics.

Patients with certain kinds of large tumors have exhibited hypoglycemic episodes. Turnover studies with glucose-^{14}C indicate that some of these patients have markedly reduced hepatic glucose production, which may relate to products of the tumor (Chap. 5). In some cases excessive glucose utilization, i.e., glycolysis, by the tumor tissue could account for the hypoglycemia. Tumor-induced hypoglycemia may be a heterogeneous disorder.

LACTIC ACID AND PYRUVIC ACID

Next to glucose the carbohydrate of highest concentration normally in the circulation is lactic acid. Its postabsorptive concentration in man is usually in the range of about 5 to 10 mg./dl. (0.5 to 1.0 mmoles/L.), thus about one-tenth that of glucose. However, its fractional turnover rate is much higher and its absolute rate of turnover not much less than that of glucose. Corresponding to its high metabolic activity, the physiological and pathological significance of lactate in the circulation is very important. The immediate oxidative product of lactate, i.e., pyruvate (Chap. 4), circulates normally at a level of about one-tenth that of lactate. The two circulating acids can be rapidly equilibrated in vivo, probably mainly by lactate dehydrogenase of red cells. However, in non-steady state conditions and some pathological conditions, the ratio of the concentrations of the two acids can be distorted from normal. It is appropriate to discuss jointly the characteristics in the circulation of these closely related acids.

Distribution, Turnover, and Fate

Single injection studies with lactate-^{14}C in dogs and humans indicate that lactate is probably distributed in a multicompartmental system, but the number and size of compartments and their turnover times cannot be elucidated by this method owing to appreciable losses of ^{14}C into nonlactate compartments prior to mixing in pools with high turnover rates. By evidence from continuous infusion of a quantity of lactate (nonlabeled) into the circulation of humans, and with the assumption of a single pool at concentration identical to that of blood lactate, the size of such a "pool" is about 25 per cent of body weight. The absolute turnover rate of lactate in this pool, as measured by techniques of intravenous infusion of lactate-^{14}C, is about 80 mg./kg./hr. in sheep and in man and about 130 mg./kg./hr. in dogs. Thus in each species the turnover rate is about two-thirds of the glucose turnover rate when measured under similar circumstances. At a lactate concentration of 10 mg./dl. in a pool of .25 L./kg., the turnover time is therefore about 20 minutes for man and sheep and 12 minutes for dogs ($T_{1/2}$ of 14 minutes and 8 minutes, respectively). From such data it is evident that the Cori cycle for lactate (Fig. 8-2) is an active metabolic process.

As already mentioned, much of the lactate turnover can come from glycolysis of glucose primarily by such tissues as brain, skeletal muscle and red blood cells. The proportion of circulating lactate derived from circulating glucose (according to ^{14}C studies) has appeared to differ with species, being about 20 per cent in dogs, 40 per cent in sheep and possibly much higher in man. The source of the rest of the lactate is not known, but could include muscle or liver glycogen, glycerol from lipolysis in different tissues, and amino acids convertible to lactate. There is also an origin from propionate in ruminants. The origin of lactate is probably quite variable, depending on existent metabolic circumstances. Glycogenolysis either in liver or in skeletal muscle can contribute large amounts under the influence of hormones, exercise, and other factors, as described below.

Circulating lactate and pyruvate are utilized by liver and kidney primarily, mostly for gluconeogenesis, and significantly also by heart muscle. Skeletal muscle uses lactate to an insignificant extent at rest. In sheep about 15 per cent, in man about 20 per cent and in dogs about 40 per cent of lactate appears to be converted to glucose. Only about 10 per cent seems to be oxidized directly to CO_2 under postabsorptive resting conditions in man. However by 2 hours after single injection 20 to 30 per cent of lactate-^{14}C is converted to $^{14}CO_2$. The fate of the remainder of utilized lactate is unknown.

The reserve capacity for utilization of circulating lactate at concentrations above that in the normal, basal state is usually high. When 5 g. of lactate is administered intravenously to a normal human (which would theoretically elevate the concentration in the lactate pool about fourfold initially), it is cleared from the blood in about 30 minutes. This is equivalent to a daily clearance rate of about 240 g., or about twice the turnover rate in the basal state. The extrahepatic tissues also have a capacity for extra utilization of lactate at elevated concentrations, but the liver is probably chiefly responsible for disposal of excess amounts of lactate. It follows that any abnormal state of blood lactate accumulation is likely to include a component of hepatic dysfunction.

Physiological Effects

Exercise

The physiological aberration which most profoundly and significantly affects the concentration and turnover of circulating lactate or pyruvate is exercise. By the priming-continuous infusion of lactate-^{14}C, the rates of formation and disposal of plasma lactic acid in exercising (running) dogs appear to be about twice those of dogs at rest. As previously noted in Chapter 6, most of the lactate production, at least for early and more intense exercise, comes from glycogenolysis in the working muscle and to a lesser extent from blood glucose. For small muscle groups (forearm) at moderately heavy exercise in man, the lactate production during the early phase (5 to 15 min.) can increase more than a hundredfold above the resting state (Chap. 4). At this time the venous-arterial difference is more than 1 μmole lactate/ml. of blood flow across the forearm. As exercise of this type is prolonged up to one hour, the lactate production diminishes. This coincides with other evidence that "white" skeletal muscle is most active during initial phases of exercise, while "red" muscle with greater aerobic capacity comes to predominate as exercise continues. For large muscle groups, e.g., human femoral muscles exercising via a bicycle, the lactate production is relatively low (< 0.1 μmole/ml. blood) for light to medium work (50 to 100 watts) but increases to 0.3 μmole at 150 watts and to 0.75 μmole or more at 200 watts, which is heavy exercise.

In accordance with these sharply increasing rates of lactate production as intensity of exercise increases, the elevations of lactate concentration in the blood are distinctly a function of exercise intensity and are less related to amount or duration of exercise. As heavy exercise continues, the lactic acid concentration usually reaches about 15 mmoles/L. at the time exhaustion occurs. At this time the concentration in muscle, as measured in human quadriceps femoris, is slightly greater than blood concentration. This helps to account for the "washout" effect, which is a momentary increase in blood lactate concentration during the early recovery phase after cessation of exercise. This accumulation of lactate in the muscle may be the limiting factor for maximal physical exercise of short duration; maximal breakdown of phosphagens (ATP and creatine phosphate) occurs early during exercise and appears not to be rate-limiting. According to Margaria and co-workers, the "alactacid oxygen debt" is that portion of oxygen utilized after exercise to restore the previous amount of ATP and creatine phosphate, whereas the "lactacid oxygen debt" is the oxygen utilized at that time to remove the accumulated blood lactate, either by its oxidation or its conversion to glucose or both. The proportion of oxidation versus resynthesis to glucose probably varies depending upon different conditions.

When exercise is less than severe or maximal, the blood lactate concentration may rise to about 3 to 5 mmoles/L. and then decline slowly with a half-time of about 20 to 30 minutes, even though the exercise continues. As previously remarked, this may be due to a switch from the use of white skeletal muscle to that of red muscle with more mitochondrial capability. However, other mechanisms are possible, e.g., increased gluconeogenesis, and more evidence is needed to clarify this phenomenon. During mild exercise, either in dogs or humans, there usually is no rise in blood lactate at all, even though an oxygen debt, which is by definition "alactacid," may be discerned. Brief, light exercise in healthy subjects can result in a lowering of the lactate/pyruvate (L/P) concentration ratio during the recovery phase relative to the initial, resting state. During moderately heavy exercise, even though concentrations of both lactate and pyruvate rise, there is usually no change in L/P ratio in normal dogs or humans. However, with severe exercise the L/P ratio usually is elevated, at least in humans, and may reach values of about 50 instead of the normal ratio of 10 at rest. Underlying chronic defects in tissue oxygenation, e.g., anemia, or abnormalities in lactate metabolism can lead to unusually early rises in L/P ratio (and formation of "excess lactate," as defined below) during exercise.

The rate of rise of blood lactate during exercise is an inverse function of an individual's aerobic oxidative capacity or "max. VO_2" (Fig. 8-5). As is to be expected, in athletic subjects the rate of rise is slower than in less fit subjects. Moreover, when exercising at the same percentage of maximal aerobic utilization, there is a lesser rise of blood lactate in trained subjects than in non-trained. Thus at 70 per cent of maximal VO_2, the non-trained human subject displays early in exercise a rise to peak levels of about 5 mmoles lactate/L. of blood. When a trained runner's oxygen consumption is less than 70 per cent of maximal VO_2, little or no increase in blood lactate will occur during prolonged running. Some highly trained runners can utilize up to 90 per cent of their maximal VO_2 with only moderate accumulations of lactate.

It might be supposed that individuals with greater constitutional capacity for avoidance of lactatemia during exercise would become athletes by natural selection. To some extent this may account for the differences described above. However, several studies have indicated that adaptive mechanisms occur in training which reduce blood lactate levels during equivalent work loads. For instance, the venous-arterial lactate difference of exercising leg muscles in adolescent humans is significantly

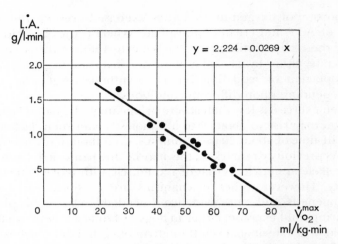

Figure 8-5. Rate of increase of lactic acid in blood of men during supramaximal exercise vs. maximum oxygen consumption. (From Margaria et al.: Europ. J. Physiol. *326*: 152, 1971.)

lower after a 6-week period of training than before. Comparison with the a-v oxygen difference indicates that the training was accompanied by a greater decrease in production of lactate than in oxygen consumption. Moreover, the respiratory quotient during exercise is lower after training than before, which implies that after training the oxidative needs during exercise are to a large extent covered by fatty acids. The bulk of the energy needs in any case during exercise, particularly when prolonged, are met by the oxidation of free fatty acids extracted from the serum.

If a period of prolonged work in humans precedes (by approximately 1 hour) a certain work load (submaximal or supramaximal), then the blood lactate rises less during the work load. This appears to be due to an increased capacity for the tissues, particularly skeletal muscle, to take up lactate from the circulation. However, the preceding, prolonged work does not change muscle lactate concentration during exercise. Therefore, an explanation other than increased blood or muscle lactate may be needed to understand why subjects are more easily exhausted after a prolonged period of exercise.

During exercise about half of the plasma lactate turnover is being taken up by the liver, which again emphasizes the importance of this organ in lactate metabolism. Probably the lactate taken up by liver is converted largely to glucose, since, at least in man, there is little extra CO_2 production by the hepatosplanchnic tissues. In running dogs studied with lactate-^{14}C, about three-fourths of the total blood lactate turnover is converted promptly to CO_2. However, a large part of this may be due to metabolism by heart muscle or relatively quiescent skeletal muscles, both of which take up lactate in proportion to the arterial concentration. So also does the renal cortex, which may, however, convert much of its lactate to glucose.

Diet

Dietary changes do not seem to have any pronounced effects on concentration or turnover of plasma lactate. After a week of starvation, obese humans show no change in either concentration of lactate or turnover of lactate-^{14}C. A larger percentage of the lactate is converted to glucose and the fraction of the glucose turnover derived from lactate increases twofold. However, the amount of glucose converted to lactate is decreased during starvation, which means an increase in formation of lactate from some other source to account for lack of change in concentration or turnover of lactate.

The production of lactate during exercise is greater when subjects have been on a preceding high-carbohydrate diet than when they have been on a high-fat diet. This suggests that glucose (or glycogen) is a more predominant fuel for exercise in the carbohydrate-adapted state. This is also indicated by the higher respiratory quotient during exercise after a high-carbohydrate diet.

A special dietary effect is that of deficiency of thiamine (vitamin B_1), since oxidative decarboxylation of pyruvate is dependent on thiamine (Chap. 4). Levels of blood pyruvate (and to a lesser extent lactate) rise during B_1 deficiency in man or animals, and capacity for oxidation of pyruvate-^{14}C is decreased in thiamine-deficient rats. The pyruvate pool size, as measured with labeled pyruvate, is increased threefold in thiamine-deficient calves. This has particular implications for metabolism in heart muscle and brain, as suggested below (see also Chap. 4).

Lactate and pyruvate levels are elevated in the blood of patients with kwashiorkor (protein deficiency), although blood glucose levels are often decreased (Chap. 5). There may be an impairment of oxidation of the 3-carbon acids due to vitamin deficiencies (thiamine or pantothenic acid), but no definite cause of the elevation of blood levels has been established.

Hormones

Epinephrine is a strong stimulus to increase in blood lactate levels, presumably via its glycogenolytic effects on both liver and muscle. On the other hand, glucagon, which stimulates glycogenolysis in liver but not muscle, does not raise the concentration of blood lactate. In patients given glucogenic adrenal cortical steroids or in women taking anovulatory steroids as oral contraceptives, the blood pyruvate levels are elevated postabsorptively and in the course of glucose tolerance tests. This may be associated with an increased production from glucose, since there is a normal rate of clearance of a lactate load from the blood. On the other hand, there is an alteration in the kind, if not rate, of blood pyruvate–lactate disposal during glucocorticoid excess. Various evidence indicates a lower rate of oxidation and a higher rate of gluconeogenesis from the 3-carbon acids. The definitive studies to determine effect of glucocorticoids on turnover rate of lactate have not been done. Growth hormone excess increases blood pyruvate levels and decreases rate of oxidation of pyruvate-^{14}C to $^{14}CO_2$ according to some studies, but there are also negative findings with some acromegalic patients concerning pyruvate levels before or after glucose loads. In the blood of hyperthyroid patients, the postabsorptive concentrations of both lactate and pyruvate are elevated without any change in L/P ratio. The blood levels of pyruvate (lactate not measured) are also high in such patients after a glucose load. Whether this indicates impaired removal or a greater flow of glucose through the glycolytic pathway is not known. In hypothyroid patients there is a lowered pyruvate and an unchanged lactate concentration, thus an abnormally high L/P ratio. It is possible that this reflects a cellular deficiency in disposal of reducing equivalents through the electron transport system.

Cellular pH and pCO_2

There is a consistent effect of alkalosis, either respiratory or metabolic, to augment the blood lactate concentration. Conversely, hypercapnia (high pCO_2) can cause a significant fall in blood lactate to less than half of the control level. The pH effect is believed to operate on the reaction catalyzed by the enzyme phosphofructokinase, so that glycolysis is increased by alkalosis and decreased by acidosis. This might be a mechanism for homeostatic control of intracellular pH. The lactatemia of alkalosis does not derive principally from muscle, liver, intestine, or brain and is not associated with tissue hypoxia. The major source of the blood lactate rise seems to be glucose glycolyzed in red blood cells. Another possible explanation for the rise in lactate during respiratory alkalosis caused by hyperventilation is the reduction in hepatic blood flow associated with the latter.

The elevation of blood lactate due to hyperventilation in man is of a minor degree (1 mmole/L. body water, i.e., about twice normal) unless there are other contributing factors. However, with prolonged passive hyperventilation in dogs, the level of lactate in the cerebrospinal fluid is seen to rise several mmoles/L. higher than the elevation of the concentration in blood. This and other evidence suggests that the blood-brain barrier does not allow blood to reflect accurately the lactate concentration in brain and that the level in the CSF is a much better indicator. The hyperventilation associated with exercise (particularly in untrained subjects) accounts for only a small portion of the total increment of lacate during and after exercise.

Barometric Changes

As expected, after rapid ascent to high altitudes (4,300 meters), exercise there is accompanied by higher pyruvate levels than exercise at sea level. Intermediate values occur after gradual ascent to high altitude. The excessive rise in both lactate and pyruvate during exercise at high altitude is presumably consequent to hypoxemia, hypocapnia and alkalemia. After 2 weeks' acclimatization at high altitude, however, both resting and exercise levels of pyruvate and lactate are lower than previously at sea level, and after return to sea level further declines are noted. Natives of high altitudes for a given amount of exercise accumulate lesser amounts of blood lactate than do natives of sea level. Whether this relates to a lesser production of lactate or a greater rate of removal is not clear.

There are changes in blood pyruvate at intervals following simulated deep excursions by divers. Although blood lactate remains essentially constant, pyruvate rises after a few hours of recovery at 35 feet below the surface and then falls below normal several hours later at the surface. The temporary rise in pyruvate coincides with a time of elevation of lactate dehydrogenase in the plasma. An interesting facet of these changes in pyruvate and LDH in association with diving is that they occur when dives are performed at 2400 hours, but not if dives are at 0800 hours. This circadian difference has been related to the evidence for a similar diurnal variation in glycolysis, which appears to be in an ascending phase of activity during the day and a descending phase (with storage of glycogen) during the night. There is a circadian cycle of LDH activity due to periodic inflow and outflow of muscle subunit (M4) form, which is a regulatory enzyme controlling glycolysis. The changes caused by dives at 2400 are a disruption of the normal cycle, which does not occur after dives at 0.800.

Thermal Changes

Under conditions of extreme hyperthermia (42° C for 2 hours) in dogs with maintenance of high arterial pO_2 and no respiratory alkalosis, there is no change in L/P ratio and not much change in either lactate or pyruvate in arterial or venous blood of muscle or brain. However, there is quite an increase in concentration of both acids (without change in L/P ratio) in the CSF. This again shows that a-v differences of lactate or pyruvate across the brain are not a good index of cerebral metabolism, and that measurements of the CSF are a better indicator of metabolism in the central nervous tissue. Plasma lactate levels in cold-exposed lambs and newborn rabbits increase markedly (by 40 mg./dl. in lambs) and to a greater extent than do glucose levels. This finding is consistent with a marked increase in the Cori cycle upon cold exposure, as previously discussed.

Pathological Changes

Rises in blood lactate and pyruvate occur in a variety of acute and chronic illnesses, including surgical shock, hemorrhage, cardiac failure, uremia, bacterial infections, alcoholism, liver disease, leukemia, neoplasms, severe anemia, glycogen storage disease, diabetes mellitus, neurologic disorders, hypertension, and others. Obviously there must be diverse causes for these rises among such a variety of diseases. The elevations of the 3-carbon acids can, of course, in general be associated with overproduction or underutilization or both. The liver is of such importance in total lactate removal that it can again be emphasized that in any abnormal elevation of lactate or pyruvate the question of hepatic dysfunction being at least contributory to the condition must be considered. Pathological increases of blood lactate, when pronounced, are designated lactic acidosis. Lactic acidosis occurs most typically, though not necessarily, in clinical situations when there has been prolonged or severe hypoxia of the tissues, as in shock with hypotension, cardiac failure, pneumonia, and other serious illnesses.

In order to express more accurately than simply by concentration of blood lactate the degree of hypoxia in various pathological or physiological states, Huckabee derived the term "excess lactate," as defined by the formula:

$$XL = (L_n - L_o) - (P_n - P_o) \times (L_o/P_o)$$

where XL = excess lactate, L_o and P_o are initial arterial lactate and pyruvate concentrations under non-hypoxic conditions, and L_n and P_n are the values of these acids in arterial blood in the hypoxic state. Sometimes the L/P ratio in the hypoxic (or pathological) state, i.e., L_n/P_n, is taken as a measure of lactic acidosis rather than the more complete calculation of XL. Further to express the excess lactate of a particular organ or body region, he developed the equation:

$$XL_{v-a} = (L_v - L_a) - (P_v - P_a) \times (L_a/P_a)$$

where subscripts v and a designate concentrations of lactate or pyruvate for venous and arterial blood of an organ or region. The XL_{v-a} for different organs will vary depending upon the etiology of the lactic acidosis. Skeletal muscle, gastrointestinal tract, and liver are more likely sources of major quantitative change in hypoxic conditions, while heart, brain, and kidney are less so.

Often the XL or L_n/P_n values have not correlated well with arterial pO_2, as earlier expected. More recently, Huckabee has suggested that local abnormal changes in the microvasculature could decrease the tissue oxidative rate in spite of normal arterial O_2 concentration and gross blood flow. Furthermore, he considers that in cases of intractable lactic acidosis (specifically those with continuing or recurrent high XL or L_n/P_n), there may be some disorder or "breakdown" of the electron transport system superimposed upon other pathology. Such a disorder may also be an underlying constitutional difference in cases of idiopathic lactic acidosis, as discussed later. In any case, the L/P ratio in peripheral blood varies in proportion to changes in redox potential in the cytoplasm of tissues (NADH/NAD$^+$ ratio), although whether the mitochondrial state of oxidation is as clearly reflected by the blood L/P ratio is questionable, since the redox states in cytoplasm and mitochondria may be dissociated (Chap. 3). Whereas tissue hypoxia can be indicated by evidence of excess blood lactate, including elevation of the blood L/P ratio, it is important to appreciate that such changes may reflect a transient, nonequilibrium state rather than a steady state.

"Spontaneous" lactic acidosis associated with various abnormalities can have in common certain physiopathologic mechanisms which combine to elevate the blood lactate concentration progressively. The mechanisms indicated in Fig. 8-6 contain the elements of vicious cycles which, once started, inexorably worsen unless the cycles are interrupted. Hyperventilation produces respiratory alkalosis (low pCO_2), which, as stated before, is a strong stimulus to glycolysis with resultant lactate production. This normally does not become excessive or persist, but if there is liver damage and therefore inadequate lactate removal, the lactic acidosis further stimulates hyperventilation via a pH effect on the breathing center. Severe acidosis leads to shock and induces tissue anoxia, which is another cause of lactate accumulation (Pasteur effect). The low blood pressure of shock is accompanied by low liver blood flow, further impairing capacity for removal of lactate. The liver itself shifts to anaerobic metabolism and contributes to the lactate production.

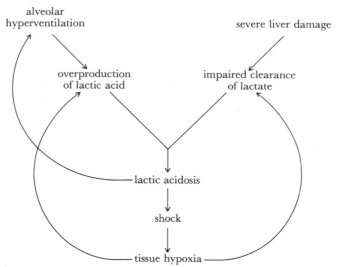

Figure 8-6. Cycles of pathological changes promoting lactic acidosis. (From Perret et al.: Helvetica Medica Acta 35:377, 1970.)

Besides correction of obvious factors (shock, hemorrhage, cardiac failure, and so forth) which perpetuate or extend lactic acidosis, the use of glucose and insulin has been clinically effective in lowering high blood lactate concentration. The administration of sodium bicarbonate may serve to interrupt the cycles of Figure 8-6. Methylene blue, which accepts reducing equivalents, has been used with limited clinical success.

Particular characteristics of blood lactate and pyruvate and lactic acidosis in different abnormalities are further discussed below.

Shock, Hemorrhage, and Injury

In these conditions with varying degrees of hypoxia, there is reflex peripheral vasoconstriction which leads to large intracellular accumulations of lactate. This is reflected by increases of serum lactate, which may temporarily increase after restoration of blood flow due to the "washout phenomenon." With reduction of cardiac output and decreased blood flow, particularly through the liver, the hepatic uptake of lactate is markedly reduced, even when oxygen consumption by the liver remains normal. Pig livers retain the capacity for metabolism of lactate even at 10 per cent of normal blood flow if oxygenation of the blood is 100 per cent. However, at 50 per cent oxygen saturation of the blood, a flow rate less than 20 per cent of normal is inadequate for metabolism of lactate. Lactate levels between 5 and 10 mmoles/L. may yet carry a good prognosis if the arterial oxygenation remains adequate and the pathological condition is temporary, but such levels in patients with continuing poor oxygen saturation imply a poor prognosis. In one study as lactate concentration increased from 2 to 8 mmoles/L. in clinical circulatory shock, the estimated probability of survival was found to decrease from 90 to 10 per cent. The level of blood lactate alone has appeared to be as useful an indicator of circulatory shock as L/P ratio or excess lactate; however, the latter may yet have prognostic value in particular situations and patients. During hemorrhagic shock in dogs, brain lactate accumulates rapidly and becomes two to three times higher than blood lactate levels. This helps to explain the vulnerability of the brain to hypotensive, hypoxic episodes. Under circumstances of hemorrhagic shock in humans, the administration of glucocorticoids appears to be accompanied by a more rapid decrease in blood lactate concentration during hypovolemic treatment. Whether this is due to circulatory effects or metabolic effects (e.g., increased gluconeogenesis) of the glucocorticoids is not clear.

Excess Ethanol

Ethanol is well known to induce hyperlactatemia acutely. Ethanol decreases the rate of disappearance of lactate from the blood after a lactate load. Studies of the fate of lactate-^{14}C infused in humans indicate that ethanol does not change the rate of production of blood lactate but decreases the rate of utilization by 50 per cent or more. Conversion of lactate to glucose is markedly inhibited to about one-third of the normal rate; oxidation of lactate is not notably decreased. The decrease in gluconeogenesis is pertinent to the mechanism of development of hypoglycemia with ethanol excess. As discussed above and in Chapter 5, the large quantity of reducing equivalents generated by ethanol oxidation effectively decreases the concentration of intermediates (pyruvate and oxaloacetate) needed for adequate gluconeogenesis. Under some circumstances or for a limited period of time, glycogenolysis or some other compensating mechanism may maintain a normal rate of glucose production, but not indefinitely.

A moderate intake of ethanol (only 24 g.) 30 minutes prior to short-term severe exercise by an adult human can effectively delay the removal of lactate from the blood to the extent that an hour after the exercise the blood lactate level is twice that which it would be without the ethanol. Even at rest the normal lactate concentration is doubled by such an alcohol intake. Among the adverse metabolic effects of this situation is the potential deficiency of glucose production needed to sustain blood glucose levels during a time of continuing heavy muscular effort.

Chronically excessive ethanol intake can lead to the type of general liver damage known as cirrhosis (which can have other causes also). In hepatic cirrhosis, the blood lactate levels are twice as high as in a control group with lesser elevations of the pyruvate level, so that the L/P ratio is slightly increased. Nevertheless, even severe chronic liver damage may not be accompanied by very high elevations of the 3-carbon acids. Seemingly, acute metabolic or circulatory alterations in the liver are more likely to disturb blood lactate levels.

Infections

Increased levels of pyruvate and lactate occur in the blood of febrile, infected patients. Various reasons for this have been advanced. These include increased glycogenolysis and glycolysis secondary to high metabolic rate, epinephrine release, alkalosis, glucocorticoid excess, and relative thiamine deficiency. Any or all of these may be contributory. It is possible that excessive glycolysis and lactate production by white blood cells during infection may be significant etiologically when coupled with impairment of lactate oxidation by other tissues. In some cases of serious infection, there is a progressive development of severe lactic acidosis, which has appeared to be spontaneous. It is likely that a combination of factors, as discussed previously in connection with Fig. 8-6, is responsible for this grave development.

Diabetes and Obesity

Patients or animals with diabetes commonly have moderate elevations of blood lactate and pyruvate and some rise in L/P ratio. Some studies have indicated no appreciable disturbance of the turnover rate, but others, both with and without the aid of labeled lactate, have provided evidence for impairment of the rate of removal of a lactate load or of endogenous lactate when the blood level is high. At the same time, the amount and proportion of lactate originating from glucose in the diabetic may be less than normal. Infusion of a glucose load can markedly increase the fraction of lactate deriving from glucose in normal, but not diabetic (pancreatectomized) dogs. Lactate metabolism in the diabetic is perhaps most notably changed in the mode of removal from the blood rather than the rate. Both cardiac and skeletal muscle take up less of the 3-carbon acids from the blood of diabetics, while utilization by the liver is likely to be increased. Probably in connection with this shift, there is typically a decrease, to as little as 50 per cent of normal, in the percentage of blood lactate oxidized to carbon dioxide and an increase by as much as two- to fourfold in the percentage of lactate converted to blood glucose.

In obese patients, even those without diabetes or notable glucose intolerance, there is often an increase in proportion of blood lactate converted to glucose and a decrease in oxidation to carbon dioxide, which may be hardly less in degree than in diabetic patients. In obese, nondiabetic patients, however, there is no decrease in total rate of lactate removal from the blood and, in fact, there is some suggestion of increased rate of production. The fasting levels of both lactate and pyruvate and the levels after glucose load are higher than normal in obese subjects.

As in the case of severe infections, some patients with uncontrolled diabetes, generally with some other superimposed problem disposing to the condition, have been seen to develop spontaneous lactic acidosis of severe degree. Whether this relates directly to their diabetic state or reflects merely the greater likelihood of recognition of lactic acidosis in diabetics is a matter of question. One facet of lactatemia in insulin-treated diabetics is the occurrence of even greater blood lactate levels following than before insulin treatment. The larger formation of lactate from glucose induced by insulin is not matched by an equal increase in utilization, partly because gluconeogenesis from lactate is reduced by insulin effect on the liver. Therefore, low blood bicarbonate levels in later stages of treatment may be due to lactic acidosis rather than to ketoacidosis. The paradoxical difference between this phenomenon in diabetics and the lactate-lowering effect of glucose and insulin in other circumstances (see above) may relate to differing quantities of substrate and hormone.

Neurological and Muscular Disorders

With certain kinds of encephalopathy, encephalomyelopathy and myelopathy, there are associated rises of both lactate and pyruvate of the blood without significant change in L/P ratio. These may include Wernicke's encephalopathy, acute peripheral neuropathy, multiple sclerosis, and progressive muscular dystrophy. In some cases thiamine treatment reduces pyruvatemia in these neuromuscular disorders and sometimes not. The nature of the impaired pyruvate metabolism in these conditions is not known, but been considered as possibly etiologic for some of the clinical symptomatology. In a congenital type of encephalopathy a deficiency of pyruvate carboxylase has been recognized.

In this general category (and possibly related in some way to one or another of the above conditions) is a rare familial disorder, infantile lactic acidosis, appearing shortly after birth or in early childhood with severe symptoms of muscular hypotonia, convulsions, mental retardation, and (curiously) obesity. The cause for elevations of blood lactate and pyruvate is not known, but a defect in pyruvate utilization is evident from abnormal pyruvate and lactate tolerance. Increase of blood glucose level during infusion of lactate indicates that gluconeogenesis occurs adequately. A similar inheritable occurrence of chronic "idiopathic" lactic acidosis has recently been recognized in patients who are not so severely affected, do not have any clinical symptoms other than those referable to muscle weakness and easy fatigability (which is exacerbated by exercise or alcohol ingestion), and in whom the disorder may not be detected until adulthood. The biochemical mechanism of these genetic abnormalities has not yet been elucidated, but difficulties in the mitochondrial function or the cytoplasmic hydrogen ion shuttle (Chap. 3) are suspected.

Some patients with anxiety neurosis or neurocirculatory asthenia are likely to have rapid elevations of blood lactate during exercise and in connection with such elevation may have symptoms of tremor, faintness, nervousness or weakness or all of these. Other studies show that infusions of lactate are more prone to produce anxiety and other signs and symptoms of neurasthenia in such patients than in normal persons. An accompaniment of these infusions in either anxious or normal persons is certain signs of hypocalcemia. When a lactate infusion is given together with Ca^{++} to susceptible persons, there can be a mitigation of the neurasthenic symptoms, though exceptions occur. From these findings a theory has been devised that a biochemical mechanism of anxiety may involve a complexing of ionizable calcium at the surface of excitable membranes of some neuronal cells by lactate produced intracellularly. Whether this implies that such persons have an excessive production of lactate (perhaps evoked by a ready supply of epinephrine) is an interesting question. In any case, treatment of patients with neurocirculatory asthenia by graded muscular exercise may lower the extent to which lactate rises after exercise as well as improve them symptomatically.

Cardiovascular Disorders

Patients with clinical myocardial infarction usually have elevated levels of pyruvic and lactic acids in the blood during at least the first 2 days after infarction. The L/P ratio is not elevated. The mean rise of the 3-carbon acids is about twofold and is greater than the degree of hyperglycemia which occurs in these patients. Moreover, in patients with chronic ischemic cardiovascular disease, the fasting blood levels of lactate and pyruvate tend to be high. Elevation of these blood acids in heart disease is suggestive of increased myocardial glycolysis. In fact, there is evidence for increased lactate production by ischemic myocardium. However, increase of lactate and pyruvate with heart disorders, particularly when acute, may be due to less specific effects of general stress.

As a group, hypertensive human subjects have about 25 per cent higher resting blood lactate levels than the normotensive population. Although some abnormal steroid effect could be suspected, there is no evidence for this or any other explanation. No difference from normal lactate level is found in rats with experimental hypertension. Low tissue oxygenation in the muscles of patients with femoral artery occlusion is evident from higher than normal L/P ratios in the venous return from affected leg muscles during exercise.

GLYCEROL

Turnover and Fate

The concentration and turnover of free glycerol in plasma have been of interest particularly because they seem to represent, better than do the concentration and turnover of free fatty acids, the amount of lipolysis occurring in peripheral tissues. The fatty acids released by breakdown of triglycerides may be reutilized appreciably for triglyceridogenesis at the site of lipolysis, but the glycerol is little, if any, reutilized before travel first through the blood stream to the liver, where a significant part may be again converted to triglycerides. The evaluation of lipolysis by glycerol turnover is probably useful only for the fasting condition, since in fed animals a significant amount of circulating glycerol is being formed from glucose. In the rat in the fed state, as much as 70 per cent of net formation of glycerol comes from glycolytic intermediates, although in the fasting animal only 17 per cent of the formed glycerol derives from glucose and 83 per cent comes from glyceride-glycerol. Possibly other species have lower fractions of glycerol deriving from glucose in the fed state, since the rat has an unusually high rate of turnover of circulating glycerol.

Concentrations of plasma glycerol in man and sheep in the early postabsorptive state are normally about 5 to 10 μmoles or 0.5 to 1.0 mg./dl. Thus the concentration is much less than that of glucose and lactate and similar to that of pyruvate. In the rabbit and dog the concentration is somewhat higher and in the rat it is higher yet (10 to 15 μmoles/dl.). Although present in very low concentration, the turnover rate of glycerol is fairly rapid. In the rabbit the early half-time of singly injected tracer glycerol is about 10 minutes and the fractional turnover rate about .06/min. The fractional rate is similar in the dog and slightly lower in man (.04/min.). The absolute turnover rate is estimated at about 1 μmole/kg./min. in postabsorptive man, i.e., 5.5 mg./kg./hr. (much less than glucose or lactate). The turnover rate when adjusted for body size is similar in sheep, about twice as high in the dog, and several times higher in the rabbit and the rat, which has a fractional turnover rate of 0.2 to 0.35/min.

The turnover rate for circulating glycerol can be correlated to a great extent with the concentration. In the dog the uptake of glycerol is generally proportional to the concentration over the usual physiological range, i.e., up to about 50 μmoles/dl. At much higher levels of plasma glycerol, the uptake of glycerol falls off markedly from proportionality with concentration (Fig. 8-7). This suggests a saturation-type kinetic relationship. Similarly, the K_m (concentration at one-half the maximum rate) for glycerol utilization by the rabbit has appeared to be about 0.33 mmoles, which is well above the normal concentration. The maximal capcity for glycerol utilization in man at high concentration of plasma glycerol is found to be about the same as that indicated for the dog by Figure 8-7.

The glycerol pool size is commonly assumed to be 60 to 65 per cent of the body weight, i.e., essentially a distribution in total body water, but aspects of the kinetics of glycerol-^{14}C during constant infusion in dogs suggest that the pool size is somewhat smaller.

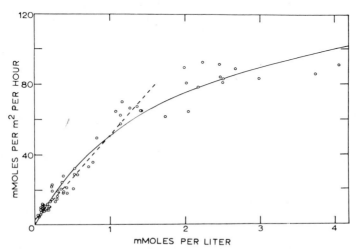

Figure 8-7. Glycerol uptake in dogs vs. plasma glycerol concentration. (From Winkler et al.: Amer. J. Physiol. *216*:191, 1969.)

Liver and kidney account largely for the bulk of the net uptake of plasma glycerol in vivo. According to some studies, about 70 per cent of utilized plasma glycerol is removed by the liver in man. In both liver and kidney, a substantial portion of the utilized glycerol is converted to glucose. In the dog it is estimated from ^{14}C studies that about one-third of the glycerol removed from the blood is converted to glucose and in sheep about one-half. Possibly in the rat it is an even higher fraction in view of the high rate of glycerol turnover. In postabsorptive dogs the circulating glycerol contributes about 7 per cent of the carbon of newly synthesized glucose. In fasting sheep the value is about 20 per cent, although in fed sheep it is only 5 per cent. In either species, when the concentration and turnover of glycerol are increased, e.g., by bulk infusion of glycerol, the contribution of glycerol to glucose output is increased. Thus in dogs the fraction of glucose derived from glycerol may rise to about 50 per cent. However, total glucose release under these circumstances increases only 20 per cent, indicating that infused glycerol substantially replaces other substrates for gluconeogenesis.

Another significant fraction of utilized glycerol is converted to lipids, primarily by the liver. The fraction so utilized is not clearly known. Of the glycerol converted to liver lipid in rabbits in vivo, about half is in neutral lipid and about half in phospholipid. The specific radioactivity of plasma lipids is very much lower than that of blood glucose after administration of ^{14}C-labeled glycerol, which may indicate much less total conversion to lipids or else a large pool of formed hepatic lipids. Relatively little circulating glycerol appears to be oxidized directly to carbon dioxide. Usually not more than 5 per cent of total CO_2 is derived by oxidation of glycerol.

Although liver and kidney take up most of the circulating glycerol, according to in vitro studies there can also be utilization of glycerol by a number of other tissues, including intestinal mucosa, brown and white adipose tissue, muscle, lymphatic tissue, nerve tissue, thyroid, lung, pancreas, and leukocytes. It has been suggested that in the fasting state, when glycerol level in plasma becomes elevated, brain might use substantial amounts for its energy needs. However, although there is some response of the brain in vivo to glycerol, utilization is evidently much slower than for glucose and concentrations of glycerol in vitro must be 10 times those of glucose to obtain the same respiratory rates.

Physiological Effects

Diet

During the first day of a starvation regimen in normal man there is little change in blood glycerol concentration, but thereafter the glycerol concentration rises to twice the normal level at 3 to 4 days of starvation. Presumably this is due to increased lipolysis. As already stated, the source of plasma glycerol can shift markedly from glucose to triglycerides. As the plasma glycerol concentration rises, so does its turnover and so does the percentage of new glucose which is formed from glycerol. In both the dog and the sheep, it rises from about 5 per cent to about 20 per cent of the total gluconeogenesis. In the rat the fraction of glucose derived from glycerol after 48 hours of fasting has appeared to be 75 per cent or more. This and other differences in plasma glycerol between rats and other species have not yet been explained. There is evidence from in vitro studies that glycerol uptake by liver or kidney is not changed by starvation or by a low-carbohydrate diet. A high-glycerol diet for more than 2 weeks in rats does not change uptake by liver, but does increase uptake by kidney.

There are some factors which, even in the rat, can limit the amount of glycerol converted to glucose. One of these is the concentration of FFA in plasma. If the level of palmitate in the circulation (of perfused rat liver) is raised to 2 mmoles/L., there is almost a complete suppression of gluconeogenesis from glycerol, though little suppression in uptake. Part of this change can be attributed to diversion of glycerol to synthesis of triglycerides in combination with the palmitate. A factor which can decrease, or prevent, further increase of plasma glycerol concentration, as well as that of FFA, is ketone bodies in the blood. Blood ketones rise considerably in normal subjects during total starvation. There may be a kind of feedback control inhibition of lipolysis by these compounds which are derivatives of products of lipolysis (FFA). The inhibition of lipolysis may occur via insulin, since ketone bodies can stimulate insulin secretion. These observations raise questions about the extent to which plasma glycerol can contribute to formation of blood glucose (and thereby substrate for the CNS) during starvation.

Conversely to the effect of fasting, the administration of glucose tends to lower, by as much as 50 per cent, the plasma glycerol concentration and turnover. This is most likely due to the elevation of blood insulin, which has a direct inhibitory effect on peripheral lipolysis.

Hormones

Generally those hormones—thyroid hormone, epinephrine, growth hormone—which promote lipolysis at peripheral sites, particularly adipose tissue, have the effect of increasing concentration and turnover of plasma glycerol. This effect is not shared by one lipolytic hormone, glucagon, possibly because the latter provokes secretion of insulin, which has an antilipolytic effect and itself decreases glycerol concentration. Clinical hyperthyroidism is accompanied by elevations of both concentration and turnover of glycerol, which are about threefold normal.

The changes in glycerol metabolism are a better indicator of the lipolytic effect of thyroid hormone than are changes in plasma FFA. However, hypothyroid subjects cannot be distinguished by plasma glycerol levels. The lipid-mobilizing effect of thyroid hormone appears to be mediated via its effect on the response of adipose tissue to catecholamines. On a regimen of bovine growth hormone, dogs show an increase of turnover of glycerol from about 8 mmoles/hr./m.^{2} body area to a maximum of about 20 mmoles/hr./m.^{2} in the first few days with a gradual decline, but an elevation remaining after 2 weeks at about 13 mmoles/hr./m.^{2}. The elevation persists longer than does the increased turnover of plasma FFA. There are similar relationships between patterns of change in plasma concentrations of FFA and of glycerol, both of which rise with growth hormone administration (glycerol to three times the normal level), but glycerol remains elevated longer.

Exercise

The level of plasma glycerol rises during exercise and this is accompanied by an increased utilization of glycerol by working muscle. After training, the glycerol, as well as the FFA, of the blood rises to a greater extent during exercise than before training. This probably indicates a greater adaptation of the trained athlete to the utilization of lipid as energy fuel during exercise, as discussed earlier. Although human growth hormone (HGH) increases in plasma concentration during muscular exercise, the time of increase of immunoreactive HGH and that of glycerol during exercise are not the same and the evidence is against the hypothesis that HGH could be intermediary in the response of lipid mobilization to exercise. Probably the release of epinephrine, which can be variable depending upon psychological factors, is an important component of lipid mobilization during exercise.

Other Physiological Effects

During cold exposure or conditions of fever, glycerol levels in the blood rise to double or more the base-line levels. The serum FFA rise also, though not so markedly. This is an expression of increased peripheral lipolysis, which is at least part of the mechanism of thermogenesis.

In the rat there is a marked change in the potential metabolic fate of glycerol according to the age of the animal as it develops through the neonatal period. Although gluconeogenesis from glycerol is low during the first 2 days of life, subsequently during the suckling period, when the bulk of calories are derived from fat, a large amount of circulating glycerol can be converted to glucose or glycogen. In the young rat after weaning, when the animal's diet has shifted largely to protein, much less glycerol is converted to glycogen and glucose. Four-week-old human infants eliminate glycerol from the blood much more rapidly than do 1-day-old infants. These observations may have significance in relation to infant artificial nutrition.

Pathological Changes

Diabetes

The plasma glycerol concentration, like that of FFA, is generally increased in diabetes. In diabetics of either juvenile or maturity-onset type, there is a greater than normal increase of plasma glycerol during exercise, which indicates excessive lipid mobilization. Presumably this indicates insufficient amount or ineffectual action of insulin at the lipolytic site. Control of juvenile-type diabetics with insulin is accompanied by a much more normal response of plasma glycerol to exercise. However, prediabetic human subjects, as defined by low and delayed insulin responses to glucose, are not different from normal subjects in their responses to epinephrine in terms of elevation of blood glycerol (or glucose) levels. It is probable that in the diabetic the fate of circulating glycerol is shifted toward the formation of glucose and away from formation of glycerol-containing lipids. In the diabetic rat, though not the normal, the feeding of glycerol can stimulate the activity of hepatic gluconeogenic enzymes. The use of glycerol as a carbohydrate fuel has sometimes been advocated for diabetics, in whom glycerol feeding may decrease ketosis and glucosuria. Evidently the conversion of glycerol to glucose is not so great as to defeat the purpose of its use as a carbohydrate which can be utilized better than glucose, at least under some circumstances by certain diabetic patients.

Other Pathological Changes

The metabolism of glycerol is strongly inhibited by the presence of ethanol. Splanchnic uptake of glycerol, presumably by liver, in humans is reduced to about one-third of control values by infusion of ethanol at only moderate concentrations (3 mmoles/L.). Nevertheless, when ethanol is introduced into the blood stream, the concentration of glycerol does not rise but falls, as does that of FFA. This is evidently due to a reduction in lipolysis, which is possibly an indirect effect of ethanol via some metabolite or metabolic change in the liver associated with oxidation of ethanol. Perhaps this is in some way related to the effect of ketone bodies to inhibit lipolysis.

As stated before, the plasma glycerol concentration is a good index of lipid mobilization, and, in various stress situations such as injury, of the general degree of sympathetic activity, i.e., catecholamine response.

The plasma glycerol concentration is higher in patients with myocardial infarction with arrhythmias than in uncomplicated myocardial infarction or angina, and this has been suggested as a way of selecting cases prone to complicating arrhythmias. Probably this proneness depends upon the levels of both epinephrine and thyroid hormone. The occurrence of markedly high concentrations of plasma glycerol in hyperthryoidism has already been mentioned.

GALACTOSE

Galactose enters the circulation as the free monosaccharide predominantly through exogenous supply. This can occur after hydrolysis of ingested lactose in milk and to a lesser extent from glycoproteins or glycolipids in food. There is a large capacity for utilization of galactose, but the occasional occurrence of genetic enzymatic defects and the general susceptibility of galactose metabolism to disturbance of certain fundamental coenzymes (pyridine nucleotides) lead to some interesting and important aberrations of galactose clearance from the circulation.

Distribution, Turnover, and Fate

The concentration of free blood galactose (discounting, of course, that in glycoproteins) in the early postabsorptive state is about 1.6 mg./dl. with 90 per cent of values less than 3.0 mg./dl. by galactose dehydrogenase assay. The values among normal human subjects show a Gaussian distribution. Judging by extrapolation to zero time of blood galactose values after intravenous injection of a load, the total volume of distribution in normal human subjects is about 20 per cent of body weight, i.e., slightly more than extracellular volume and similar to volume distribution of glucose. In rats which have been eviscerated and nephrectomized, thus essentially eliminating metabolism of galactose, the distribution has appeared to be about 30 per cent of body weight.

In normal humans during the initial 10 to 20 minutes after intravenous injection of a large load of galactose (350 to 500 mg./kg. body wt.), the galactose is mainly equilibrated into its distribution volume. Subsequently, when the blood level still remains above 40 to 50 mg./dl., the liver removes most of the galactose and does so at a constant rate independent of concentration, i.e., the removal system is saturated above that level. Above saturation levels, the normal hepatic elimination capacity for galactose is of the order of 450 mg./min., which is about one-third of the V_{max} for glucose disappearance at high levels (see above). Renal excretion, which is generally less than 10 per cent of a galactose load, is a non-threshold phenomenon and is a constant fraction at both high and low blood galactose concentrations. During the hepatic saturation period, the renal excretion and further extravascular diffusion followed by increasing degree of re-entry combine to make the rate of decrease of blood concentration more exponential than rectilinear, as would be expected by hepatic uptake alone. Later at concentrations below maximal hepatic uptake, all elimination processes are concentration-dependent and decrease exponentially at different rates. The clearance curve represents the sum of re-entry diffusion of galactose from the extravascular pool, of renal excretion, and of hepatic utilization. The latter remains the major mechanism of galactose clearance. Although the parameter of rate of exponential decline is an undefined mixture of these component processes, there is a characteristic normal half-life in adult humans for galactose disappearance of about 10 to 12 minutes, when measured over a time span including hepatic saturation as well as subsaturation galactose concentrations.

At high plasma levels, galactose-^{14}C is relatively little utilized by the non-splanchnic organs, i.e., skeletal muscle, heart, lung, adipose tissue, and brain, compared with utilization by the splanchnic organs, i.e., liver, kidney, spleen, pancreas, and gastrointestinal tract. At low concentrations in the plasma, however, a considerable fraction of circulating galactose may be utilized and oxidized by remaining organs of the eviscerated rabbit. The extent of utilization of galactose by the kidney has hardly been investigated in mammals, but in chicks it is quite active. Utilization of galactose by human liver is twice as fast as by rat liver, according to in vitro study with organ slices.

Physiological Effects

Little is known concerning the effect of various physiological influences on the removal of galactose from the circulation. The identity of galactose with glucose in configuration of the first three carbon atoms makes its transport into cells susceptible to insulin, phlorizin, and so forth (Chap. 3). The effect of insulin to increase the galactose space in eviscerated animals has been a mainstay of the theory of insulin action which considers the effect of insulin on glucose transport as primary. Present evidence indicates that the effect of insulin on galactose transport is pronounced for heart and skeletal muscle but is not discernible for adipose tissue. Whether insulin or other hormones exert any significant effect physiologically on hepatic removal of plasma galactose has not been studied. However, possible effect via action on hepatic intermediary metabolism of galactose should not be discounted. Theoretically, the thyroid hormone may affect the rate of metabolism of galactose and possibly its clearance from the blood, in view of the importance of thyroxine in regulation of electron transport and hence the redox potential in the liver. A more reduced state of redox pairs, including serum L/P ratio, is known to occur in hypothyroidism. This would be especially inimical to the disposal of galactose load in the circulation.

There is no evidence that ordinary changes of diet, other than the amount of galactose itself, have any effect on turnover or fate of galactose in the circulation. The only influence known is the curious effect of choline-deficient diets to offset the inhibitory effect of ethanol on galactose utilization (see below).

Age has some effect on the galactose tolerance. In the first three neonatal hours, the $T_{1/2}$ values for galactose disappearance after intravenous injection are higher than subsequently in human infants or in later life. After the first day or two, the mean $T_{1/2}$ gradually increases from about 7 minutes in infants or very young children to the adult values of approximately 10 to 12 minutes in children of age 10 to 15 years. These differences are about as expected on the basis of activity of intracellular galactose metabolism at different ages (Chap. 4).

Pathological Changes

Galactosemia

The enzymatic defects and some clinical characteristics of the rare but serious condition of hereditary galactosemia in human infants have already been discussed in Chapter 4. When there is total genetic loss of UDP-Glu; Gal-1-P uridyltransferase, as in homozygotes for "true" galactosemia, the ingestion of milk by infants may cause elevations of the blood galactose to 100 mg./dl. or more, and usually even in the fasting state levels greater than 10 mg./dl. are seen. Since normal infants even after milk have less than 10 mg. of galactose/dl., analysis of blood for galactose (by enzyme assay) can be clearly indicative of this homozygous state. Recognition of heterozygotes presently depends upon enzymatic assay of blood cells for uridyltransferase. Neither the level of galactose in the blood nor oxidation of galactose-1-^{14}C by red blood cells is adequate to detect heterozygotes. This is probably due to the occurrence of alternate pathways of galactose metabolism, i.e., via UTP; Gal-1-P pyrophosphorylase or galactose dehydrogenase (Chap. 4). Further studies with differently labeled galactose, either in vitro or in vivo, may provide more sensitive and distinctive isotope methods for detection of heterozygotes or evaluation of possible change in galactose tolerance of homozygotes.

Galactosemia is now known to include a "Duarte variant," which is also recessive and associated with a gene allelic to the gene for true galactosemia. Homozygotes for the Duarte variant have 50 per cent of normal activity of uridyltransferase, as do heterozygotes for true galactosemia, and like the latter, they are asymptomatic and show no abnormal elevations of blood galactose. Heterozygotes for Duarte variant have approximately three-fourths normal enzyme activity; this group may comprise as much as 10 per cent of the population. Heterozygotes for both Duarte variant and true galactosemia have one-fourth normal activity and may have galactose intolerance and perhaps pathological consequences. Whereas heterozygotes for true galactosemia have an apparent population incidence of 1 to 2 per cent, this number may include homozygotes for Duarte variant. The incidence of homozygous true galactosemia is therefore hard to establish but appears to be about 1 per 25,000 births. In the black race, there are further variants in whom galactose intolerance is less severe, owing evidently to higher activity of alternate metabolic pathways.

The occurrence of galactokinase deficiency is now recognized as another genetic, autosomal recessive type of galactosemia. Homozygotes have near zero or absent enzyme activity and heterozygotes have intermediate amounts. The heterozygous incidence appears to be about 1 per cent of the population and that of homozygotes, therefore, about one per 40,000. In this condition, the early symptoms are absent or less pronounced than in uridyltransferase deficiency, probably because the highly toxic galactose-1-phosphate does not accumulate. Cataracts may be the first manifestation. Because of the difficulty of clinical recognition, screening procedures may be even more important in galactokinase deficiency than in uridyltransferase deficiency. Since galactokinase is the rate-limiting enzyme for normal galactose metabolism (Chap. 4), even heterozygotes are at risk for galactose intolerance and have been observed to develop cataracts in the third or fourth decade of life.

Pregnant women who are heterozygous for deficiency of galactokinase or uridyltransferase of certain types (see above) could have abnormal elevations of blood galactose if they consume large amounts of milk. If galactose were to pass readily through the placenta, the level of galactose in the fetal circulation could also be high. It has been suggested that this could be detrimental to the developing embryonic tissue.

It is questionable whether there is an increased galactose tolerance in galactosemic individuals as they grow older. Probably there is a reduced tissue sensitivity and possibly alternate pathways may develop gradually.

Excess Ethanol

Ethanol readily shifts the redox potential of the liver to a more reduced state. As already discussed in Chapter 4, it thereby interferes with galactose utilization because of the high sensitivity of UDP-Gal 4-epimerase to an increased $NADH/NAD^+$ ratio. Oral administration of ethanol (300 mg./kg. body wt.) 15 minutes prior to an intravenous galactose tolerance test in normal human subjects increases $T_{1/2}$ for galactose disappearance approximately threefold. The larger the dose of alcohol administered, the greater is the inhibition of galactose elimination from the blood. As little as 10 ml. of ethanol in a human (hardly an excess) can substantially decrease the rate of oxidation of galactose-^{14}C to ^{14}CO$_2$. In normal rats, ethanol (250 mg./100 g. body wt.) decreases the rate of oxidation of galactose by 50 to 90 per cent.

In alcoholics with fatty liver, the $T_{1/2}$ of galactose disappearance from the blood without ethanol is essentially normal; moreover, in such patients the administration of ethanol prior to the test does not decrease the galactose elimination rate as it normally does. Likewise, rats on a high-fat, choline-deficient diet show no decrease in $T_{1/2}$ of galactose disappearance after ethanol, in contrast to normal rats, and, indeed, the rate of galactose elimination without ethanol is twice as high on the abnormal diet as for rats on a normal diet. Administration of choline together with the high fat corrects the abnormalities of galactose utilization. The fatty liver in choline deficiency (and perhaps chronic alcoholism) may possess a greater than normal capacity to oxidize the NADH or else the control of the hepatic cytoplasmic redox state is otherwise different. The study of fatty infiltration of the liver, as well as other liver diseases, may be facilitated by the use of galactose tolerance tests with and without ethanol.

Liver Diseases

In about 85 per cent of cases of liver cirrhosis, the $T_{1/2}$ for blood galactose elimination after intravenous administration of a load is prolonged (Fig. 8-8). In one study which compared other liver function tests with intravenous galactose tolerance, the latter was deemed superior for

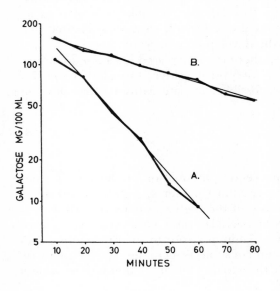

Figure 8-8. Rate of blood galactose elimination in a normal subject (A) and in a patient with cirrhosis of the liver (B). A, $T_{1/2}$ = 13 minutes; B, $T_{1/2}$ = 47 minutes. (From Tengstrom, B.: Scand. J. Clin. Lab. Invest. 18:Suppl. 92: 118, 1966.)

detection of the existence of hepatic cirrhosis. The abnormal $T_{1/2}$ value is said to be generally correlated to the severity of cirrhosis in patients. On the other hand, poor correlation is found between galactose tolerance and the extent of hepatic fibrosis in Eck fistula dogs. Using hepatic arteriovenous uptake of galactose-1-^{14}C as a measure of intrahepatic arterial shunting in human subjects, the amount of shunting is four times higher in patients with hepatic cirrhosis than in normal subjects. Hepatic uptake of the labeled galactose is also lower than normal in patients with chronic hepatitis and those with primary or metastic carcinoma. The $T_{1/2}$ of galactose load disappearance is prolonged in about 60 per cent of patients with infectious hepatitis and in about 50 per cent of patients with cancer metastases to the liver. The galactose tolerance test has prognostic value in hepatic disorders and is useful diagnostically in the differentiation of parenchymatous and obstructive jaundice, since the test is positive in less than 10 per cent of cases of the latter condition.

Diabetes and Obesity

Although galactose transport is known to be affected by insulin, there has been little study of galactose tolerance in diabetic humans or animals. The transport effect may not be significant in the main organ of utilization, the liver. However, the hepatic intermediary metabolism of galactose may be disturbed in diabetes, since the redox potential in the liver by various evidence is shifted in the same direction as it is by administration of ethanol. A variety of patients with galactose intolerance do not show corresponding glucose intolerance, but the reverse study has not been made. In obese subjects, who bear metabolic similarities to diabetics, the mean value of postabsorptive blood galactose concentration is 2 to 5 mg./dl., i.e., about 60 per cent higher than normal.

FRUCTOSE

The monosaccharide fructose is found in significant amount in the circulation only after ingestion of food containing free fructose or sucrose, the disaccharide from which fructose is derived by intestinal hydrolysis (Chap. 2). Many people throughout the world now eat large amounts of sucrose, so the effects of large amounts of fructose in the circulation (particularly portal) on metabolism of lipids and other constituents in the liver and other tissues become important. Fructose itself is sometimes used clinically for carbohydrate nutrition instead of glucose or its polymers.

Distribution, Turnover, and Fate

The endogenous postabsorptive level of serum fructose is about 0.5 to 1.0 mg./dl. After oral ingestion of fructose or sucrose (1 to 2 g./kg.) by normal human subjects, the serum concentration of fructose reaches a peak at about 30 to 90 minutes and at a level of about 10 to 15 mg./dl. The time curve of concentration is normally a resultant of two exponentials, one for absorption and one for disappearance of fructose from the blood. Fructose appears in the peripheral circulation of humans even more rapidly after ingestion of sucrose than after equivalent amounts of glucose and fructose. This is probably due to absorptive differences (Chap. 2).

Fructose probably has a larger volume of distribution after intravenous injection than does glucose, since fructose distributes unchanged into the extracellular fluid of muscle, which utilizes fructose slowly. This can explain why plasma fructose level rises less after fructose injection by peripheral vein than does glucose after glucose injection. However, it may also be due to rapid hepatic utilization before mixing. After intravenous injection of 0.5 g. fructose/kg. body wt., the half-life for fructose in the blood was shorter in male than in female baboons. This may relate to other sex differences noted below.

Removal of a large fraction by the liver during the first portal passage is an important factor in disposal of an oral fructose load. In dogs systemic blood levels of fructose are only half as high when the portal vein is infused as when the same amount of fructose is given via a systemic vein. From 50 to 80 per cent of fructose infused into a peripheral vein in humans is taken up by the liver according to arteriovenous differences. About 50 per cent of the hepatic uptake of fructose is converted to glucose normally. This is evident from findings with ^{14}C-labeled fructose in humans which suggest that about 30 to 40 per cent of injected fructose is converted to glucose rather promptly. Besides the liver, there are probably other sites of lesser conversion, e.g., kidney and intestinal mucosa.

The mechanism by which fructose is rapidly metabolized to lactate by the liver with elevations of blood lactate has been discussed in Chapter 4. Upon intravenous infusion of 0.6 g. fructose/kg./hr., human subjects (without diabetes or liver disease) seemingly fall into two groups in terms of a bimodal distribution of equilibrium concentrations of lactate or pyruvate in the blood during the infusion. The lower lactate level is apparently a function of metabolism of the fructose, since lactate tolerance was not found to be different between the two groups defined by the fructose administration. Definite correlation with any other differences between the groups has not been made.

A portion of an intravenous load of labeled fructose in man is oxidized directly and rapidly. Besides early conversions to glucose, pyruvate, lactate, and CO_2, carbon from fructose may be converted significantly to triglycerides and possibly glycogen. More labeled fructose than glucose becomes incorporated into triglycerides; this is due mainly to much more extensive formation of glyceride-glycerol from fructose than from glucose. Whether a significant fraction of circulating fructose is transformed to glucosamine and other constituents of heterosaccharides in the liver or elsewhere is not known.

To what extent fructose is utilized by extrahepatic tissues in normal man remains unsettled. Muscles of dog and man use fructose in vivo, according to arteriovenous differences, and formation of muscle glycogen is demonstrable in man. However, the rate of replenishment of glycogen in a muscle that has been depleted of glycogen by work occurs more rapidly during an infusion of glucose than of fructose. This could be due to a difference between the two monosaccharides in their evocation of insulin. Adipose tissue may utilize fructose considerably at high concentrations, as suggested in Chapter 4.

Physiological Effects

Acute Loads of Fructose

An intravenous load of fructose in rats causes acute depletion of total liver adenine nucleotides (mainly ATP) and of inorganic phosphate (Pi). This is due to rapid phosphorylation and glycolysis of fructose. The amount of total nucleotides diminishes because of increased activity of AMP deaminase. The depletion of ATP is accompanied by a temporary acute decrease in protein synthesis. Also the depletion of Pi limits oxidative phosphorylation, so there is a general inhibition of respiration of liver cells. A lack of ATP interferes with hepatic glycogenolysis, since there is inadequate conversion of phosphorylase b to a form (Chap. 6). This effect, along with inhibition of phosphohexose isomerase by Fru-1-P (Chap. 4), can lead to hypoglycemia in response to a fructose load in newborn infants and even in normal adults with large doses. Another effect also discussed in Chapter 4 is the production of transient hyperuricemia by a fructose load. This is probably related to the increased AMP deaminase activity. Infusions of fructose at rates of 1.0 to 1.5 g./kg./hr. are needed to produce elevations of serum uric acid in normal adults, but children may require less.

Uptake of fructose by the liver enhances esterification of circulating free fatty acids to form triglycerides. Also, the secretion of triglycerides into the circulation as very low-density lipoproteins is increased by perfusion of the (rat) liver with fructose. On the other hand, the absorption of fructose in normal subjects or patients with hypertriglyceridemia is followed by a transient decrease in concentration of circulating triglycerides, a phenomenon which is as yet unexplained.

A consequence of acute intravenous injection of a load of fructose (0.5 g./kg.) is a prompt and considerable increase in plasma concentration of growth hormone. This is, of course, opposite to the effect of glucose injection. Whether this is a direct effect of fructose on metabolism of the hypothalamus or pituitary or some indirect effect, e.g., via a circulating hepatic metabolite, is not known. Since fructose and glucose are metabolized by different pathways and to some extent by different organs, fructose does not interfere with the over-all utilization of glucose. However, in some particular sites there might be interference, and the possibility of competitive inhibition by fructose of uptake of glucose by the hypothalamus needs to be considered.

High Fructose or Sucrose Diets

Several studies have demonstrated that diets chronically high in fructose or sucrose in several species cause an increase of serum lipids, more frequently the triglycerides and sometimes cholesterol. In humans this effect can be noted at levels of 15 to 20 per cent of total calories as sucrose in the diet, but it is more consistent at higher levels. Even the higher amounts elevate triglycerides more in men and postmenopausal women than in premenopausal women. This may have significance for the well-known higher incidence of coronary thrombosis (and probably atherosclerosis) in men than in women. After several weeks on diets high in sucrose, the conversion of sucrose-^{14}C to serum glycerides is increased from the prediet level in male or female baboons and in male rats. The male baboons show a higher rate of conversion than the females. In adult humans of either sex (mostly obese) there has been found a doubling or more of the conversion within one week on diets containing 40 per cent of total calories as sucrose instead of equivalent starch. Adaptive enzymatic mechanisms for this effect have been discussed in Chapters 2 and 4. In baboons, serum fructose levels are higher after a sucrose meal when the animals have been adapted to high sucrose feeding. The males show higher serum fructose than females and the higher values correlate in individual cases with higher incorporation of sucrose-^{14}C into serum triglycerides (Fig. 8-9). The mechanism for higher peripheral blood fructose values is not yet clear, but increased rate of absorption due to adaptive increase in intestinal sucrase could be one explanation. The effect of diets high in sucrose or fructose to cause elevation of the serum triglycerides is age-dependent; whereas the triglyceride level in mature rats may be almost doubled by a diet of 70 per cent sucrose (by weight), there are no significant effects on the level in weanling rats. There is some evidence from the human studies of partial deadaptation while on continuing high sucrose diets, i.e., a return toward prediet values of blood triglyceride levels and hepatic lipogenic activity.

The effect of high sucrose diets on glucose tolerance is controversial, with impaired tolerance found in rats and increased tolerance in man. The effect on plasma insulin may also be different in the two species; in rats, plasma insulin is lower and in man it is higher (relative to

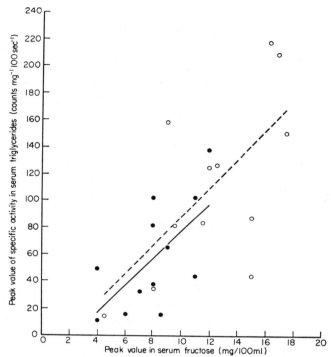

Figure 8-9. Peak value of serum fructose in baboons after sucrose meal (with sucrose-^{14}C) vs. peak value of specific activity of serum triglycerides. Solid regression line is for values before 75 per cent sucrose diet (dots) and broken regression line is for values after 13 weeks on 75 per cent sucrose diet (open circles). (From Coltart and Crossley: Clinical Science. *38*:427, 1970.)

the glucose concentration) after a sucrose meal during a high sucrose diet.

Hormones

As mentioned earlier, a distinctive feature of fructose metabolism is its independence from insulin for transport into cells and intracellular utilization, at least in the initial enzymatic steps. Also unlike glucose, fructose only weakly, if at all, stimulates the production or release of insulin by the pancreas. Glucagon from that organ interferes with the elevation of blood pyruvate and lactate during fructose infusion. On the other hand, glucagon is known to promote formation of glucose from fructose. Evidently, this hormone changes the balance of glycolysis and gluconeogenesis from fructose after its initial hepatic utilization. The effect of fructose to elevate plasma growth hormone has been mentioned above.

Altitude

There are some remarkable differences between sea-level and high-altitude residents in their handling of fructose loads injected into the circulation. Young men living at sea level show utilization of fructose primarily by the liver, but high-altitude residents seem to utilize fructose substantially via extrahepatic tissues. Whereas sea-level residents display considerable rises in lactate, pyruvate, and glucose of blood following the injection of fructose, highlanders (4,500 m.) have lesser elevations of lactate and pyruvate and no rises in blood glucose. These differences may in part be due to the difference in organ site of utilization of fructose and also probably to the greater capacity of high-altitude residents for oxidative utilization of the 3-carbon acids.

Pathological Changes

Diabetes

Because insulin is not needed for metabolism of fructose, the latter has been tried often as carbohydrate nourishment in acute or chronic diabetes, with mixed results (Chap. 4). The strong tendency of fructose to form lactate poses a hazard of aggravation of acidosis, when fructose is tried as a glucose substitute for patients with uncontrolled diabetic ketoacidosis. Moreover, in such patients an inordinate amount of fructose is converted to glucose. As a carbohydrate food in mild diabetes, fructose has found better acceptance, since glucosuria and hyperglycemia are diminished in some cases. The use of fructose in diabetics needs further to be weighed in respect to its effect on metabolism of lipids and other constituents in the liver.

Liver Diseases

Although cirrhotic patients as a group have a lower fructose tolerance than normal, there is a large overlap between the two groups and marked disturbances of portal circulation occur in the cases of significantly altered fructose tolerance. Fructose metabolism may be more sensitive to acute inflammatory changes in the liver. In patients with viral hepatitis, there are additional exponential functions to the rate of blood fructose disappearance after oral sucrose, and the time of maximum concentration of fructose is later than normal even in some mild cases of the disease. Lactic acidosis has been precipitated by administration of fructose in a case of acute liver necrosis due to infectious hepatitis.

Ethanol

The administration of fructose simultaneously with alcohol in man results in faster disappearance of alcohol from the blood. This may be due to provision of substrates, glyceraldehyde and fructose, which can be reduced by NADH to glycerol and sorbitol, respectively. These reductions would facilitate the oxidation of ethanol by making available more NAD^+. The fact that fructose infusions tend to produce excess lactate (increased L/P ratio in the serum) seems not consistent with this explanation, since excess lactate should signify accumulation of hepatic NADH. However, the system is not in equilibrium and the pyruvate in reduction to lactate is probably acting as yet another electron acceptor which further increases the rate of turnover of the NAD^+-NADH couple. Although there are claims that intravenous administration of fructose is particularly efficacious in improving symptoms of delirium tremens, this view is contradicted by others who find no superiority of fructose over glucose in treatment of this acute alcoholic condition.

An effect of ethanol on fructose utilization is an increase of the extent of formation of alpha-glycerol-phosphate from fructose in the liver, again dependent on the decrease in redox potential of the pyridine nucleotide system. This facilitates conditions for exaggerated triglyceridogenesis. The same can be said for the accelerating effect of fructose on the lipogenic potency of ethanol.

Gout

There is evidence that gouty patients and children of gouty parents respond to oral loads of fructose with greater than normal rises in serum uric acid. Moreover, after limitation of sweets in the diet of such gout-prone individuals, the fructose load test is followed by much less rise of serum uric acid. This both indicates the adaptive nature of the mechanism which relates fructose utilization to acute hyperuricemia and suggests that the finding has application in the dietary management of gout.

Hereditary Fructose Intolerance

In this rare condition with deficiency of fructose-1-P aldolase (Chap. 4), the oral administration of a fructose load (1.0 g./kg.) is usually followed by elevation of blood fructose to levels of 25 to 100 mg./dl. or more. Other effects may be marked hypoglycemia, excessive lactic acidemia and fructosuria. This is a hazardous diagnostic test, but otherwise the diagnosis must be made by liver biopsy and assay for fructose-1-P aldolase. Test of the metabolic fate (formation of blood glucose or lactate or breath CO_2) by use of fructose labeled with a carbon isotopic tracer in a relatively small load of fructose could provide a safer test for the existence of this genetic defect.

Use of the non-radioactive, stable carbon isotope, ^{13}C, in this situation as well as those involving other more common disorders of carbohydrate metabolism could be quite advantageous for pediatric diagnosis and clinical research. The same can be said for studies in pregnant women and indeed populations at large if there develops sufficient indication for widespread use of carbon isotope tracer tests in vivo in human subjects.

REFERENCES

For Glucose

Ballard, F. J., Hanson, R. W., and Kronfeld, D. S.: Glucogenesis and lipogenesis in tissue from ruminant and nonruminant animals. Federation Proceedings 28:218–231, 1969.
Bergman, E. N., and Hogue, D. E.: Glucose turnover and oxidation rates in lactating sheep. Am. J. Physiol. 213:1378–1384, 1967.
Cahill, G. F., Jr., and Owen, O. E.: Some observations on carbohydrate metabolism in man. In: Carbohydrate Metabolism and Its Disorders, Vol. I, pp. 497–522, ed. by F. Dickens, P. J. Randle and W. J. Whelan, Academic Press, New York, 1968.
Dunn, A., Chenoweth, M., and Schaeffer, L. D.: Effect of adrenalectomy on glucose turnover, the Cori cycle and gluconeogenesis from alanine. Biochim. Biophys. Acta 177:11–16, 1969.
Felig, P., and Wahren, J.: Influence of endogenous insulin secretion on splanchnic glucose and amino acid metabolism in man. J. Clin. Invest. 50:1702–1711, 1971.
Forbath, N., and Hetenyi, G., Jr.: Glucose dynamics in normal subjects and diabetic patients before and after a glucose load. Diabetes 15:778–789, 1966.
Gatewood, L. C., Ackerman, E., Rosevear, J. W., Molnar, G. D., and Burns, T. W.: Tests of a mathematical model of the blood-glucose regulatory system. Computer and Biomedical Research 2:1–27, 1968.
Goodner, C. J., Conway, M. J., and Werrbach, J. H.: Control of insulin secretion during fasting hyperglycemia in adult diabetics and in nondiabetic subjects during infusion of glucose. J. Clin. Invest. 48:1878–1887, 1969.
Hermansen, L., Pruett, E. D. R., Osnes, J. B., and Giere, F. A.: Blood glucose and plasma insulin in response to maximal exercise and glucose infusion. J. Applied Physiology 29:13, 1970.
Hetenyi, G., Ninomiya, R., and Wrenshall, G. A.: Glucose production rates in dogs determined by two different tracers and tracer methods. J. Nuclear Med. 7:454–464, 1966.
Issekutz, B., Jr., Issekutz, A. C., and Nash, D.: Mobilization of energy sources in exercising dogs. J. Applied Physiol. 29:691–697, 1970.
Kronfeld, D. S., Ramberg, C. F., Jr., and Shames, D. M.: Multicompartmental analysis of glucose kinetics in normal and hypoglycemic cows. Am. J. Physiol. 220:886–893, 1971.
Long, C. L., Spencer, J. L., Kinney, J. M., and Geiger, J. W.: Carbohydrate metabolism in normal man and effect of glucose infusion. J. Appl. Physiol. 31:102–109, 1971.
Moorhouse, J. A., Smithen, C. S., and Houston, E. S.: A study of glucose transport kinetics in man by means of continuous glucose infusion. J. Clin. Endocr. Metab. 27:256–264, 1967.
Sanbar, S. S.: Alternations in glucose turnover following single intravenous injections of epinephrine and glucagon in dogs. Metabolism 17:631–637, 1968.
Shen, S-W., Reaven, G. W., and Farquhar, J. W.: Comparison of impedance to insulin-mediated glucose uptake in normal subjects and in subjects with latent diabetes. J. Clin. Invest. 49:2151–2160, 1970.
Shipley, R. A., Chudzik, E. B., Gibbons, A. P., Jongedyk, K., and Brummond, D. O.: Rate of glucose transformation in the rat by whole-body analysis after glucose-^{14}C. Amer. J. Physiol. 213:1149–1158, 1967.

Shreeve, W. W.: Effects of insulin on the turnover of plasma carbohydrates and lipids. Am. J. Med. *40*:724–734, 1966.
Shreeve, W. W., Hoshi, M., Oji, N., Shigeta, Y., and Abe, H.: Insulin and the utilization of carbohydrates in obesity. Am. J. Clin. Nutr. *21*:1404–1418, 1968.
Steele, R., Bjerknes, C., Rathgeb, I., and Altszuler, N.: Glucose uptake and production during the oral glucose tolerance test. Diabetes *17*:415–421, 1968.
Waterhouse, C., and Kemperman, J. H.: Changes in oxidative metabolism with glucose ingestion. J. Lab. Clin. Med. *68*:250–264, 1966.
Whichelow, M. J., and Butterfield, W. J. H.: Peripheral glucose uptake during the oral glucose tolerance test in normal and obese subjects and borderline and frank diabetics. Quart. J. Med. *40*:261–273, 1971.
Young, D. R., Pelligra, R., Shapira, J., Adachi, R. R., and Skrettingland, K.: Glucose oxidation and replacement during prolonged exercise in man. J. Applied Physiol. *23*:734–741, 1967.

For Lactate

Doar, J. W. H., and Cramp, D. G.: The effects of obesity and maturity-onset diabetes mellitus on L(+)-lactic acid metabolism. Clinical Science *39*:271–279, 1970.
Forbath, N., and Hetenyi, G., Jr.: Metabolic interrelations of glucose and lactate in unanesthetized normal and diabetic dogs. Canad. J. Physiol. and Pharmacol. *48*:115–122, 1970.
Huckabee, W. E.: Hyperlactatemia. Helvetica Medical Acta *35*:363–376, 1970.
Keuel, J., Doll, E., and Keppler, D.: The substrate supply of the human skeletal muscle at rest, during and after work. Experientia *23*:974–979, 1967.
Kreisberg, R. A., Pennington, L. F., and Boshell, B. R.: Lactate turnover and gluconeogenesis in normal and obese humans: effect of starvation. Diabetes *19*:53–63, 1970.
Kreisberg, R. A., Owen, W. C., and Siegel, A. M.: Ethanol-induced hyperlacticacidemia: inhibition of local utilization. J. Clin. Invest. *50*:166–174, 1971.
Margaria, R.: Anaerobic metabolism in muscle. Canad. Med. Assn. J. *96*:770–774, 1967.
Margaria, R., Aghemo, P., and Sassi, G.: Lactic acid production in supramaximal exercise. European J. Physiol. *326*:152–161, 1971.
Oliva, P. B.: Lactic acidosis. Am. J. Med. *48*:209–225, 1970.
Perret, C., Poli, S., and Enrico, J. F.: Lactic acidosis and liver damage. Helvetica Medica Acta *35*:377–405, 1970.

For Glycerol

Bergman, E. N., Starr, D. J., and Reulein, S. S., Jr.: Glycerol metabolism and gluconeogenesis in the normal and hypoglycemic ketotic sheep. Am. J. Physiol. *215*:874–880, 1968.
Himms-Hagen, J.: Glycerol metabolism in rabbits. Canad. J. Biochem. *46*:1107–1114, 1968.
Tibbling, G.: Glycerol turnover in hyperthyroidism. Clinica Chimica Acta *24*:121–130, 1969.
Winkler, B., Steele, R., and Altszuler, N.: Relationship of glycerol uptake to plasma glycerol concentration in the normal dog. Am. J. Physiol. *216*:191–196, 1969.
Winkler, B., Rathgeb, I., Steele, R., and Altszuler, N.: Conversion of glycerol to glucose in the normal dog. Am. J. Physiol. *219*:497–502, 1970.

For Galactose

Condon, R. E.: Blood galactose clearance in normal and Eck fistula dogs. J. Surg. Research *11*:135–139, 1971.
Hansen, R. G.: Hereditary galactosemia. J. Am. Med. Assn. *208*:2077–2082, 1969.
Hsia, D. Y.: Clinical variants of galactosemia. Metabolism *16*:419–437, 1967.
Salaspuro, M. P.: Application of the galactose tolerance test for the early diagnosis of fatty liver in human alcoholics. Scand. J. Clin. Lab. Invest. *20*:274–280, 1967.
Tengstrom, B.: An intravenous galactose tolerance test and its use in hepatobiliary diseases. Acta Med. Scand. *183*:31–40, 1968.

Waterman, F. K., and Hetenyi, G., Jr.: The effect of insulin on the distribution space of galactose in eviscerated normal and diabetic rats and their selected organs. Canad. J. Physiol. and Pharmacol. *44*:923–932, 1966.

For Fructose

Atwell, M. E., and Waterhouse, C.: Glucose production from fructose. Diabetes *20*:193–199, 1971.

Bergstrom, J., and Hultman, E.: Synthesis of muscle glycogen in man after glucose and fructose infusion. Acta Med. Scand. *182*:93, 1967.

Coltart, T. M., and Crossley, J. N.: Influence of dietary sucrose on glucose and fructose tolerance and triglyceride synthesis in the baboon. Clinical Science *38*:427–437, 1970.

Kaufmann, N. A., Kapitulnik, J., and Blondheim, S. H.: Studies in carbohydrate-induced hypertriglyceridemia. Aspects of fructose metabolism. Israel J. Med. Sci. *6*:80–85, 1970.

Maenpaa, P. H., Raivio, K. O., and Kekomaki, M. P.: Liver adenine nucleotides: Fructose-induced depletion and its effect on protein synthesis. Science *161*:1253–1254, 1968.

Poznanska, H.: Blood fructose and inorganic phosphorus after oral saccharose load as index of fructose metabolism. Polish Medical J. *6*:576–580, 1967.

Reynafarje, B., Oyola, L., Cheesman, R., Marticorena, E., and Jimenez, S.: Fructose metabolism in sea-level and high-altitude natives. Am. J. Physiol. *216*:1542–1547, 1969.

GLOSSARY OF SOME ABBREVIATIONS

BGS	= blood-group substances
cAMP	= cyclic- 3,5-adenosine-monophosphate
CD	= ceramide dihexoside
CH	= ceramide hexoside
CM	= ceramide monohexoside
CS	= chondroitin sulfate
C-4-S	= chondroitin-4-sulfate
C-6-S	= chondroitin-6-sulfate
CT_3	= ceramide trihexoside
CT_4	= ceramide tetrahexoside
DHAP	= dihydroxy-acetone phosphate
Dr	= 2-deoxy-D-ribose
DS	= dermatan sulfate
FDP	= fructose-1,6-diphosphate
Fr or Fru	= fructose
Fu	= L-fucose
GA-3-P	= glyceric acid-3-phosphate
GAl-3-P	= glyceraldehyde-3-phosphate
Gal	= D-galactose
Gal-1-P	= galactose-1-phosphate
GGM	= glucose-galactose malabsorption
Gl or Glu	= D-glucose
Gl-6-P	= glucose-6-phosphate
GM	= monosialoganglioside
G6PDH	= glucose-6-phosphate dehydrogenase
GSD	= glycogen storage disease
HA	= hyaluronic acid
Hex	= hexose
HFI	= hereditary fructose intolerance
HS	= heparan sulfate
2-KGA	= 2-keto-glutaric acid
KS	= keratan sulfate
LDH	= lactate dehydrogenase
L/P	= lactate/pyruvate
Man	= D-mannose
Man-6-P	= mannose-6-phosphate
3-O-MG	= 3-oxy-methylglucose

NAcGal	= N-acetyl-D-galactosamine
NAcGl or NAcGlu	= N-acetyl-D-glucosamine
NGal	= D-galactosamine (2-amino)
NGl or NGlu	= D-glucosamine (2-amino)
NAD	= nicotinamide adenine dinucleotide
NANA	= N-acetylneuraminic acid
NGNA	= N-glycolylneuraminic acid
OAA	= oxaloacetic acid
PAPS	= phospho-adenosine-5-phosphosulfate
PC	= pentose cycle
P-C	= pyruvate carboxylase
PEP	= 2-phospho-enol-pyruvate
PEP-CK	= phospho-enol-pyruvate carboxykinase
PFK	= phosphofructokinase
6PGDH	= 6-phospho-gluconate dehydrogenase
Rbu	= ribulose
Rib	= D-ribose
SA	= sialic acid
TA	= transaldolase
TK	= transketolase
UDPG or UDP-Glu	= uridine-diphospho-glucose
UDP-GA	= uridine-diphospho-glucuronic acid
Xyl	= xylulose

INDEX

Note: Page numbers in *italics* refer to illustrations.
Page numbers followed by *t* refer to tables.

Absorption, of carbohydrates, 38–52
    monosaccharide, 45–48
        rates of, 47*t*
    deficiencies of, 48–52
Acetyl CoA, oxidation of, 114–125, *116–117*
    pyruvate utilization and, 110, *111*, 113
N-Acetyl-D-glucosamine, in human milk, 10
Acetylhyalobiuronic acid, 18
α-D-N-Acetylneuraminic acid, *8*
Acid α-1,4-glucosidase, deficiency of, 200*t*, 202
Aconitase, 115, *116*
ACTH, 64, 68
Adipose tissue, fructose utilization in, 126
    gluconeogenesis and, 151
    glycogen metabolism in, 199
    glycolysis in, 99
    hexokinase activity in, 89
    mannose utilization in, 134
    monosaccharide transport in, 67
    pentose cycle in, 107
    tricarboxylic acid cycle in, 124
    uronic acid pathway in, 171
Adrenal gland, glycogen in, 199
    pentose cycle in, 106
Adrenal steroids. See also *Steroids*.
    glucose circulation and, 269
Adrenocorticotrophic hormone, 64
Age, galactose tolerance and, 306
    glucose circulation and, 276
Alanine, hepatic gluconeogenesis and, 267
Alanine-glucose cycle, *256*
Aldohexoses, 6
Aldopentoses, 6
Aldose, cyclic, 5
Alkaline phosphatase, in glucose absorption, 46
Alkalosis, and lactate turnover, 290
Altitude, fructose metabolism and, 314
    pyruvate circulation and, 291
Amino acids, effects of on gluconeogenesis, 156

Amino acids (*Continued*)
    glucose tolerance and, 264
    insulin action and, 62
Amino sugars, biosynthesis of, 172–178
α-Amylase, in glucose hydrolysis, 39
    glycogen degradation with, 12
Amylo-(1,4→1,6)-transglucosidase, deficiency of, 200*t*, 203
Amylo-1,6-glucosidase, deficiency of, 200*t*, 202
Anderson's disease, 203
Anoxia, glucose uptake and, 65
Anthrone, 6
Antigen-antibody specificity, heteroglycans and, 15
Aorta, pentose cycle activity in, 109
    proteoglycan metabolism and, 211
Arterial tissue, glycoproteins of, 36
    pentose cycle activity in, 109
    proteoglycan metabolism and, 211
Arthritides, mucopolysaccharide disorders in, 216
Ascorbic acid, 8
Asialoganglioside, 235
Asthenia, neurocirculatory, lactate levels and, 297
Atheroma, proteoglycan metabolism and, 211
Atherosclerosis, glucose tolerance and, 283

Barometric changes, fructose metabolism and, 314
    pyruvate circulation and, 291
Bence-Jones protein, 32
Bladder, glucose transport in, 76
Blood, glycolipid disorders involving, 251
    types of. See *Blood-group substances*.
Blood cells, glycolysis in, 97
    glycoproteins in, 227
    proteoglycan metabolism in, 215
Blood vessels, pentose cycle activity in, 109

321

Blood vessels (*Continued*)
  proteoglycan metabolism and, 211
Blood-brain barrier, 69
Blood-group substances, glycolipid function and, 238
  glycoproteins and, 25, 227
  oligosaccharide units in, structure of, 27
  urine oligosaccharides and, *232*
Bone, glucose transport in, 76
  glycoproteins of, 36
  proteoglycan action in, 209, 210
  tricarboxylic acid cycle in, 124
Bone marrow, glycolysis in, 97
Bovine gland glycoproteins, 28
Brain, fructose synthesis in, 181
  fructose utilization in, 126
  glycogen metabolism in, 196
  glycolipids in, 235, 237, 238
  glycolysis in, 97
  glycoproteins in, 230
  hexokinase activities in, 86
  insulin sensitivity of, 86
  lactate turnover and, 285
  monosaccharide transport in, 69
  pentose cycle in, 106
  proteoglycans in, 214
  pyruvate utilization and, 112
  tricarboxylic acid cycle in, 122

Canine submaxillary gland glycoprotein, 28
Carbohydrates, body turnover of, 253–318
  cellular translocation of, 53–80
  chemistry of, 1–37
  dietary, constituent characteristics of, 40*t*
    fate of, 38–52
    glucose tolerance and, 265
    lactate turnover and, 289
  digestion and absorption of, 38–52
  intracellular translocation of, 76–79, *78*
Carbon-14, glucose tracing with, 254, *261*
Cardiac muscle, glucose uptake in, 65
  glycogen metabolism in, 195
  glycolysis in, 96
  hexokinase activity in, 87
  in thiamine deficiency, 112
  pentose cycle in, 109
  pyruvate utilization in, 113
Cardiovascular disorders, lactate levels and, 297
Carnivores, dietary carbohydrates in, 38
  glucokinase activity in, 88
Carrier, monosaccharide, concept of, 54–57, *55*

Cartilage, glycoproteins of, 36
  glycosaminoglycans in, 209
Cell, carbohydrate translocation within, 76–79, *78*
  glycoprotein synthesis in, 220, *221*
Cell membrane, insulin effects on, *67*
  transport and, 58–61
Cellobiose, 40*t*, 42
Cellulose, hydrolysis of, 39, 40*t*
Ceramide galactoside, 235, *236*
Ceramide hexosides, chemistry of, 235
  in brain, 238
  sulfated, 237
  synthesis of, 246
Cerebroside, 235
  degradation of, 248
Ceruloplasmin, 34
Cervical mucus, 35
Chediak-Higashi syndrome, 252
Cholinesterase, 35
Chondroitin, 21
Chondroitin sulfate, in proteoglycan synthesis, 205*t*, 206
Chondroitin-4-sulfate, 20
  location of, 21
  properties of, 21
  units of, *19*
Chondroitin-6-sulfate, units of, *19*
Chondrosin, 20
Choroid plexus, glucose transport in, 75
Chromium, dietary, glucose tolerance and, 266
Circulation, fate of carbohydrates in, 253–318
Cirrhosis, hepatic, 295
  fructose metabolism in, 314
  galactose turnover in, *308*
Citrate synthase, activity of, 115, *116–117*
Citric acid cycle, 114–125
Cold exposure, effect of on gluconeogenesis, 157
  glucose circulation and, 277
  glycerol levels in, 302
  glycogen metabolism in, 194
Collagen, glycoproteins in, 228
Connective tissue, glycoproteins in, 36, 228
  proteoglycan functions in, 208
  synthesis, 205
  uronic acid pathway and, 171
Cori lactate-glucose cycle, *256*
Cornea, glycoproteins of, 37
  pentose cycle activity in, 109
Cortisone, gluconeogenesis and, 152
  glycogen synthesis and, 192
Crabtree effect, 97
Cystic fibrosis, 48
  mucopolysaccharide disorder in, 219
  urinary glycoprotein and, 33

INDEX

2-Deoxy-2-amino-D-galactosamine, 8
2-Deoxy-2-amino-D-glucosamine, 8
Deoxyribose, synthesis of, 162, *163*, *164*
2-Deoxy-D-ribose, *4*
Dermatan sulfate, 21
  location of, 22
  units of, *19*
DHAP, 3
Diabetes mellitus, 72
  connective tissue glycoproteins in, 228
  fructose metabolism and, 314
  galactose transport in, 309
  glucose circulation and, 277
  glycerol levels in, 303
  hepatic fructokinase in, 127
  lactate turnover and, 295
  uronic acid pathway in, 172
Diaphragm, glucose uptake by, 66
  ribose biosynthesis in, 161
Diet, carbohydrates in, 40t
  fate of, 38–52
  glucose tolerance and, 265
  galactose turnover and, 306
  glycerol levels and, 300
  high-fructose or high-sucrose, 312
  lactate turnover and, 289
Diethylstilbestrol, *60*
Digestion, of carbohydrates, 38–52
  deficiencies of, 48–52
Dihydroxyacetone, *2*, 3
Dihydroxyacetone monophosphate, 3
Disaccharidases, 41t
  deficiency of, 48
Disaccharides. See also specific disaccharides.
  hydrolysis of, 40–45
Disease, carbohydrate digestion deficiencies in, 48–52
  glycolipid disorders and, 249–252
  lactate/pyruvate levels and, 292
  mucopolysaccharide-related, 216
Dog, brain glucose in, 70
  glycerol uptake in, *299*
Durand syndrome, 50

Embden-Meyerhof-Parnas glycolytic pathway, *84–85*, 89, *90–91*
Encephalopathy, lactate turnover in, 296
Endometrium, proteoglycans in, 214
Enzyme glycoproteins, 34
Enzymes, digestive. See also specific enzymes.
  nomenclature of, 41
Epinephrine, 66, 68
  blood glucose and, 273
  gluconeogenesis and, 152
  glycerol turnover and, 301

Epinephrine (*Continued*)
  glycogen metabolism and, 188, 191t, 194
  lactate turnover and, 289
Erythrocytes, glucose transfer in, 57, *58*
  glycogen metabolism in, 197
  glycolipids in, 240
  glycolysis in, 98
  glycoproteins in, 227
  hexokinase activity in, 89
  hexosamine synthesis and, 177
  monosaccharide transport in, 71
  pentose cycle in, 108
  tricarboxylic acid cycle in, 124
Erythropoietin, 34
D-Erythrose, *2*, 4
Essential pentosuria, 172
Estradiol, *60*, 63
Estrogen, and uterine glycogen, 196
  glucose tolerance and, 271
  proteoglycan metabolism and, 214
  vascular proteoglycans and, 213
Ethanol, excessive, galactose levels and, 307
  glucose circulation and, 282
  lactate turnover and, 294
  fructose metabolism and, 314
  gluconeogenesis and, 157
  glycerol levels and, 303
  hepatic glycoprotein synthesis and, 225
Exercise, blood glucose and, 275
  glycerol turnover and, 302
  glycogen synthesis and, 193
  insulin-like effect of, 65
  lactate turnover and, 286, *288*
Eye, glucose transport in, 75
  tricarboxylic acid cycle in, 124

Fabry's disease, 251
Familial amaurotic idiocy, 250
Fasting, gluconeogenetic effect of, 155
  glucose circulation in, 266
Fat tissue, fructose utilization in, 126
  gluconeogenesis and, 151
  glycogen metabolism in, 199
  glycolysis in, 99
  hexokinase activity in, 89
  mannose utilization in, 134
  monosaccharide transport in, 67
  pentose cycle in, 107
FDPase, activation of, in gluconeogenesis, 146, *147*
Feeding, acute effects of, 261
  gluconeogenesis and, 155
Female genital tract glycoproteins, 35
Fetuin, 30
Fibrinogen, 32

Follicle-stimulating hormone, 33
Forbe's disease, 202
Free fatty acids, gluconeogenesis and, 153
  glucose tolerance and, 264
Freezing. See *Cold exposure.*
Fructokinase, 126, 127
Fructose, biosynthesis of, 180–183
    sorbitol pathway for, *181*
  body turnover of, 309–316, *313*
    in acute loading, 311
  erythrocyte uptake of, 71
  hereditary intolerance to, 128, 315
  utilization of, 125–129
β-D-Fructose, 7, 8
FSH, 33
Fucose, biosynthesis of, 178–*179*
L-Fucose, in human milk, *10*
L-Fucosyl-1,2-D-galactosyl-1,4-D-glucose, *10*
Fucosyl-lactose, *10*
Fumarate, 119
Furanose ring, 5
Furfural, 6

Galactokinase, 129, 130
  deficiency of, 307
β-D-Galactopyranosyl-1,4-D-glycopyranose, *9*
D-Galactosamine, *8*
Galactose, absorption of, disorder of, 51
  biosynthesis of, 180
  body turnover of, 304–309
  utilization of, 129–134
D-Galactose, in human milk, *10*
Galactosemia, congenital, 132, 306
  ganglioside disorder and, 250
Gangliosides, biosynthesis of, *243*, 246
  catabolism of, *244*
  chemistry of, 235, *236*
  in brain, 238
Gangliosidosis, generalized, 250
Gargoylism, glycolipid disorder in, 250
Gastrointestinal tract, disaccharidases in, 42, 43
  fructose utilization and, 125
  galactose utilization and, 129
  glucose transport in, 43, *44*
  glycolipids in, 242
  glycoproteins in, 35, 229
  hexokinase activity in, 89
  monosaccharide absorption capacities of, 47
  mucus secretions of, 35
Gaucher's disease, glycolipid disorder in, 251
Genital tract, female, glycoproteins of, 35
Gentiobiose, 40*t*, 42
GGM (glucose-galactose malabsorption), 51

Globoid leukodystrophy, 249
Globoside, 235
  degradation of, 248
  in red blood cells, 240
Globulin, β_1-metal-combining, 31
Glucagon, 68
  blood glucose and, 273
  fructose metabolism and, 313
  gluconeogenesis and, 153, 154
  glycerol turnover and, 301
  glycogen metabolism and, 191*t*, 192
  lactate turnover and, 289
Glucogen, storage of, 14
Glucocorticoids, 63
  gluconeogenetic effect of, 154
  glucose circulation and, 269
  glycogen synthesis and, 192
  lactate turnover and, 290
  plasma glycoproteins and, 226
  uronic acid pathway and, 172
  vascular proteoglycans and, 212
Glucokinase, activity of, 82, 88
Gluconeogenesis, 139–158
  functions of, 148
  in diabetes mellitus, 278
  physiological changes and, 152
  reactions of, 139, *141*, *142*, *144–145*
  renal, in starvation, 267
  tricarboxylic acid cycle and, 122
D-Gluconic acid, *8*
Glucosamine, 12
D-Glucosamine, *8*
Glucose, absorption of, 45, 47*t*
    disorder of, 51
  biosynthesis of, 139–158. See also *Gluconeogenesis.*
  body distribution and turnover of, 253–284
    evaluation of, 253, *255*, *256*
    in early postabsorptive state, 257, *258*
  circulating, hormonal effects on, 268
  labeling of, 254
  pathological conditions involving, 277
  physiological effects of, 261
  conformations of, 5, 7
  entry rates of, *258*
  hydrolysis of, in man, 39
  isomers of, 5, 7
  measurement of, 6
  phosphorylation of, 82–89
Glucose, tolerance to. See *Glucose tolerance.*
  transport of, 43, *44*
D-Glucose, 5
  in human milk, *10*
  reactions of, 6
α-D-Glucose, 5
β-D-Glucose, 5
Glucose oxidase, 6
Glucose-6-phosphatase, deficiency of, 200

Glucose tolerance, definition of, 262
  factors influencing, 262
  in diabetes mellitus, 280
Glucosuria, 72
  familial, 73
α-D-Glucosyl-1,4-D-fructose, 12
Glucuronic acid, 8
α-Glutamyl transpeptidase, 35
D-Glyceraldehyde, 2, 3
Glycerol, body turnover of, 298–304
  reactions of, 90–91
α-Glycerol-P, 3
Glycolaldehyde, 2
Glycogen, degradation of, 12, 188
  functions of, 14
  hydrolysis of, 40t
  linkages in, 12
  liver, 14
  metabolism of, 185–200, 187
    disturbances of, 200–205
  molecular weight of, 12, 14
  muscle, 14
  structure of, 11, 13
  synthesis of, 186
Glycogen storage diseases, 200–204
  classification of, 200t
Glycogen synthetase, deficiency of, 200t, 204
Glycogenoses, 200–204
  classification of, 200t
Glycolipids, chemistry and metabolism of, 234–252
  degradation of, 248
  disorders involving, 249–252
  functions of, 238
  synthesis of, 242, 243, 244, 245
Glycolysis, 84–85, 89–100, 90–91
Glycopeptides, 25
Glycopolypeptides, classification of, 16t
Glycoproteins, 15, 24–37
  classification of, 16t
  degradation of, 223
  enzyme, 34
  functions of, 224
  hormone, 33
  in blood-group substances, 25
  metabolism of, 219–230
    disorders of, 230
  monosaccharides in, 24
  of arterial tissue, 36
  of bone, 36
  of bovine glands, 28
  of canine submaxillary gland, 28
  of cartilage, 36
  of connective tissue, 36
  of cornea, 37
  of female genital tract, 35
  of gastrointestinal tract, 35
  of ovine submaxillary gland, 28
  of plasma, 28

Glycoproteins (Continued)
  of procine submaxillary gland, 28
  of respiratory tract, 35
  of reticulin, 36
  of skin, 36
  of synovial fluid, 36
  of tendon, 36
  of vitreous body, 37
  synthesis of, 220
  urinary, 32
β_A-α_2-Glycoprotein, 30
Zn-α_2-Glycoprotein, 30
Glycosides, 6
Glycosidic bonds, 6, 9
Glycosoaminoglycans, 14
Gonadal steroids, blood glucose and, 271
Gout, fructose metabolism in, 315
Growth hormone, glucose circulation and, 270
  glycerol turnover and, 301
  lactate turnover and, 290
Guinea pig, brain glucose in, 70

Haptoglobin, 30
HCG, 33
Heart, disorders involving, glycerol levels in, 304
  lactate levels and, 297
  glucose uptake by, 65
  glycogen metabolism in, 195
  glycolysis in, 96
  hexokinase activity in, 87
  pyruvate utilization in, 113
  in thiamine deficiency, 112
  ribose biosynthesis in, 161
Hemiacetal, internal, 5, 6
Hemorrhage, lactate turnover and, 294
Heparan sulfate, 23
  Hurler's syndrome and, 24
  in proteoglycan synthesis, 205t, 207
  linkages in, 24
  molecular weight of, 24
Heparin, functions of, 23
  molecular weight of, 23
Heparin monosulfate. See Heparan sulfate.
Heparitin sulfate. See Heparan sulfate.
Herbivores, dietary carbohydrates in, 38
  glucokinase activity in, 88
  gluconeogenesis in, 151
Hereditary fructose intolerance, 128, 315
Her's disease, 204
Heteroglycans, 11, 14–18
  classification of, 15, 16–17t
  degradation of, 15
  functions of, 15
Heteropolysaccharides, 4, 18–24
  acidity of, 18

Hexokinases, activity of, 82, *84–85*
Hexosamines, 8
 biosynthesis of, 172–178
Hexose phosphate, formation of, 146, *147*
Hexoses, 5
 phosphorylation of, 82, *84–85*
D-Hexoses, isomers of, *5, 7*
Hibernation, blood glucose during, 267
Homoglycans, 11–14
Hormones. See also specific hormones.
 effects of, on blood glucose, 268
  on cardiac glycogen, 195
  on fructose metabolism, 313
  on fructose synthesis, 182
  on gluconeogenesis, 152
  on glucose transport, 61–64
  on glycerol circulation, 301
  on lactate turnover, 289
  on ribose synthesis, 161
  on uronic acid pathway, 171
 follicle-stimulating, 33
 interstitial cell-stimulating, 33
 lipolytic, 68
 pituitary, 64
 renal gluconeogenesis and, 150
 thyroid, storage form of, 33. See also *Thyroid hormone.*
 thyroid-stimulating, 34
Hormone glycoprotein, 33
Human chorionic gonadotropin, 33
Hurler's syndrome, dermatan sulfate and, 22
Hyaluronic acid, 18
 properties of, 20
 units of, *19*
Hyaluronidase, in proteoglycan degradation, 207
Hydrocortisone, 63
 gluconeogenesis and, 152
 glycogen synthesis and, 192
Hydrolysis, of disaccharides, 40–45
 of polysaccharides, 38–40
Hydroxymethylfurfural, 6
Hypercapnia, 290
Hyperglycemia, stress-induced, 282
Hyperglyceridemia, glucose tolerance and, 264
Hypertension, carbohydrate abnormalities and, 67
 lactate levels in, 298
Hyperthermia, lactate turnover and, 291
Hyperthyroidism, glucose tolerance in, 274
 glycerol levels in, 301
Hyperventilation, lactate levels and, 290
Hypoglycemia, glucose tolerance tests and, 264
Hypophysectomy, and skin proteoglycans, 213
Hypothalamus, 64
 blood glucose and, 274

Hypothyroidism, glucose tolerance in, 274
Hypoxia, lactic acidosis and, 292, *293*

ICSH, 33
IgA, 31
IgG, 31
IgM, 31
Immunoglobulins, 31
Infants, galactose utilization in, 130
 starch hydrolysis in, 39
 lactase and, 42, 48
Infection, systemic, lactate turnover in, 295
Injury, glucose circulation in, 282
 lactate turnover and, 294
Inositols, 8
Insulin, 268
 brain sensitivity to, 86
 fructose metabolism and, 313
 galactose transport and, 305
 glucose uptake and, in brain, 70
 glycogen metabolism and, 188, 191*t*
 gluconeogenesis and, 153, 154
 in glucose transport, 62, 66, *67*
 vascular proteoglycans and, 213
Interstitial cell-stimulating hormone, 33
Intestine. See *Gastrointestinal tract.*
Intrinsic factor, 35
Islet cells, pancreatic, glycolysis in, 100
 pentose cycle in, 109
Isocitrate dehydrogenase, 118
Isomaltase, deficiency of, 50
Isomaltotriose, 12

Keratan sulfate, 22
Ketose, cyclic, 5
Kidney, fructose utilization in, 126
 gluconeogenesis in, 149
 glycerol uptake and, 300
 glycogen metabolism in, 198
 glycolipid disorder involving, 251
 glycolipids in, 239
 glycolysis in, 99
 hexokinase activity in, 89
 monosaccharide transport in, 72, *74*
 pentose cycle in, 108
 proteoglycans in, 214
 pyruvate utilization in, 113
 tricarboxylic acid cycle in, 123
 uronic acid pathway in, 171
Krebs cycle, 114–125, *116–117*
Kwashiorkor, glucose tolerance and, 265, 284
 lactate turnover in, 289

Lactaldehyde, 3
Lactase, 41*t*, 42
 deficiency of, 42, 48

Lactate, in body circulation, 284–298
    turnover of, 284
        pathological changes involving, 292, 293
Lactation, pentose cycle and, 106
    prolactin effects and, 272
Lactic acid, in body circulation, 284–298
Lactic acidosis, 292, 293
    infantile, 297
Lactose, 9, 40t, 41
    absorption of, 44
    production of, prolactin and, 272
    synthesis of, 230
Lactose tolerance tests, 49
Leukocytes, glycogen metabolism in, 197
    glycolipids in, 242
    glycolysis in, 97
    pentose cycle in, 108
    proteoglycan metabolism in, 215
Leukodystrophies, 249
Lithium, in glucose uptake, 69
Liver, 295
    disease of, fructose metabolism in, 314
        galactose turnover in, 308
    fructose metabolism and, 126, 310, 311
    fructose synthesis in, 181
    gluconeogenesis in, 148, 149t
    glucose tolerance and, 263
    glycerol uptake and, 300
    glycogen metabolism and, 191t
    glycolysis in, 100
    glycoprotein metabolism in, 224
    hexokinase activity in, 87
    hexosamine synthesis in, 177
    in mucopolysaccharidoses, 218
    lactate turnover and, 285, 288
    mannose utilization in, 134
    monosaccharide transport in, 74, 79
    pentose cycle in, 107
    proteoglycan metabolism in, 214
    pyruvate utilization in, 114
    tricarboxylic acid cycle in, 121
    uronic acid pathway in, 171
Liver phosphorylase, deficiency of, 200t, 204
Loop of Henle, glycolysis in, 100
Lung, glycolysis in, 100
    proteoglycan metabolism in, 215
Lupus erythematosus, mucopolysaccharide disorder in, 217
Lymphatic organs, proteoglycan metabolism in, 215
Lymphocytes, glycolysis in, 97
    glycoprotein synthesis in, 227

γ_1-Macroglobulins, 31
Malabsorption, of carbohydrates, 48–52
    disaccharides and monosaccharides, classification of, 49t
L-Malate, 119

Maltulose, 12
Mammary gland, gluconeogenesis and, 151
    mannose utilization in, 135
    pentose cycle in, 106
    tricarboxylic acid cycle in, 123
Man, dietary carbohydrates of, 38
    intestinal disaccharidases in, 43
Manganese, glucose tolerance and, 266
Mannose, utilization of, 134–135
Maple syrup disease, glycolipid disorder and, 249
Marrow, glycolysis in, 97
McArdle's syndrome, 203
Membrane, cellular, insulin effects on, 67
    transport and, 58–61
    mitochondrial, carbohydrate translocation and, 76, 78
Metachromatic leukodystrophy, 249
β_1-Metal-combining globulin, 31
Milk, glycolipids in, 241
    human, oligosaccharides in, 10, 230
Mitochondria, brain hexokinase and, 86
    carbohydrate transport in, 76, 78
    in gluconeogenesis, 140, 141
    tricarboxylic acid cycle and, 115, 118, 120
"Mobile carrier," in monosaccharide transport, 54, 55, 56
Monosaccharide carrier, concept of, 54–57, 55
Monosaccharides, 2–9. See also specific monosaccharides.
    absorption and transport of, 45–48
    biosynthesis of, 139–184
    malabsorption of, 49t
    transport of, 64–76
        in adipose tissue, 67
        in brain, 69
        in kidney, 72, 74
        in liver, 74, 79
        in muscle, 64
        in red blood cells, 71
    utilization of, intracellular, 81–138
MPS, metabolism of, 204–219. See also Mucopolysaccharides and Proteoglycans.
Mucins, 15
    gastrointestinal, 229
Mucoids, 15
Mucopolysaccharides, 18. See also Proteoglycans.
    classification of, 16t
    chemistry of, 205t
    metabolism of, 204–219
Mucopolysaccharidoses, 217t
Mucoproteins, 15
Mucus, cervical, 35
    of gastrointestinal tract, glycoproteins of, 35
Multiple sclerosis, glycolipid disorder in, 249

Muscle, disorders involving, lactate turnover and, 296
  fructose metabolism and, 126, 311
  gluconeogenesis in, 150
  glycogen metabolism and, 191t, 193
  glycolysis in, 96
  hexokinase activity in, 87
  lactate turnover and, 285, 286
  monosaccharide transport in, 64
  pentose cycle in, 109
  pyruvate utilization in, 113
  ribose biosynthesis in, 161
  tricarboxylic acid cycle in, 122
Muscle phosphorylase, deficiency of, 200t, 203
Mutarotation, 5
Myelopathy, lactate turnover in, 296
Myocardial infarction, lactate levels in, 297
Myo-inositol, 8, 168

Na$^+$, in glucose absorption, 45
Na$^+$ pump, 56
NADPH, in pentose cycle, 104
NANA, *8*, 9
  in human milk, 11
Nerves, gangliosides in, 239
Nervous system. See also *Brain*.
  disorders of, lactate turnover in, 296
  glycolipids in, 238, 249
Neuramine-lactose, 11
Neuraminic acids, biosynthesis of, 172–178, *174–175*
Neuraminidase, 25
Neuropathy, lactate turnover in, 297
Neurosis, lactate levels and, 297
Nigerose, 12
Notatin, 6

Obesity, galactose turnover in, 309
  glucose circulation and, 280
  lactate/pyruvate levels in, 295
Oligosaccharides, 9–11, 230–232
  in blood-group specific substances, structure of, *27*
  in human milk, *10*
Orosomucoid, 29, 225
Osatriazole, 6
Osazones, 6
Osteoarthritis, mucopolysaccharide disorder in, 216
Ouabain, effects of, 61, 66, 71
Ovine submaxillary gland glycoprotein, 28

Palatinose, 40t, 42
Pancreas, glucose uptake in, 76
  islet cells of, glycolysis in, 100
    pentose cycle in, 109

Pasteur effect, 93, 97, 98
pCO$_2$, cellular, lactate turnover and, 290
Pentose cycle, 100–110
  functions of, 104
  reactions of, 101, *102–103*
Pentoses, *4*
  biosynthesis of, 158–164
Pentosuria, 4
  essential, 172
PEP, formation of, 140, *141*
PFK, 92, 93
pH, cellular, lactate turnover and, 290
Phenolphthalein, *60*
Phloretin, 59, *60*
Phlorizin, sugar transport and, *59*
Phosphoenolpyruvate, formation of, 140, *141*
Phosphofructokinase, in glycolysis, 92
Phosphoglucomutase, in glycolysis, 92
Phosphorylase, deficiencies of, 200t, 203, 204
Phosphorylase, in glycogen degradation, 188
Phosphorylation, of glucose, 82–89
Pituitary gland, 64
  glycogen in, 199
Placenta, glucose transport in, 75
  glycogen metabolism in, 198
  glycolipids in, 241
  glycolysis in, 100
  pentose cycle activity in, 109
Plasma, glycerol circulation in, 298, *299*
  glycoproteins in, 28, 225
  lactate turnover and, *288*
Platelets, glycogen metabolism in, 197, 198
  glycoproteins in, 228
  proteoglycan metabolism in, 215
PMSG, 33
Polymorphonuclear leukocytes, glycogen metabolism in, 197
  glycolysis in, 97
  pentose cycle in, 108
Polysaccharides, 11–37
  hydrolysis of, 38–40t
  metabolism of, 185–233
Pompe's disease, 202
Porcine submaxillary gland glycoprotein, 28
Potassium, glucose tolerance and, 266, 283
Pregnancy, glucose tolerance in, 272
Prolactin, blood glucose and, 272
Propanediol, 3
Proteins, cellular transport of carbohydrates and, 58
  dietary, glucose tolerance and, 265
Proteoglycans, chemistry of, 205t
  degradation of, 207
  metabolism of, 204–219
  pathological conditions involving, 216

Proteoglycans (*Continued*)
  synthesis of, 205
Prothrombin, 32
Pyranose ring, 5
Pyruvate, dietary deficiency and, 289
  in body circulation, 284-298
  turnover of, 285
  utilization of, 110-114
Pyruvate kinase, in glycolysis, *91*, 94
Pyruvic acid, fate of, 110, *111*
  in body circulation, 284-298

Rabbit, erythrocyte glucose uptake in, 71
Rabbit serum atropinesterase, 35
Rat, brain glucose in, 70
Red blood cells, glucose transfer in, 57, *58*
  glycogen metabolism in, 197
  glycolipids in, 240
  glycolysis in, 98
  glycoproteins in, 227
  hexokinase activity in, 89
  hexosamine synthesis and, 177
  monosaccharide transport in, 71
  pentose cycle in, 108
  tricarboxylic acid cycle in, 124
Renal medulla, glycolysis in, 100
Renal tubule, glucose transport in, 72
  glycogen metabolism in, 198
  glycolysis in, 100
  hexokinase activity in, 89
  pentose cycle in, 108
  tricarboxylic acid cycle in, 124
"Reorienting pore," in monosaccharide transport, 54, *55*
Reproductive organs, glycogen metabolism in, 198
  proteoglycans in, 214
Respiratory tract glycoproteins, 35
Reticulin, glycoproteins of, 36
Retina, glycolysis in, 97
  hexosamine synthesis in, 177
Rheumatoid arthritis, mucopolysaccharide disorder in, 216
Ribonuclease, 34
Ribose, biosynthesis, of, 158, *159*
  functions of, 160
D-Ribose, 4, *4*
Ruminants, gluconeogenesis in, 151

Salivary glands, glycoprotein metabolism and, 26, 229
  pentose cycle activity in, 109
Scylloinositol, 8
Sedo-heptulose-7-phosphate, 9
Semen, fructose in, 181, 182
Seromucoid, 29
Serum, glycolipids in, 241
Shivering, gluconeogenesis during, 157
Shock, lactate turnover and, 294

Sialic acids, *8*, 9
  biosynthesis of, 172-178
  in gangliosides, 235
  in glycoproteins, 24, 25
Skin. See also *Connective tissue*.
  glycoproteins of, 36
  proteoglycan metabolism in, 213
  pyruvate utilization in, 114
  uronic acid pathway and, 164, 171
Smooth muscle, glycogen metabolism in, 195
Sodium ion, 56
  in glucose absorption, 45
Sorbitol pathway, in fructose biosynthesis, *181*, 182
D-Sorbitol, 8
Spermatogenesis, glycogen and, 199
Spermatozoa, glycolysis in, 100
Sphingoglycolipids, 235
Sphingosine, in glycolipids, 234
Spleen, ceramide hexosides of, 241
  glycolipid disorder involving, 251
  glycolysis in, 100
Starch, hydrolysis of, 40t
Starvation, gluconeogenesis and, 155
  glucose circulation in, 266
  glucose intolerance and, 267
  glycerol levels in, 300
  lactate turnover and, 289
  ribose synthesis and, 161
Stereospecificity, of monosaccharide translocation, 57-*58*
Steroids, gluconeogenesis and, 152
  glucose circulation and, 269
  gonadal, blood glucose and, 271
  in carbohydrate metabolism, 60, 63
  lactate turnover and, 289
Stress, environmental, glucocorticoids in, 270
  glucose circulation and, 277
Strophanthidin, *61*
Succinate, 118
Succinyl CoA, 118
Sucrase, deficiency of, 50
Sucrose, 40t, 41
  dietary loading of, 312, *313*
Sugars, eight-carbon, 9
  five-carbon, *4*
  four-carbon, 4
  nine-carbon, 9
  seven-carbon, 9
  six-carbon, *5*
  three-carbon, 3
  two-carbon, 2
Sulfatides, biosynthesis of, 242, *245*
  degradation of, 248
Synovial fluid, glycoproteins of, 36

T and H glycoprotein, 32
Tay-Sachs disease, ganglioside disorder in, 250

TCA cycle, 114–125, *116–117*
　functions of, 120
Temperature, environmental, lactate levels and, 291
Tendon, glycoproteins of, 36
Testosterone, fructose synthesis and, 182
　glycogen metabolism and, 194
Thermal changes. See also *Cold exposure*.
　lactate turnover and, 291
Thiamine, deficiency of, lactate turnover and, 289
　ribose synthesis and, 161
　in pyruvate utilization, 112
Thyroglobulin, 33
Thyroid gland, tricarboxylic acid cycle in, 124
Thyroid hormone, glucose circulation and, 274
　glycerol turnover and, 301
　lactate turnover and, 290
　storage form of, 33
Thyroid-stimulating hormone, 34, 64
Thyrotropin, 34
　and skin proteoglycans, 214
Thyroxine, gluconeogenesis and, 153
　vascular proteoglycans and, 212
Transaldolase, 3
Transferrin, 31
Translocation, definition of, 53
　cellular, of carbohydrates, 53–80
　stereospecificity in, 57–*58*
　intracellular, of carbohydrates, 76–79, *78*
Transport, 53–80
　hormonal effects on, 61–64
　monosaccharide, 45–48, 64–76. See also *Monosaccharides, transport of*.
　rates of, 47*t*
　non-active, 54, 56
　parameters of, 57, *58*
　of glucose, 43, *44*
Trehalase, in renal function, 74
Trehalose, 40*t*, 42
Tricarboxylic acid cycle, 114–125
　functions of, 120

Triose phosphate, formation of, 143
TSH, 34, 64

Uremia, glucose tolerance and, 283
Urine, glycoproteins in, 32, 230
　in mucopolysaccharidoses, 217*t*
　mucopolysaccharides in, 215
　oligosaccharides in, 231, *232*
Uronic acid pathway, 164–172, *166–167*
　functions of, 164, 171
Uterus, glycogen metabolism in, 195, 196
　proteoglycans in, 214

Vagina, glycogen metabolism in, 198
Vascular system, pentose cycle activity in, 109
　proteoglycans and, 211
Veins, pentose cycle activity in, 109
Vitamin A, in glycoprotein synthesis, 222
　mucopolysaccharide metabolism and, 219
Vitamin B_1, deficiency of, lactate turnover and, 289
　ribose synthesis and, 161
　in pyruvate utilization, 112
Vitamin C, 8
Vitreous body, glycoproteins of, 37
Von Gierke's disease, 200

White blood cells, glycolipids in, 242
　glycolysis in, 97
　pentose cycle in, 108

Xylose, 21
D-Xylose, *4*
Xylulose, in uronic acid pathway, 168, *169, 170*
L-Xylulose, *4*

Zinc, in glucose uptake, 69